MALAY FISHERMEN

Their Peasant Economy

I   LINE FISHERMEN AT WORK

Catching small cuttle-fish; part of each man's catch lies in front of him.   The round pedestal-boxes contain food, the rectangular boxes hold betel materials, hooks, etc.   The Photograph was taken soon after dawn, about 10 miles off the Kelantan coast.

MALAY FISHERMEN

Their Peasant Economy

by

RAYMOND FIRTH

PROFESSOR OF ANTHROPOLOGY
IN THE
UNIVERSITY OF LONDON

The Norton Library

W·W·NORTON & COMPANY·INC·

NEW YORK

Books That Live
The Norton imprint on a book means that in the publisher's
estimation it is a book not for a single season but for the years.
W. W. Norton & Company, Inc.

Library of Congress Cataloging in Publication Data
Firth, Raymond William, 1901–
  Malay fishermen.
  (The Norton library)
  Reprint of the ed. published by Archon Books, Hamden
Conn.
  Bibliography: p.
  1. Fish Trade—Malay Peninsula. 2. Fishermen—
Malay Peninsula. I. Title.
[HD9466.M32F5   1975]   338.3'72'709205951   75-8523
ISBN 0-393-00775-8

Printed in the United States of America
1 2 3 4 5 6 7 8 9 0

CONTENTS

TABLES

FIGURES

PLATES

PREFACE TO FIRST EDITION

This book is a study of some Far Eastern peasant problems, based mainly on field research. As originally planned, it was intended to demonstrate four main points. The first is the need for much more attention to the " native " fishing industry in tropical regions ; though forming the livelihood of large numbers of people, by comparison with the " native " tropical agriculture it has suffered from neglect by both scientists and governments. The second point is the need for more studies of the economics as distinct from the technology of these peasant systems ; before the war the investigation and handling of peasant economic affairs tended to fall between the two stools of the administrative and the technical departments—each was interested but its major job lay elsewhere, with the result that the economic aspects of the problems were not approached in an integrated way. The third point is the need for basing generalizations about a peasant economy on systematic, planned research of an intensive kind. Just as a sound policy in matters of public health or of agriculture must be founded on a great deal of factual knowledge, carefully accumulated by trained personnel working on the spot, so also policy aimed at improving the economic conditions of peasant communities should rely to a large extent on a body of factual inquiries made by people whose special job is to make such inquiries and who have been trained for the work. The fourth point is the importance in such work of collaboration between two or more spheres of interest or scientific disciplines. Research into peasant economic systems—for which even ordinary statistics are usually not obtainable—demands special techniques for collecting the information. Some kind of fusion between the theoretical apparatus of the economist and the field techniques of the anthropologist seems called for, as one type of attack on the problems. This book is an essay in this kind of approach. Whatever may be its short-comings, I do think that it shows the need for more work of a similar nature.

At the present time much attention is being given to the ways in which anthropology can be of help in dealing with administrative problems. I think it will be generally agreed that no scientist should have his work dictated to him by considerations which appear to be purely " practical " ; that he should be free to choose what he regards as the most significant lines of research to elucidate general principles in his field of study. At the same time it is reasonable to ask in these days that the scientist should indicate what, if any, are the practical implications of the work that he is doing. I have accordingly attempted to point these out in the first and last chapters of this book. But a warning is necessary here. " Science " has now acquired a kind of mystical value for the layman. There is a tendency to put to the scientist almost any type of problem, ask him to take a quick look at the conditions, and to give an answer which will help to solve it almost at once. Often it is not realized that to give an answer which is worth anything—particularly on a social question—a mass of basic information has to be systematically collected, and this may take months of inquiry. It may demand a study by several people in a number of different areas. If therefore the results of one man's work may seem inconclusive and disappointing from the practical point of view, this is often not an argument for abandoning scientific inquiry, but for having much more of it. Moreover, even a large-scale scientific investigation cannot necessarily result in the solution of a pressing social problem— some of the conditions for settling it may lie outside the scientist's province. To raise the standard of living of a peasant society, as in Malaya, for instance, involves major questions of social and economic policy—on capital expenditure ; on the relations which it is desired to maintain between Malays and Chinese ; on the plantation system and on industrialization ; on the aims of colonial government itself. The scientist may have his ideas about what should be done, but he is usually in no position to implement them, even if he were competent to do so.

The first draft of this book was written in 1940–1, before the entry of the Japanese into Malaya ; its completion has been delayed by war conditions. The bulk of the book has therefore been left as a description of how a Malay fishing economy ordinarily functions. It is probable that it is not simply of historical interest, but will serve as a general guide to the kind of conditions that will operate when a peace-time regime is once more established. It will, I hope, give some idea of the adaptative nature of these peasant economic systems, of the value of their traditional forms of coöperation, and of the claims of such types of society to survival in the face of pressure from forces which threaten to disrupt them while offering them no alternative forms of communal existence. I hope also that it may be possible to follow this work by another now in preparation, on Malay peasant agriculture, which will carry the economic analysis into the complementary field.

In carrying out the field-work on which the greater part of this book is based, assistance was received from many sources. Acknowledgement is due in the first place to the Committee of the Leverhulme Research Fellowships, whose award to me made the investigation possible, and to whom I am deeply indebted for their help. I am indebted also to the Governors of the London School of Economics for their liberal help in giving me a year's leave of absence from teaching duties. To Messrs. Alfred Holt, Ltd., and to the management of the Federated Malay States Railways thanks are also due for the assistance they gave in travel facilities to my wife and myself.

Of the many people from whom we received help in Malaya it is impossible now to think without emotion—some are missing ; others are in Japanese hands ; and others are scattered through many countries. To mention by name all who gave us hospitality, information and other help is not possible, but acknowledgement, however inadequate, must be made to a few : to Sir Alexander Small, formerly Colonial Secretary, Singapore ; to Captain A. C. Baker, M.C.S., formerly British Adviser to the Government of Kelantan ; to G. A. de C. de Moubray, Esq., M.C.S., formerly Acting British Adviser to the Government of Trengganu ; to S. W. Jones, Esq., M.C.S., formerly Colonial Secretary, Singapore ; to G. S. Rawlings, Esq., M.C.S., District Officer, Kota Bharu ; to Anker Rentse, Esq., of the Kelantan Civil Service ; to Tĕngku M. Mahyiddeen, then Director of Education, Kelantan ; to Nik Mustapha bin Nik Mahmud, then Assistant District Officer, Bachok ; to Dato' Jaya Perkasa, O.B.E., then Principal Officer of Customs, Trengganu ; to David Somerville, Esq., M.C.S., then Assistant Adviser, Besut ; to Mrs. Anthony Walker, of Kota Bharu ; to W. A. Bangs, Esq., of Kuala Pergau ; to W. Birtwistle, Esq., Director of Fisheries, S.S. etc. ; to Dr. F. N. Chasen, Director of the Raffles Museum ; to Noel Ross, Esq., M.C.S., formerly Assistant Adviser, Besut ; to Dr. J. L. Strachan, Medical Officer, Tampin ; to Dr. H. J. Lawson, Chief Medical Officer, Kelantan. Lastly, but by no means least, to Dr. H. D. Noone, Director of the F.M.S. Museum, Taiping, and to D. W. Le Mare, Esq., Assistant Director of Fisheries, S.S., I am indebted for much technical and other help.

I am also indebted, for advice before going to Malaya, to A. S. Haynes, Esq., C.M.G., and to W. W. Skeat, Esq., from whose long experience of the Malay peasant I derived great help. My thanks are also due to Sir Richard Winstedt, K.B.E., C.M.G., unrivalled in his knowledge of Malay history and language, for reading the original manuscript of this book and making various helpful suggestions.

Work of this detailed kind would have been impossible without the active coöperation of the people of the community studied. To Awang Lung, Awang Muda, Awang-Yoh and Pa' Che Mat of Pantai Damat ; Awang-Me' Sari of Paya Mengkuang ; the late To' Mamat Mindo' of Kubang Golok, and many others I owe much for patient explanation and help in gathering information. To the assistance of my wife at every stage of the investigation no adequate acknowledgement can be made ; all that can be said here is that we worked together on many aspects of the inquiry, that her special studies of the local domestic economy covered a field that I could not have entered alone, and that my friendly relations with the village people owed a very great deal to her presence.

RAYMOND FIRTH.

Cambridge,
June, 1944.

PREFACE TO SECOND EDITION

In issuing a second, revised edition of this book my aim is much the same as before—to offer a contribution both to economic anthropology and to the economic history of Malaya. As far as I know, though the major field research was carried out in 1939–40, this is still the only detailed analysis of production, marketing and distribution in a Malay fishing community, related to community structure and values. But modern anthropologists have become increasingly interested in studies of conservatism and change, linked with a clearer recognition of the significance of the time dimension of their subject. When my study was made this Malay fishing industry and community had just begun to be affected by the War, though the country had not yet been invaded and come under Japanese occupation. Even before the War some changes had seemed imminent, with the development of more advanced technology and increase of contact with the world outside. As I emphasized at the end of the first edition of this book, radical economic, social and political modifications in the society of Malaya were inevitable, with consequent effects upon the conditions of the fishermen.

In 1947 I revisited Malaya for three months to examine generally the social and economic conditions of the Malay peasantry, particularly with a view to suggesting priorities in problems for research. The results of this survey were published as a *Report on Social Science Research in Malaya* (see Bibliography). Unfortunately, the onset of the Communist " Emergency " set back most of these plans for a number of years, and in the Malay field only the studies by Djamour and to some degree Swift were the result. As part of my own survey in 1947 I was able to live for about a fortnight in the area my wife and I had investigated formerly, and to see how the local fishing industry and the community had accommodated themselves to the Japanese occupation and then to the complex conditions of re-establishment of ordinary civilian life.

In 1963 my wife and I were able to return to the same area. After a quick look at conditions on some parts of the Kelantan–Trengganu coast, we worked for about six weeks in the same village as before, living in a Malay house rented from the widow of Po‘ Su (p. 175), a fisherman well known to us. With friends of twenty-odd years before again as neighbours, and using again the vernacular (if rather awkwardly), we found it relatively easy to collect a great deal of information on recent local conditions, and to note the main social and economic changes which had taken place since our earlier stay. In particular, while my wife studied again the Malay domestic economy, and the broad changes made by education and the social services, I concentrated on detailed observation of the local fishing industry and fish marketing. To back up all this material with some quantitative data we repeated a sociological census, surveying a set of households occupying about half of the same area as we had examined in 1940. This part of the work was carried out with the very effective help of Abdul Aziz bin Ismail and Wan Yusoff bin Wan Ismail, undergraduates from the University of Malaya and both from Kelantan; to them our most grateful thanks are due.

For the new edition of this book the central chapters (III–X) dealing with the particular community studied, in the vicinity of Perupok, have been kept in their original form. They constitute a body of empirical record and a mode of analysis which can be used for comparison by any future student. But many of the general observations in Chapters I and XI, and some at the beginning of Chapter II, have now become well out-of-date and have been revised. The present version is designed to bring out more clearly the structural aspects of a fishing economy and to examine their relation to the modern social situation in Kelantan, with particular reference to the rural Malay society of the coastal area which we studied intensively. In the course of the revision Chapter XI has been considerably expanded. The original appendices have been retained ; the bibliography has been modified and enlarged.

Other claims have interfered with the completion of the work on Malay peasant agriculture referred to in the original preface. With the bulk of new material now available on the subject there is little point now in producing the work in the form then drafted, though I hope it will be still possible to publish some of my field data as a contribution to Malay social and economic history. The domestic side of the Malay peasant economy, together with other important social and economic aspects of the life of the people studied by us in 1939–40, was dealt with by my wife—Rosemary Firth, *Housekeeping Among Malay Peasants*, London School of Economics on Social Anthropology, no. 7,

London, 1943. This work, which she has now revised, should be read in conjunction with the present analysis. (For printing convenience, this work is referred to unchanged in the body of the text of my book ; the new edition of 1965 should be consulted for further details.)

For facilities in carrying out my study in 1963 I am very much indebted to grants from the Research Committee of the London School of Economics and Political Science, and from the Central Research Fund of the University of London. Our work was also greatly helped by a grant to my wife by the London Committee of the London-Cornell Project for Social Research in Southeast Asia.

In carrying out my further studies I have as before received much help from people in Malaya. Many of those named in the original preface are no longer living. But I would like to pay tribute to the late Anker Tentse and to the late Tengku M. Mahyiddeen, Kelantan friends who were both of great assistance to me in 1947. I am also glad here to reaffirm my obligation to D. W. Le Mare, former Director of Fisheries, Malaya and Singapore ; to Noel Ross, M.C.S. ; to Haji Mubin Sheppard and to Dato' Haji Yusoff Bangs who did much to smooth my path and provide me with information on various occasions.

For field research in 1963 I and my wife owe a particular debt to Professor Ungku A. Aziz, head of the Department of Economics, University of Malaya, Kuala Lumpur, and to Inche' Mokhzani bin Abdul Rahim, Lecturer in Rural Economics in that Department, who helped us greatly in the planning of our programme and in many other ways. I am greatly obliged also to the Director of Fisheries, Malaysia, for kindly supplying me with copies of statistical returns for 1963 from his Department. For more local advice and assistance I was indebted also to : Dato' Hashim bin Haji Mohamed, State Secretary, Kelantan ; Nik Hussein bin Nik Ali, President, Sessions Court of Kelantan ; Inche' Nasir bin Yusoff, Assistant State Secretary, Kelantan; Inche' Abdul Rahim bin Abdul Kadir (then Secretary Town Council, Kota Bharu), now Assistant State Secretary, Kelantan ; Inche' Admad Noordin bin Wahab, District Officer, Bachok ; Tengku Adlin bin Tengku Mahmood (then Assistant District Officer, Bachok), now Assistant State Secretary, Kelantan ; Inche' Ayoub bin Zakaria, Social Welfare Officer, Headquarters, Social Welfare Department, Kelantan ; Inche' Abdul Salam bin Haji Ibrahim, Coöperative Marketing Officer, Kelantan ; Inche'

Abdul Malim Baginda, State Welfare Officer, Trengganu ; Syed Mansur bin Syed Salim, Coöperative Marketing Officer for the East Coast, Kuala Trengganu.

I also benefited considerably from the opportunity to present sections of my material for discussion in a field seminar on social research in Malaya and Indonesia which took place in Kuala Lumpur, September 9th–11th, 1963. This seminar I organized with the help of A. R. Mokhzani under the auspices of the Cornell Committee of the London-Cornell Project for Social Research in Southeast Asia, to whom I am indebted for the invitation and facilities.

I wish to acknowledge also the kindness of Inche' Abdul Rahim bin Abdul Kadir in reading and commenting upon parts of my revised text, and the help of Mrs. D. H. Alfandary generally in the preparation of this new edition.

Finally, as before I am greatly indebted to my wife, with whom I had constant fruitful discussion of the problems examined in this book, and whose friendships in the village lightened immeasurably the fieldwork task. Without the help of our Malay friends—our housekeeper Che' Minoh, her son Che' Nur, her niece Che' Limoh, Che' Awang Lung, Che' Awang bin Awang Hitam and his wife Che' Yoh, this book could not have been re-issued in its present form. Parallel to it is a new edition by Rosemary Firth of her *Housekeeping Among Malay Peasants*.

RAYMOND FIRTH.

London,
    December, 1964.

MALAY FISHERMEN

Their Peasant Economy

CHAPTER I

FISHING ECONOMY AND MALAY RURAL SOCIETY

For Malays in both town and country rice and fish are staple foods. The primary producers of these commodities are therefore of great importance to the nation in which Malays are a major component, and at the present time a high proportion of Malays are still engaged in these occupations. Proper studies of their economy and society should therefore be available, both as a matter of scientific interest and as a basis for planning any attempts to help them to take fuller advantage of their resources and improve their position. It is hoped that this study will be a contribution to a better understanding of Malay fishing economy and Malay rural society.

The field chosen for study has been the coastal Malays of Kelantan, the north-easternmost state of the Malay Peninsula, in the Federation of Malaysia. As a general background the first two chapters of this book discuss some of the salient characteristics of Malay rural society, especially in the coastal area, in Kelantan ; the main features of Malay marine fishing ; and the particular situation of the fisheries in Kelantan and Trengganu. The body of the book then deals with what is in effect an historical case study in economic anthropology, of one fishing community analysed in detail in 1939–40. Finally, an account is given on a comparative basis of recent developments in the same community, to bring out some of the underlying social and economic forces that have been at work during the past generation.

MALAY FISHERMEN OF KELANTAN

Ethnically the structure of the population of the Kelantan coast is overwhelmingly Malay. In the whole state, according to the census of 1957, the last taken, the " Malaysians ", i.e. primarily Malays, numbered over 460,000, or over 90 per cent., out of a total population of just over half a million. The greater part of this population was a rural one ; out of the half a million, only 115,000 people lived in (gazetted) areas of 1,000 population and over, and less than 50,000, or less than 10 per cent., in urban areas of 10,000 population and over. The capital and largest town, Kota Bharu, had in 1957 a population of only just over 38,000.[1] In the thickly populated coastal district of Bachok (where our particular study was located), with an area of 100 sq. miles and a total population of 51,500, Malays numbered 50,000 or about 97 per cent., nearly all locally born ; the rest were mainly Chinese.

Kelantan is predominantly an agricultural state,[2] with rice as the principal crop and with some production of rubber and copra for export as well as of fruit and vegetables and forest products for local consumption. But about 6,500 Malays, or 3 per cent. of the Malay male population, is engaged in the fishing industry as fishermen, and many others are employed as dealers, curers or carriers of fish.

The economy of a fishing community, while it shares many of the general characteristics of an agricultural economy, has some special features arising from its specific technical conditions. By contrast with agriculture they may be summarized as follows.

Whereas agricultural yield is largely seasonal, with long gaps during which no direct income is received,[3] the yield from fishing is largely one of daily increments. Since each day's labour commonly gives its return on the spot, with no question of waiting as is inherent in the growing of crops, the fisherman's planning of production can take different forms. Long-term planning is still essential for the accumulation of technical equipment and other capital and for preparation against seasonal changes. But there is more room for short-term planning. In particular, there is more opportunity for the entry of the marginal worker whose main interests lie elsewhere. Fishing can be used like casual day labour, to give an agriculturalist a little extra food or cash immediately, or to fill a gap in another task.

The problems of planning are affected in other ways also. The agriculturalist, receiving the majority of his crop in bulk at one time, can plan in advance, decide what he will retain and what he will sell, estimate his margin of saving against his con-

[1] Federation of Malaya, *Year Book* (1962), pp. 37, 39, 463.

[2] See Raymond Firth (1943), pp. 193–205. For some modern accounts of Kelantan, see E. K. C. Dobby (1957), pp. 3–42 ; R. E. Downs (1960). A useful summary of the major features of Malay society in general is given by M. Freedman and M. G. Swift (1959) ; a detailed study of an agricultural Malay society and economy is given by M. G. Swift (1965).

[3] This refers primarily to cereal cropping ; the situation is different for example with rubber. See M. G. Swift (1957), p. 336.

sumption months ahead, and narrow or expand the one at the expense of the other in terms of his bulk supplies in hand. The fisherman, with his daily income, often very irregular, must calculate against greater uncertainty. He must think of saving in smaller increments ; he cannot set aside so much in bulk and divide for daily consumption the remainder into appropriate fractions. For both, saving lies in abstention. But while for the agriculturalist it is the problem of abstention from drawing on a store already there, for the fisherman it is the problem of abstention in order to accumulate a store. Each has its difficulties, but the type of foresight demanded differs. Again, the question of storage is different. The agriculturalist's seasonal crop normally needs more space, but the fisherman's catch, if it is to be stored, needs more labour and outlay in equipment for its preservation. Hence the tendency to a greater development of middlemen who take these matters off his hands.

The nature of the yield in relation to nutrition and food habits also tends to give a different economic bias to each occupation. The agriculturalist's main crop is usually also his staple food, but the fisherman does not live mainly on fish. He must also have rice or similar vegetable food as his staple. Hence for him exchange of his product, or part-time agriculture, is a necessity ; full-time fishing, therefore, tends to be more definitely associated with an exchange economy than does full-time agriculture. The nature of the production unit is different also. In agriculture there is more scope for complete family activity at all stages ; in fishing the work at sea, partly by tradition but mainly by physical necessity, is primarily restricted to men— though women and children can participate in the secondary processes on shore. Again, there is more scope in fishing for permanent day-to-day coöperation in moderately large groups, with this there is more tendency for complex systems of distributing the earnings to arise. Finally, there are differences in the opportunities for investment. Investment in agricultural land has a permanency not found in fishing enterprises ; and fishing-boats and gear, though perhaps as durable as agricultural implements and cattle, are, on the whole, more liable to sudden damage and loss. Hence the provision of capital is apt to involve risks of a different order, and to attract investors of a different type.

In Kelantan, the people directly associated with fishing, especially the actual fishermen, are known as *orang ka laut*, the folk who go to sea ; they distinguish themselves from the *orang darat*, the inland folk, who from their point of view are land-lubbers. The fisherfolk regard themselves as specialized in the life of boats, nets and fishing ; they contrast themselves with the rice-planters and say they are not adept in agriculture. They own neither buffalo nor plough, the symbols of the culti-vator, and only rarely an ox or cow. There is some notion here akin to the mediaeval idea of a " calling " to which one is bound. One fisherman complained to me of his hard lot, but when I asked him why he did not go and tap rubber answered, " If I did how would the inland folk get fish ? When I die my children too will go to sea, just as when *orang darat* die their children too will work rice."

Yet while technologically different the fisherfolk are not rigidly separated economically from the agriculturalists. Rural Malays live in small aggregates, rather than in isolated dwellings. *Kampong*, meaning literally a grouping or gathering together, is the term commonly applied to a village or hamlet, and includes not merely the cluster of buildings but also the mass of coconut palms, fruit trees and other appurtenances of the settlement. Our detailed study in 1963, as in 1939-40, comprised major sectors of three contiguous *kampongs*, and though the boundaries of our survey area do not correspond with these, the location is referred to for convenience by the name of one *kampong*, Perupok.[1] A *kampong* is not normally an administrative or religious unit, but it is a social unit with some degree of solidarity and neigh-bourly feeling. In the area we worked 70 per cent. of the adults and practically all the children were locally born.

In this small area of about 180 households, more than 110 had no rice lands, but 60 odd did, getting from them annually in the region of 12,000 gallons of rice. (Comparable figures for an earlier period are given on p. 284.) About two-thirds of this was gained from leasing the land to be worked by agri-culturalists, especially in cases where the fields were not local, but some distance away, usually inherited. But about one-third of the rice was produced personally by about a score of house-holds. People outside were hired to do the ploughing, but all the planting, weeding and harvesting was done, in nearly every case, by men—usually with their wives' help—who were for the greater part of their time fishermen. In a couple of cases where

---

[1] Our survey area is thus not the same as " Perupok Village ", a local government unit. Perupok is also the name of a *daerah*, a larger administrative sub-division.

the rice planter did not go to sea, his or her son was a fisherman. Moreover, the fishermen had economic relations with agriculture in another way—through the coconuts from the palms growing alongside their houses. The nuts for these were collected periodically, five or six times a year, by copra makers, using monkeys to climb the palms for them. Some fishermen also were vegetable planters for the market.

Thus the fisherfolk of the Kelantan coastal area live side by side with people of other occupations, including agriculturalists ; have economic relations through leasing land or its product to them ; and do in some limited cases plant rice themselves. Moreover, they have elaborate and intimate social relations with the agricultural sector of the population. Not only have they in nearly every case close kin in the inland rice plain, whom they visit occasionally and whom they invite to weddings and other gatherings ; they may intermingle fishermen and agriculturalists in their own families. For instance, the wife of the man quoted above came from a fisherman's family and two of her mother's brothers were also fishermen ; yet another, their elder sibling, was a rice planter and vegetable planter of ordinary agricultural style.

It was for this reason that in the original edition of this book I represented these fishermen as forming part of a peasant economy : with relatively simple, non-mechanical technology ; small-scale production units ; and a substantial production for subsistence as well as for the market. The major features of such an economy may be outlined.[1]

Such a peasant economy is not necessarily either a closed economy or a pre-capitalist economy in the literal sense of these terms. It commonly has external market relationships. There is production of a limited range of capital goods, with some degree of individual control over them ; there is some lending of them out to people requiring them, and interest in commodity or money form may exist as an economic category. There may even be some persons whose major economic rôle is the provision of such capital goods for the processes of production. But the economy does not function mainly by its dependence on foreign markets, nor do its providers of capital constitute a separate class, nor has its elementary capitalism developed any concomitants of extensive wage labour and complete divorce of the worker from control of the means of production.

[1] See also Raymond Firth, *Elements of Social Organization* (3rd ed., 1961), pp. 87–92.

The technical factors in this peasant production—whether tools and traction in agriculture, or boats and gear in fishing—are relatively simple. The equipment is not mechanized, and it is not integrated with any complex scheme of ideas in which mechanization is envisaged as a desirable and attainable end. The actual production units are small, even though the community as a whole may be large. Though in function the agents of production can be recognized according to ordinary economic analysis, in form they are often not clearly separable into the categories which commonly appear in a capitalist organization. Instead of the familiar pattern of capitalist-rentier, organizer of production, and worker, in fairly clear-cut divisions, the one set of persons often fulfils all three functions. The capitalist works with his hands ; the labourer provides part of the capital ; and both participate in organizing the activity. Their functions grade almost imperceptibly into one another, as far as personnel is concerned. Linked with this, the system of distribution tends to take on a form quite different from that of a capitalist economy, where, institutionally, wage relationships assume great importance. In a peasant economy the manner of apportioning the product of the economic process is in some cases not very clearly defined in an overt way—as when the producing unit is an individual family ; in other cases it may be laid down by definite rules of custom, and be quite complex (see pp. 236–54). In all cases the broad principle operates, that units of resources are evaluated, and receive direct or indirect equivalent for their participation in production. But the difficulty which is recognized in the analysis of a capitalist economy, of dividing these equivalents into clear-cut categories of rent, interest, profits and wages, is much greater in that of a peasant economy. Again, inequalities in the possession of capital goods are often levelled out or at least lessened by free borrowing or the exercise of communal rights, on a scale or of a kind not ordinarily operative in a capitalist economy.

Economic ties between peasant producers tend to reach out into spheres beyond their immediate common interest in the act of production and its associated rewards ; similarly also with the ties between producers and consumers. The factors which they take into account in making their decisions include some of a very long-term character—such as permanent kinship obligations, or the network of village coöperative services. In one sense their economic relationships may be said to have a more

personal content than do those of parties to an economic transaction in a Western capitalist society ; in another sense they
may be said to have a more social content. The choices of the
parties are governed by their total relationship to one another,
not simply by their immediate relationship in the exchange
situation. They are thus often willing to forgo some elements
of present satisfaction in favour of future satisfactions, even though
these be less well-defined. Such behaviour is sometimes regarded
by a Westerner as "non-economic", but the term is misapplied.
The economic aspect is still there, but the decisions are taken on
a wider range of preferences than is apparent to the observer.

But a classification that could be justified twenty years ago
is more difficult to apply today. One of the reasons for this is
political : " peasant " has come to be regarded in some of the
new nations as a term of disrespect, implying not merely a low
economic level and small-scale semi-subsistence production, but
also a low cultural, even intellectual position. It should hardly
need to be said that in the extensive literature by historians,
anthropologists and other social scientists no such meaning was
intended by the term. But leaving this issue aside, there are
more objective reasons for modifying this categorization as
applied to the Malay fishermen. As will be shown later, modern
Malay fishermen in Kelantan have to a considerable extent
adopted mechanization of their fishing fleets, and side by side
with this has come a change in the capital structure of the local
fishing industry which has affected also the relation of the
ordinary fisherman to the control of the enterprise in which he is
engaged. The result is a closer approximation of much Malay
fishing on the Kelantan coast to business enterprises of other
kinds rather than to Malay agriculture. It is no longer so appropriate therefore to label the economy a peasant one now as it
was in 1940. But the label has been retained in the title of the
book as a description of a situation that obtained a generation
ago.

But what about the rural coastal society in general ? Whether
or not one uses the description of " peasant ", how far may it
be characterized in terms of " folk-society " or " folk-culture " ? [1]

[1] One of the most sensitive analyses of this is by Robert Redfield, *Peasant Society
and Culture* (Chicago, 1956) ; cf. also the very useful categorization by Eric R. Wolf,
" Types of Latin American Peasantry ", *American Anthropologist*, vol. 57, pp. 452–71
(1955).

RURAL SOCIETY OF COASTAL KELANTAN

The coastal community is still very much of a rustic community, remote from modern urban facilities such as a reticulated water supply and a corresponding sewage system. Electric light has been proceeding through the countryside, but has not yet reached the Bachok area, which uses kerosene pressure lamps for illumination on public occasions, as in the celebration of the Prophet's birth month. Modern transport in the form of buses, taxis and pedal-propelled trishaws have come to the area, and a school system has developed with most striking effect. But domestically and in their ordinary social contacts the Malays of the Kelantan coastal region still retain a great deal of their traditional culture.

Market relationships are of prime importance to these coastal people and are one of the most significant elements in their incorporation in a larger society. They rely on the market to dispose of the greater part of their catch of fish, and to supply them with not only their fishing equipment but also a range of consumers' goods such as cloth, tobacco, matches, crockery, lamp oil and many foods. It would be possible of course for these fishermen and their families to have such essential market relations and yet remain relatively isolated from the wider society, if they were served completely by an alien group of middlemen, who were the unique intermediaries with the outside world. But this is not the case with the Kelantan fishermen. They are served with consumer goods by local shopkeepers of whom a large proportion are alien Indians and Chinese, and even the local Malay shopkeepers are supplied by alien wholesalers. But the fish dealers are almost without exception Malays, and local residents at that, so that they spread a network of communications, of a social as well as of an economic order, over a great part of the more accessible inhabited areas of the state. In answering the question as to how far a coastal community composed largely of fishermen is to be regarded as having a society and culture of its own, clearly the character of the middlemen is highly significant.

Again, the fishing community does not rely solely on middlemen for its economic relations with the outside market. No coastal village in Kelantan is very far from a town—either Kota Bharu or Pasir Puteh (Fig. 2)—and though fishermen and their wives may go there only infrequently, these urban contacts for purchases or visits to kin are important to their social perceptions. They play no part in the sophisticated society of the Court officials and businessmen—though they may have relatives therein, with whom some contact may be maintained. But they go into the coffee-shops, the cloth shops, the goldsmiths and, above all, they frequent the market-place. For the market —the physical market—depends on the country-folk. It is normally the centre of the town—only with the growth of the towns and the raising of standards of municipal sanitation and traffic improvement is there a tendancy to shift the market away from the centre. There the country-folk bring their fruit, vegetables, eggs, basketwork and other products, the dealers bring their fish, fresh and dried, and there both urban and rural buyers assemble from an early hour in the morning, streaming in by bus, taxi, trishaw and on foot to the number of several hundreds or it may be thousands. In this urban environment of the market square and its surrounding shops fishermen and their wives may not be completely at ease. But it is an environment in which they can move freely, among people many of whose ways of life, language and jokes they share, so relations with the town are in supplement rather than in opposition to their own social sector.

Kelantan fishermen and their families are then enmeshed in a wider society. They have their own technological subculture, with which landsmen, particularly urban dwellers, are almost totally unfamiliar. But this is only in the technological field ; economically, politically, socially and religiously they are part of a larger universe. Their economic relationships will be described in some detail later ; here a brief outline will be given of the structure of their relationships in the other fields.

Politically, the Kelantan fishermen are enclosed within a state structure, itself a part of the larger unit of Malaysia. Kelantan, with the Sultan as Ruler, is divided for administrative purposes into eight districts (*jajahan*), and further into sub-districts (*daerah*) and parishes (*mukim*). The districts are controlled by officers of the Federation or State Civil Service, posted and transferred in the ordinary steps of a career. Under them the sub-districts and parishes are in the charge of salaried local officials of the Government (the *penggawa*). Assisting them in the control of affairs in the *mukim* are the *penghulu*, who are not Government officials, but are government-nominated and receive a small annual income. The *penghulu* do their work part-time, combined with some other occupation. The *penghulu* for the area where our study was conducted, for example, was an agriculturalist with a large amount of rice land, which he cultivated himself, also two buffaloes, six cattle and two wives. The post was not hereditary ; his father had not been a *penghulu*, though a rich man. His functions include registration of cattle, hunting up defaulters on land tax, making arrangements for permission to hold shadow-plays, spirit-medium performances and other entertainments, investigation of assault or other trouble in the villages. For these and many other public services he received in 1963 a salary of $240 a year, and he regarded himself as greatly underpaid—he alleged that $70 or $100 a month would be more equitable. A *penggawa*, and to a greater degree a *penghulu*, is the official with whom the ordinary rice cultivators and fishermen come most into contact. He is the intermediary between them and the District Officer, often a comparative stranger ; he has power of arrest if need be. He represents in one aspect the amorphous and mysterious *Kerajaan*, the authority of Government, but in another aspect he is a fellow-villager and perhaps kinsman. He is thus treated with a mixture of respect, fear, confidence, jocularity and even (behind his back) with that tinge of disdain apt to be felt for those who have gone over to the other side.

Outside the immediate administrative framework the Kelantan fisherman's acquaintance with political institutions is as yet not great. Party politics have developed in Malaya since the war, and the fisherman is able to vote in both State and Federal elections, as also for a local Council where such exists. But the functioning of government in Kelantan has been complicated in recent years by the fact that the majority party which controls the affairs of the State, the Pan-Malayan Islamic Party (PMIP or PAS), has had a very different policy in some of its aspects from that of the Alliance Party which supplies the Government of Malaya. This has resulted in some difficulties in regard to allocation of federal funds for economic development, and perhaps at times some divided loyalties for officials of the Federation stationed in Kelantan. But little of this has reached the ordinary fishermen, for whom the issues involved are often oversimplified or obscure. Political sophistication varies greatly. Typical of many of her sex, perhaps, was the woman who while very shrewd in ordinary affairs said only half-laughingly that all parties were alike to her—she did not understand them. Their respective graphic symbols for identification by the illiterate in the countryside were a boat under sail (Alliance), bull's head (Socialist Front), bunch of rice in grain (Party Negara) and crescent and star (PMIP), and this woman voted (she said) for the one whose symbol most appealed to her—the bundle of rice, which represented her staple food. But there are others, like the line fisherman who, nearly illiterate, could argue with political knowledge and skill, on the pros and cons of the setting up of the Federation of Malaysia, defending the Kelantan government's policy against what he conceived to be the dangerous inclusion of alien elements in the Federation.

What the political parties have done in Kelantan, as else-where in Malaya, is to provide new foci of social as well as political leadership. For the most part the Kelantan social system lacks some of the institutional framework for leadership, notably any general system of corporate descent groups. Negri Sembilan has its named matrilineages and clans, in one or other of which every Negri Sembilan Malay is a member. In Kelantan there is a royal lineage, that of the Ruling House, all the members of which bear the title of Tengku, by patrilineal descent. There are also what may be termed aristocratic lineages, not named and not of a highly corporate character, in which membership is patrilineal and which exist primarily for the perpetuation of title and status, and in some cases the transmission of office. These include the families of men who in the *ancien régime* would have been territorial chiefs holding their fiefs from the Ruler, and who in modern times have titles of honour conferred upon them and exercise ceremonial functions at the Court or play a prominent part in the councils of the State.[1] They also include families of Syeds, descended from the Prophet. They may be regarded as including too the families of men with the title of Raja, putatively descended patrilineally from an ancient royal house long since ousted from power, the members of which now live and work as ordinary agriculturalists and fishermen ; as also the bearers of other titles such as Nik and Wan, deemed to have originated in hypogamous marriages of women of royal blood with non-royal husbands. But the great body of Kelantan people, the *orang kebanyakan*, the multitude, have no such descent groups ; they operate a bilateral kin system alone. They are thus without one important source of built-in patterns of leader-ship. In former times the Court, the territorial chiefs and other

[1] For discussion of such " lineages " in the Western States in a wider political context see J. M. Gullick (1958).

aristocrats holding positions of power provided stimulus and incentive, aided by a system of patronage and clientship, of which the remnants were still visible in the coastal areas in 1940.

But now that the administrative apparatus of a modern state has replaced the system of government focused on the Court, the hereditary aristocracy tend to exercise power and provide leadership side by side with men and women of non-aristocratic origins, operating largely through the bureaucratic machine. Whereas in 1940 the number of children receiving education in Kelantan was still very small, by 1963 schools of various grades had been established throughout Kelantan, as elsewhere in Malaya, and literacy both in English and in Malay was rapidly growing. Social mobility is still perhaps not very marked, but there are distinct opportunities for bright boys and girls from rural areas to obtain higher education and governmental posts in a manner almost impossible twenty years before. The rise of these has been greatly facilitated by spectacular modern developments in education.

The State Legislative Assembly and even the local councils, through their control of expenditure, also offer avenues to power and leadership, which the search for votes provides incentive to use. One aspect of this is sometimes a tendency to division of authority, between the District Office, whose officials were formerly responsible for much local development work, and the central and local elective councils, whose sanction is now needed before the proposals can be put into effect and who in their turn may have conflicting proposals of more popular appeal. Hence if a State Councillor or a local councillor be resident in a coastal village the inhabitants are provided with a social personality whose opinions tend to have weight. Incidentally, somewhat unexpected in a state governed by an orthodox Islamic party, the performance of shadow-plays and of spirit-medium séances, both anathema to scholarly pious Muslims, has been allowed to continue—presumably because if they were forbidden the elected representatives of the people would lose votes, both locally and state-wise.

But if the secular leadership of the Kelantan fisherman has become more diffused, the spiritual leadership has remained firmly in the hands of the Muslim hierarchy. True, the spirit-medium performances have not been prohibited, as was rumoured they would be, but Islam seemed to play in 1963 a more obvious and formal part in the life of the fishermen than in, say, 1940.

Kelantan is not a theocracy, but as in the other Malay States, Islam is the State Religion, and the Sultan is the Head of Religion, his authority being delegated to the Majlis Ugama, the Religious Board.   This body is in charge of the religious treasury, the collection of tithes and religious head tax, is trustee of all property dedicated for religious or charitable purposes, is in charge of all mosques in Kelantan and gives rulings on points of religious doctrine.   For religious purposes the state is divided into districts and parishes (*mukim*, but not identical with the administrative units of the same name).   Each district has its central mosque (*masjid*) and each *mukim* one or more local mosques (*surau*). Each set of mosque officials is in charge of an Imam, whose duties include leading the Friday prayers and having some pastoral care of the congregation.

Islam is one of the most powerful bonds linking the fishermen together with other members of their local community into a social unit.   Not only do they attend the mosque regularly, and keep the fasting month of Ramadan, but they gather together at religious festivals, publicly or in private houses, and they may support a prayer-house (*surau*) or house for teaching the Koran (*madrasa*) in the immediate vicinity, going there to pray and helping in its upkeep.   Moreover, the bond of Islam links the members of the local community with other sections of Kelantan society and even further afield, as may be seen when visiting religious dignitaries come from outside the state to participate in formal chanting of the Koran at a coastal mosque to celebrate the birthday of Mohammed.   Islam, like politics and commerce, has benefited from the coming of the radio and the amplifier ; religious chants compete over the air with popular songs and pulpits wired for sound allow the words of the preacher to reach an almost unlimited audience of the Faithful.

In one respect, however, in the religious field there is still a division of authority—between the Imam, representative of orthodoxy and the hierarchy, and concerned for the care of souls ; and the spirit-medium healer (*bomor*) concerned for the care of the body, but operating through the soul.   The relation between them is asymmetrical, in that whereas the Imam as a rule opposes the efforts of the spirit-medium and declares him to be trafficking with evil powers, the medium respects the Imam and declares himself a good Muslim, assisting the work of Allah. The Kelantan fishermen are not alone in supporting both. Like the Kelantan country-folk in general, they believe firmly

in the existence and the power of spirits of various types, some to
hurt and others to heal. In time of bodily or mental stress
then, either before trying Western medicine or after it has
proved unsuccessful to deal with some refractory illness, they
resort to the spirit-medium and follow his injunctions. Conflict
with the orthodox religious authorities does not appear to develop
very often, possibly because these authorities themselves in the
villages are either not completely convinced of the falsity of the
medium's claims or are unwilling to move frontally against public
opinion, which for the most part tolerates and shares the medium's
views. It is in this particular field above all that the traditional
elements of Kelantan folk-culture still display themselves.

GENERAL CHARACTER OF MALAY SEA FISHING

I turn now to a consideration of Malay sea fishing in general,
as an introduction to the analysis of the situation in Kelantan.

In Malaya, according to the census of 1931 (the last before
our original study was made) there were over 36,000 Malays
employed in fishing. By the census of 1947 the number had
risen to about 41,000. By 1963 their number seemed to have
slightly decreased, the official figure being nearly 36,200.[1] But
in each case the figure does not include the considerable number
of men who divide their time between fishing and agriculture,
nor, despite the fact that it includes a small number of women,
does it take acount of the fishermen's families, fish dealers and
carriers of fish, who depend on the industry for a livelihood.

The kinds of equipment and technique used in fishing in
the waters around Malaya show considerable variety.[2] Nets
include : seines (hauling-nets, including purse-nets) for use along
the shore or at sea ; drift-nets ; gill-nets and lift-nets (ground-
nets). There are several varieties of all these, with different size,
weight, cord and mesh. Though usually much smaller than
analogous types in Europe, these nets may be 100 fathoms or
more in length. Small hand-nets include casting-net and push-
net (scoop-net). In Malaya in 1963 there were about 500 lift-
nets, 2,200 seines, 5,000 gill and drift-nets, 2,300 bag-nets and
500 push or scoop-nets. Netting is most highly developed on
the east coast, which is exposed to the force of the north-east

[1] Federation of Malaya, *Year Book* (1962), p. 281 ; statistical return, Director of
Fisheries, Malaysia (1963).
[2] For description see M. L. Parry (1954) ; also T. W. Burdon (1955), pp. 10–23 ;
G. L. Kesteven (1949), pp. 47–57 ; D. G. Stead (1923).

IIA   HAULED UP DURING THE MONSOON

Fishing boat of kolek lichung *type.   The bow, with its crutch* (bangar) *to hold spars, is towards the camera.*

IIB   " PRAYERS OF HOPE " (*Sěmbahyang Hajat*)

The service of prayer in the boats before the new fishing season begins, to promote good catches. Those taking part are men with some religious learning (orang lebai)*; they are rewarded with a meal and a few cents each.*

IIIa   A MASTER FISHERMAN

Awang Lung, a lift-net expert (jurusĕlam takur) *of Pantai Damat, Kelantan.*

IIIb   AT THE END OF THE DAY

A fisherman goes home with his kit and a pair of Spanish mackerel (tĕnggiri) *on the end of his paddle.   The woman has small fish in the baler.*

monsoon and has few good harbours, though long stretches of clear sandy beach. Traps of many kinds are used, from small portable ones (Plate VIIA) to fixed palisaded constructions of stakes and rattan. These last, which are sometimes very large, are especially common around Singapore and in the Straits of Malacca, where there is less exposure to the monsoon than on the east coast. In 1963 there were in the States of Malaya (excluding Singapore) about 5,000 small traps and pots, over 1,200 small stake-traps and about 670 large stake-traps. These last were mainly in Johore. Hand-line, rod and line, and long-line (both baited and unbaited) are also used. The long-lines, like the nets, are much smaller than in Europe, being only about 200 fathoms long instead of several miles. In 1963 there were about 3,000 lines licensed in Malaya.

Most of these methods necessitate the use of boats, which until recently were mostly shallow undecked craft, sailed to and from the fishing grounds and then handled by paddles. While the large areas of shallow seas around the coasts facilitated fishing, the boats used to depend very much on the wind and had a comparatively limited range. But the fisheries of Malaya have for some years been in a transitional state. Until about 1950 there were very few powered fishing craft in Malaya, but by 1963 out of a total of about 22,700 registered fishing boats, about 10,500 were powered by either outboard or inboard engines. In the early stages of mechanization outboard engines, which could be used in the traditional craft with little modification, were installed mainly, but these are being replaced by inboard diesel engines, which are more reliable, more powerful and more economical. Whereas in 1955 inboard engines made up only about 13 per cent. of the total number of engines in use, by 1960 the percentage had risen to 44 per cent., and by 1963 to 60 per cent.[1] (In 1963 whereas boats with outboard engines (*prahu sangkut*) were still in use at places such as Beserah in Pahang, inboard engines (*prahu motor*) were generally in vogue to the north.) These modern developments are likely to convert much of Malayan fishing from operations in near coastal waters to much more far-ranging efforts.

Each type of fishing demands knowledge of local conditions, considerable skill, and often arduous work for small return. The element of chance plays a large part in the result, but over

[1] Federation of Malaya, *Year Book* (1962), p. 28 ; statistical return, Director of Fisheries, Malaysia (1963).

and above this the margin between success and failure is frequently not great ; a small error in judgement or skill, and the shooting of a net, the setting of a trap or the operation of a line is in vain.   Malay fishing is a traditional occupation, and faced with these risks of nature's variability or man's fallibility the industry has developed a respect for certain specialist accomplishments and rules of behaviour in the work, which are believed to minimize the dangers.   They are especially noticeable in forms of coöperative fishing such as the handling of a large net, where a slip by one man may threaten the success of the whole crew. In this work the direction of affairs is commonly entrusted to an expert, often given a special title, who is apt to combine technical knowledge and ability with some knowledge of the ritual procedure believed to be necessary in order to propitiate spirits and attract fish.   Associated with this may be certain ritual observances by the crew, as, for instance, the avoidance of the use of the names of some specified land animals when at sea—a custom which obtained not only in Malaya but also, for example, among the Achehnese and on the north coast of Java.

The manufacture of some fishing equipment is commonly a local activity, forming a set of ancillary crafts which yield a considerable income to women and other members of the community who do not go to sea.   But small and isolated as they often are, the various fishing communities are not autonomous economic units.   The coastal areas provide few of the materials they require, and they import many supplies, in all stages from the raw form to the fully made-up article.   Take, for example, the east coast of Malaya, which until recently has been comparatively remote from the large commercial centres.   Bamboo for fish-curing trays, baskets and net-drying rests ; rattan for occasional fish-traps and many kinds of lashing ; pandanus leaf sheets for covers ; resin for caulking boats ; timber for boat-building, are all brought from inland districts.   Mangrove bark for dyeing nets is imported from the west coast.   Ramie twine for certain kinds of nets, and even some ready-made nets of this material, were imported before the war, ultimately from China, and from China too came large amounts of cotton yarn to be manufactured locally into nets and hand-lines.   Nowadays both ramie and cotton nets, especially drift-nets, are increasingly being replaced by nylon or other synthetic materials, which are imported ultimately from abroad.   The outboard and inboard engines are also imported, as is their fuel and paraffin for torches, or, latterly, pressure lamps for night fishing.   Paint, copper nails and various other accessories all come from outside.   Though all the boats are made on that coast their building is concentrated mainly in recognized centres, and there is much buying and selling of boats all along the region ; Kelantan fishermen even import some boats from Patani in Siam.   And though there is still some local manufacture of fish-hooks from brass and iron, and casting sinkers from lead, the raw materials are imported, and nearly all fish-hooks now are brought in ready-made from Europe.   In other fishing regions the balance between local manufacture and ready-made import, between use of local materials and of foreign materials, may differ.   But the general picture of a busy intra-regional trade in supplies, and of considerable dependence upon external sources, is much the same. This means that to fill the technological requirements of the fishermen, trading middlemen are necessary, with contacts in the importing centres outside.   It also means that quite apart from the marketing of their fish, the fishermen are affected by external price fluctuations in certain important classes of commodities.   This is all the more true if the demand of the fishermen and their households for consumers' goods such as rice, clothing, crockery and lamp oil is taken into consideration.

One may be tempted to think of these Malay fishing communities as requiring little capital and having their labour as their main investment.   This would be a grave misconception. The amount of fixed capital involved was very considerable, even in 1940, if it be measured in relation to local income levels. A large boat and net might well cost in 1940 up to $500, a sum far beyond the means of most ordinary fishermen, and demanding perhaps six months' to a year's earnings, even from a successful man already in possession of much large-scale equipment. By 1963 prices had multiplied several times even for the same equipment.   But making allowance for changes in technology, an analogous investment might then well cost in the region of $40,000 (see Chapter XI, p. 312).   For the period 1938–9 I had put the investment in fishing boats and gear through Malaya (including Singapore) at perhaps $2½ million ; this included a large amount of investment by Chinese fishermen.   Nowadays, the total comparable investment, even by Malays alone, must be many times more.   (For a rough calculation on a sector of the Kelantan coast see pp. 312–13.)   Apart from the initial outlay on equipment, there is also the cost of overheads, as for engine fuel and oil, net dye, twine for repairs, paint and caulking materials for boats and upkeep of the vans which carry the fish to market. To meet these considerable liquid capital is needed. All such investment, even by Malay fishermen alone, and even taking account of the fact that much of the equipment is bought on hire purchase, must certainly amount to millions of dollars. Even nowadays too little information is available to enable one to generalize precisely on the ways in which all this capital is found, or to say just how far and where there is inadequacy or shortage of capital. It is clear that some capital is the result of personal saving, some is borrowed from kinsfolk and other Malays, often in complex transactions. But shortage of capital among the fishermen is illustrated by the fact that nowadays, as in 1940, many of them work with equipment owned or financed by Chinese merchants.

The variety of fish that can be taken in these waters is considerable. Maxwell gives a round figure of 250 species of valuable marine food fishes for Malaya, and if species of crustacea and mollusca were added as well, the total would be much larger.[1] Of the pelagic fish, feeding near the surface, mostly in shoals, the most important are types of herring (such as dorab or wolf-herring, shad, sprats and anchovies) and of mackerel, horse-mackerel and tunny. Of demersal fish, normally feeding at the bottom of the sea, the most important include jewfish, sea-bream, sea-perch, snapper, grey mullet and flat-fish. Sharks and rays are also taken in quantity. In addition to fish, there are also prawns, shrimps and crabs. Many kinds of fish are caught in only small quantity. But from the landings of marine fish recorded from Malaya in 1963 there were more than twenty kinds of which over 1,000 tons each were landed. Among these the most prominent were mackerel, small horse-mackerel and herring, of which more than 10,000 tons each were landed that year, while anchovy, tunny, Spanish mackerel, snapper were also taken in quantity.

Production of fish by Malayan fishermen, though much below that of European fishermen or Chinese or Japanese fishermen, was very considerable even before the war. The amount of " wet " fish recorded as landed in Malaya in 1938 was about 87,500 tons, a figure which did not include a great deal of the fish consumed locally on the east coast, but which did include a

[1] C. N. Maxwell (1921), pp. 3, 9 ; cf. also D. G. Stead (1923), pp. 230–5 ; G. L. Kesteven (1949), pp. 70–3 ; W. Birtwistle (Singapore, 1939), pp. 20–7.

larger number of small fish used as pig and duck food and manure. In 1960, the total landings of fish in the Federation of Malaya amounted to nearly 140,000 tons, and in 1963 to 183,600 tons (3 million piculs) of a total value of well over M$150 million.

FISH MARKET RELATIONSHIPS

Malay fishing is not a purely subsistence occupation : it depends for its prosperity on market conditions. The market is of several kinds. First there is the fresh-fish market, which nowadays is of major importance both locally and in inland centres. Its radius was formerly governed closely by the range of available means of transport—which varied from the basket carried on a pole at the trot, to bicycle, motor vehicle and railway. Since fish will not keep fresh long in the tropics, speed is essential in supplying this market or fish must be frozen or packed in a liberal supply of ice. Before the war ice and brine had been used, but only to a small extent and mainly by Chinese and Japanese fishermen. The result was that the first stages of decomposition had often set in before the fish was sold to the consumer. Nowadays ice is in plentiful supply on most parts of the west coast of Malaya, and its use has been spreading rapidly on the east coast, especially with the setting up of ice manufacturing plants. Nowadays then most of the Malayan fish of better quality is sold as fresh fish.

A second type of market is that for cooked fish. This allows the period during which the fish are handled to be prolonged, but even then the fish do not keep for many days, and this market is of only limited scope. More important to most areas is the market for cured and dried fish, with which is associated the market for the shrimp paste commonly known as *belachan*. Curing and drying allows the fish to be transported great distances and held in reserve against periods of bad weather or seasons when fish are scarce ; it also provides the means of dealing with temporary surpluses. Before the war, and even quite recently, processes connected with this branch of the industry employed large numbers of people, especially dependants of fishermen. Nowadays, however, partly owing to the much extended use of ice and partly to the fact that fish drying requires a great deal of labour, this side of the industry has lost some of its importance. Before the war there was little canning of fish and there is still no fish canning industry of any importance in Malaya. Small quantities, however, of canned fish are produced in Penang irregularly for local consumption and export.[1]

A remarkable feature of the whole market situation is the extent of its ramifications. One might expect from the simple character of these fishing communities that their primary trade would be with their own hinterland. Where there is a dense agricultural population, or there are large inland centres, this is the case ; the fisheries of the west coast of Malaya dispose of large quantities of their catches, both fresh and cured, in this way. But often the hinterland is comparatively undeveloped, giving an insufficient local market, and the unsatisfied demand of large consuming centres at a distance is keen ; this has led to extensive exports, especially from regions such as the east coast of Malaya. Until recently this trade has been mainly in dried and cured fish, but dealers in fresh fish have been quick to take advantage of improved communications. Fresh fish is now carried by van or motor truck for long distances, even across the Peninsula.

This fish trade, until recently at least, has taken little account of political boundaries, except when export or import taxes may have tended to affect its volume and when war, or measures of the " confrontation " type, have severely dislocated normal trade relations. The significance of such fish trade is seen by the fact that in 1960 25,000 tons of fresh fish and more than 5,000 tons of dried fish, of a total value of nearly $20 millions, were exported from the Federation of Malaya, mainly to Singapore. Conversely, more than 8,000 tons of fresh fish and 9,000 tons of dried fish, of a total value of $17 million, were imported into Malaya, mainly from Singapore, Thailand and Sumatra. Moreover, in addition, there is an important trade in salt from Thailand and elsewhere as an essential element in the dried fish industry. In these, as in so many other types of commercial transaction, Singapore has served as a nodal point and great entrepôt for Malaya and neighbouring countries. The Malayan fishing industry then cannot be regarded simply as made up of a set of independent economic units, each with its producers serving only local customers.[2]

The scope of the whole system of fish marketing has led to the development of various types of middlemen who are often

[1] Federation of Malaya, *Year Book* (1962), p. 284.
[2] Federation of Malaya, *Year Book* (1962), p. 284. A very useful general study of the fishing industry and fish marketing of Malaya has been made by Ward (1964).

specialists—as wholesalers or retailers, as handlers of fresh fish or of cured and dried fish. They conduct their transactions with much haggling, and employ complicated methods of finance and credit. They are usually a separate group from the actual fishermen, though they often work at the secondary processes of preparing the fish for market. The composition of the employment category varies from one area to another. In some places they are locals, even kinsfolk of the fishermen. In others they are " outsiders ", in particular Chinese operating among Malays. (The reverse is almost completely unknown.) Women play an important rôle in some areas, especially on the north-east coast of Malaya. But there is one tendency noticeable in many areas—for the fresh fish trade to be in the hands of the locals, and the trade in cured and dried fish to be in the hands of Chinese. One reason for this would seem to be that the handling of the latter commodities, particularly in large-scale export trade, is facilitated by distant business connections and the capacity to lay out capital and wait some time before it returns. Malays are not necessarily lacking in the business ability to handle such trade, but the Chinese commercial pattern is better adapted to take advantage of the situation.

The Chinese have entered so far into the industry in some areas that, as in the west and south of Malaya, they play a very large part in the actual fishing. Before the war indeed the Chinese ring-net fishermen at Pangkor, using power craft, were the most efficient producers in Malaya. But apart from this the Chinese for a long time have extended their rôle as middle-men to embrace the control of the whole of the large-scale fish market and even to finance Malay fishermen. The *tauke* (towkay), as he is called in Malaya, works in a variety of ways. He advances money or more often goods such as rice and cloth to the fishermen in the slack season against the security of their coming catches. He lends money for the purchase of boats and nets, and may even supply such equipment without overt charge. In return he contracts with the fishermen to take their fish at an agreed price or at a price of his own setting, usually below the free market rate. In this rôle he performs important functions. He shoulders a great part of the market risks ; he supplies considerable capital in a liquid form in consumers' goods or in purchasers' goods, and he saves the fishermen much trouble in seeking buyers. On the other hand, the system has obviously great dangers both for economic development among

the fishermen themselves and for the maintenance of equable
social relations between the parties. It can give the *tauke* a
monopoly with little prospect of the fishermen reaping the
advantages of the rising market or of even getting much more
than bare subsistence earnings, and it tends very often to place
them firmly in his debt with little chance of extricating themselves
and building up their own capital.

INCOMES OF FISHERMEN

Before the war no adequate information of any general kind
was available on the incomes of Malay fishermen. Some records
from haphazard observation and some more general estimates
had been made (see 1st edition of this book, p. 15). But little of
systematic value could be extracted from the materials published
or on file. A very rough estimate of the average fishermen's
incomes in Malaya could be made from production figures.
Calculations made for the time when my original study was
carried out yielded the following. If the number of fishermen
in Malaya just before the war could have been put at a mean
figure of 47,500 and their annual output at a value of $7 million,
and the cost of maintaining equipment and of taxation were
put at an annual figure of $500,000 (on a capital value of boats,
nets and other gear at about $2½ million), the average net
income per head would have been approximately $135 per
annum, or just over $11 per month. This may be compared
with the average wage for Malay estate labour, which varied
between $12 and $15 per month. From general impressions,
fish incomes were higher on the west and south coasts, nearer
the large towns, than on the east coast, and higher also on the
whole among Chinese than among Malays. This was borne out
by a similar calculation to the above for the Malay fishermen
of Kelantan and Trengganu, whose average annual income per
head would seem to have been in the region of $8 per month.

But though these calculations were made with some care,
they were inevitably subject to a large margin of error. Figures
for the number of fishermen and for the value of their annual
output at wholesale prices were not exact, and the allowances
for repairs involved a number of rough estimates of value and
life of equipment. It was important to have more precise knowledge of total fishermen's incomes and of average income per
head in the country as a whole and in different parts of it, as an

element in the study of the national income, and as a contribution towards any planned development programme. But this knowledge could only have been obtained as the result of more adequate census and general production statistics on the one hand, and of more intensive regional surveys on capital, production costs and methods of securing incomes on the other. Moreover, figures of average income were useful only up to a certain point ; one wanted to know also the range of incomes. From such indications as were available, including the results of my own intensive study on the east coast of Malaya, it was clear that the income position was complex. It depended not only on local resources, seasonal conditions, marketing facilities and type of equipment used, but also on different institutionalized methods of putting capital into the work and dividing the proceeds. Whereas some fishermen lived on a marginal subsistence, barely able to make both ends meet, others were able to earn a comfortable income in terms of the standards to which they were accustomed, and a few could save considerable sums. This range of fishing income was an important feature of the economy and had been often overlooked when general statements were made about the industry.

What is the position today ? I doubt if we can estimate at all precisely even the average income of fishermen in Malaya at the present time. But based on figures (for 1963) of approximately 60,000 fishermen (of all ethnic groups) with annual landings of wet fish valued at some M$150 million, average gross productivity would be in the region of $200 per man per month. From this, maintenance of equipment and other costs—much heavier than in pre-war days—would have to be subtracted, and this might well reduce the figure of net individual income to half or less than half of the productivity figure. Compared with an average actual earning of a rubber tapper of $86 per month,[1] this means that fishermen as a whole are in a low income bracket. Considering their probable range of earnings indeed, some of them may well find themselves among the lowest income groups in Malaya. (The position with regard to one Kelantan area is examined in Chapter XI.)

During the last twenty-odd years a great deal of knowledge has been accumulated by officials of the Fisheries Department, by Coöperative Marketing Officers, by Social Welfare Officers, and others in close contact with fishing communities. From

[1] Federation of Malaya, *Monthly Statistical Bulletin* (July 1963), p. 137.

their accumulated experience very much closer estimates of fishermen's incomes could be made. But we still await a comprehensive study which will reveal the incomes of fishermen of different types on different sectors of the coast, having regard not only to average earnings but also to individual differences and to daily and seasonal fluctuations. Moreover, after twenty-five years there is still no study of any single fishing economy elsewhere in Malaya comparable to that described in this book. The need for detailed socio-economic research in this field is evident.

PROBLEMS OF MALAY FISHING INDUSTRY AND FISHING COMMUNITIES

In its original form one section of this book examined the practical problems of the fishing industry in Malaya as they appeared before the war. It is of interest now to review the statements then made, to see how far these problems have persisted or have changed their shape, how far they have continued to affect the social life of the fishermen and how far predictions then made have been borne out.

The modern aim of public policy in regard to Malayan fisheries is the same as before—to make more fish available to the mass of the people, to improve its quality, and to raise the level of fishermen's incomes. The demand for fish, especially for fresh fish, remains very keen, and there is still room for improvement in the conditions of production and of marketing. Even before the war these problems had been energetically tackled in Malaya by the Fisheries Department. New fishing grounds had been explored, particularly for mackerel and other valuable types of food fish, and the results communicated to the fishermen. Experiments had been made with new or improved types of nets, and demonstrations of their utility given. On a limited scale on the west coast the introduction of motor boats had been sponsored to replace the less mobile sailing craft. Experiments were made in various types of refrigeration, in smoking and canned fish, and on the quality of salt, to help to maintain fish in fresh conditions as long as possible or to provide better methods of dealing with surpluses. A school of instruction for young Malays, preferably sons of fishermen, was set up at Singapore to teach navigation and the use of powercraft and of new types of nets and better methods of handling and preserving fish for market.   Much of this work, then at an early stage, was interrupted by the war, but has been prosecuted more vigorously and expanded greatly since that time.   Moreover, assistance has been given to fishermen in other ways as well. As part of general national development, a greatly improved road system has facilitated the transport of fish, allowing much larger quantities to be got fresh to inland markets.   Moreover, the much more widespread availability of ice, due in great measure to administrative assistance, including assistance from abroad under the Colombo plan, has meant that much greater quantities of fish now reach these inland markets in a tolerably fresh condition instead of being either partly decomposed or dried.   Parallel developments with motor transport have helped in the same direction.

In terms of general averages, it would seem that some increase of productivity has followed these measures.   Before the war a very rough estimate gave average annual output per head of fishermen in Malay as probably rather less than two tons.   By 1960 the average output per head had risen to just over $2\frac{1}{2}$ tons and by 1963 to about 3 tons per fisherman.   But what is of significance here is probably a much increased difference in productivity between high-grade and low-grade producers. Fishermen pursuing the traditional techniques of hand-lining and small netting are probably landing much the same quantity of fish as twenty-five years before.   Landings from newer or more broadly diffused technical equipment, however, such as the purse seine used from motor boats, probably have meant greatly increased output per head from that sector of fishermen.

With the problems of raising productivity have been associated the problems of raising incomes and levels of living of the fishermen.   These have proved to be in some ways more refractory.   The difficulties of the fishermen have remained fairly clear.   Many of their activities are small-scale.   Though power-driven boats have become much more common, their nets, often surprisingly large, are still hauled by manpower.   The result is that the yield per individual fisherman tends to be relatively light.   Their low *per capita* level of production is not due to lack of skill ;   the methods they use are ingenious, showing close study and intimate knowledge of fishing conditions.   Twenty years ago I wrote that, " Their difficulties were due partly to lack of acquaintance with modern mechanical equipment, partly to conservatism (which was not blind but often based on the proved value of past experience) and partly to lack of the capital needed to lift their enterprises to a much higher technical level." Nowadays the technical level has risen considerably and the capital available become much more plentiful. But, as I pointed out in my earlier comments, their low productivity per head was due also to certain more general economic and social factors.

It is often assumed that the two sets of general problems mentioned above are in fact directly linked : that an increased quantity and quality of production will by itself bring better conditions to the primary producers. Though in a general way this may be true, it will not necessarily be so in specific cases. Where, as often happens, middlemen are in virtual control of the wholesale market, an increased turnover may well be reflected in their higher profits rather than in an increased return to the fishermen with whom they deal. Yet the notion of exploiting middlemen, organized in monopolistic " rings " is not borne out fully. There is much competition between dealers in selling fish in the markets, and the arrangements between them and the fishermen help to some degree to protect both parties.

With the development of Malayan fisheries since the war, elaborate efforts have been made to evade what has been thought to be control of the market by wholesaling middlemen, as by the institution of coöperative marketing organizations. With this, however, the whole question of capital relations was bound up, since in many cases the primary producers, the fishermen, were indebted to the middlemen for their supplies of capital either in the form of equipment or of cash, and hence did not regard themselves as really free to enter into coöperative arrangements which would bypass their patrons. This problem was faced by the provision of capital facilities to fishermen, largely in the form of loans of boats and gear. This system also has met with difficulties, one reason being that the state was often obliged to take the brunt of the losses made by the less efficient fishermen.

Technical and economic developments of the kind described above since the war were foreshadowed in the earlier edition of this book (1946, pp. 16–21, 300–5). It was also stressed that quite apart from the special mechanisms inherent within the fishing industry itself, other economic and social factors would be concerned in any radical attempt to improve the lot of the fishermen. As I pointed out, the success of such technical and

economic changes must involve and be dependent upon pro-
gressive social changes.   These can be promoted by economic
developments but their precise content and impact are as much
the product of enlightened ideas as of an improvement of resources.
The problem is to understand the forces at work.   The aim here
must be to safeguard and stimulate those community ties and
values which give meaning to individual and social life and a
basis for coöperation, yet enlarge the basis of loyalty of the
peasant community to develop new foci of interest and wider
scope for common activity with the world outside.   The attainment of independence and the political transformation of Malaya
will, it is to be hoped, provide such new creative social ties.

FIG. 1.—Malay Peninsula—showing rail and road
communication with Kelantan and Trengganu.

CHAPTER II

ECONOMICS OF THE FISHING INDUSTRY IN TWO MALAY STATES (KELANTAN AND TRENGGANU)

My detailed examination of a single Kelantan fishing community needs as preface a brief description of the main features of the fishing industry in Kelantan as a whole. For comparison I also give material on Trengganu, since its much more extensive coast-line offers interesting contrasts.

To set the account of conditions in 1939–40 in modern perspective I first outline the situation in 1963. The population of Kelantan, in 1940 about 400,000, was in 1957—at the last Census—506,000 ; that of Trengganu, in 1940 about 200,000, had risen correspondingly to 278,000. Both States remained predominantly Malay in composition. In 1963 these two States together provided nearly one-quarter of the landings of sea fish in Malaya and maintained on their coasts upwards of one-third of the fishing population (Figs. 2, 3, 4). In 1963 the number of fishermen in Kelantan was recorded as just over 6,500 and in Trengganu nearly 11,500, almost exactly the same figures as for twenty-five years before (p. 37). In 1963 there were nearly 2,000 fishing boats in Kelantan and about 3,700 in Trengganu. Despite increasing mechanization (see Chapter XI) most of these craft were still non-powered. Whereas in Malaya as a whole the proportion of powered craft was then over 45 per cent., in Trengganu it was 25 per cent. and in Kelantan only 20 per cent. In both these States, however, the proportion of inboard powered craft was more than ten times that of outboard-motored, whereas in Malaya as a whole it was only about 50 per cent. greater.

The main methods of fishing adopted in these two States are described later (pp. 49–54, 83–100).[1] In 1963 Trengganu had more than twice as many nets as Kelantan—1,300 as against 640, the superiority being most manifest in seines. As regards catches, in 1963 landings of fish were more than five times as great in Trengganu (38,000 as against 6,700 tons) ; anchovy, mackerel, small horse mackerel and herring especially were

---

[1] Since the first edition of this book appeared, an excellent detailed description of fishing methods of Kelantan and Trengganu has been given by M. L. Parry (1954).

produced in great quantity.   The total value assigned to landings of marine fish in 1963 in Trengganu was nearly $35,000,000
and in Kelantan was over $8,000,000.   Thus, while the fishing

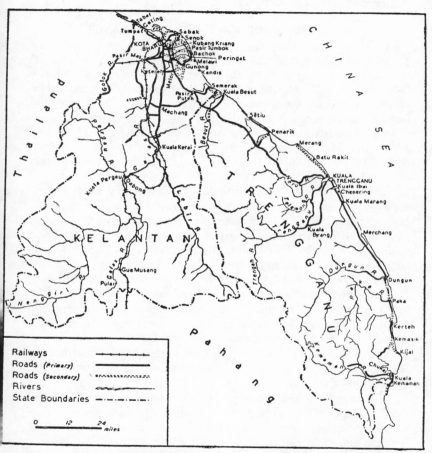

Fig. 2.—Communications and Principal Settlements in Kelantan and
Trengganu.

industry is far greater in the Trengganu economy than in that
of Kelantan, its importance to the latter is still considerable.

The following general account now summarizes the main
features of the fishing industry of Kelantan and Trengganu in
1939–40, particularly as regards personnel, trade, types of equip-

ment, and capital invested in this equipment. (Further details on numbers of boats and nets, schemes of distribution, and relations between fishermen and fish-buyers, will be found in Appendices III, IV and V.)

GENERAL CONDITIONS AND IMPORTANCE

In these two States the actual fishing is done entirely by Malays, in contrast to the situation in the south and west of the Malay peninsula, where there are large numbers of Chinese fishermen as well and where (before the war) Japanese fishermen also competed in some markets, though obtaining their catches from deep-sea grounds not fished by Malays or Chinese. Conditions in both these east coast areas favour the types of fishing which the Malays have developed, and which are adapted to their comparative lack of capital, and to their particular type of social organization.

The extension of the continental shelf for many miles out to sea gives them a large area suitable for coastal fishing, from which they can return by nightfall to their homes. Frequent river-mouths, with their characteristic sand-spit formation, afford useful harbours for their comparatively small craft ; even where such harbours are lacking the easy slopes of the many long sandy beaches allow the boats to be hauled up by manual labour, with no mechanical appliance, but only a few rough skids. The land and sea breezes in morning and afternoon are constant enough for most of the year to carry them to and from the fishing grounds by sailing, with the minimum of paddling. And fish are sufficiently abundant and are of such varied types that despite seasonal variations catches may be obtained in one way or another throughout the year.

These geographical factors are important. Malays, though excellent fishermen with a variety of techniques, do not like spending the long periods away from home which are common—indeed, almost necessary—in true pelagic fishing. (A reason they sometimes give is that they do not know what their wives may be doing in their absence.) Nor with the limited capital at the command of most of them can they afford unaided the larger boats and more expensive equipment required for fishing in more distant ocean waters. In both these respects they differ some-what from many of the Chinese fishermen, and still more from the Japanese, who are willing to spend long periods at sea and who have larger capital resources.

IVa AN EAST COAST MALAY FISHING VILLAGE
Part of the Tanjong area at Kuala Trengganu, 1963.

IVb A FISH-DEALER CUTTING UP A RAY-SHARK
The fins, greatly prized for soup by Chinese, are expensive.
(Perupok, 1947.)

Vᴀ   HAULING A BOAT UP THE BEACH
The fishermen, soaked to the waist, strain to slide their heavy craft (a kolek buatan barat)
up over the skids.

Vʙ   SCRUBBING THE PLANKS OF A BOAT
The crew use sand and water, and the thick soles of their bare feet.

But in one way the geographical conditions act as a brake on the east coast fishing industry. For at least a month and often for more, usually in December and January, the strong steady winds of the north-east monsoon, sometimes rising to gale force, tend to block all major fishing activities off the coast. The

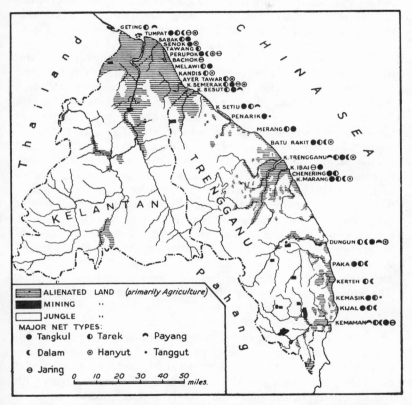

Fig. 3.—Land Utilization and Fishing in Kelantan and Trengganu

seas are considerable, and the surf heavy on the beach. The larger nets can rarely be used, and boats often cannot be launched for weeks at a stretch. Only intermittent work with hand-lines or small nets is possible at sea, and at the height of the monsoon fishing is restricted to the use of scoop-nets or other hand-nets for mullet in the breakers off the beach. The effects of this on the industry can be gauged from graphs showing variations in

monthly output (Figs. 6 and 21). In terms of consumption, this
seasonal drop in production is important since it means that every
fishing household has to save a certain proportion of its income
to tide it over the monsoon period, or else borrow, and thus in
effect mortgage future income. The monsoon period is, however,
not a total loss. During this time a great deal of work is possible
in the repair of equipment, particularly boats, and in the preparation of new equipment, particularly nets. The monsoon also
affords opportunity for carrying on secondary occupations.

The similarity between Kelantan and Trengganu in the
productive methods and organization of fishing does not extend
so completely to the marketing system. Geographical and
political factors here distinguish them. The areas of the two
States are not very different. But whereas the coastline of
Kelantan is only about 60 miles in length at the most, that of
Trengganu stretches for over 130 miles (Fig. 2). Since the
proportionate distribution of river-mouth harbours and other
fishing facilities is much the same, with, if anything, a balance
in favour of Trengganu, it is to be expected that the output of
fish is greater in the latter State. Moreover, the population of
Kelantan is practically double that of Trengganu, due mainly to
the great plain in the former State, with its eminent suitability
for rice-growing (Fig. 3). In Trengganu the plains associated
with the lower courses of the Besut and the Trengganu rivers
are very much smaller, and support less population (Fig. 4).
The effect is that Kelantan has a large agricultural hinterland
serving as a market for the products of the fishing section of the
population, whereas Trengganu has not.

Communications in each State have naturally tended to fall
into line with the topography and the distribution of the population (Fig. 2). Kelantan has a moderately good system of arterial
and interconnected roads to the east of the Kelantan River, with
the capital Kota Bharu as the centre, and extensions run more
than forty miles inland to the secondary town of Kuala Kerai.
The system of Trengganu is essentially lateral, with a single
major road running parallel to the coast. South of Kuala
Trengganu, the capital, which is about two-fifths of the distance
from the northern to the southern border, stretches of this road
are linked only by frequent ferries, till Kemaman is reached.
Before the war communication from there to the town of Kuantan,
in Pahang, was of a primitive type, running for many miles along
the open beach. The only inland road of any consequence was

that to the small centre of Kuala Brang, at the apex of the plain through which the lower Trengganu River runs, and extending for a little over 20 miles. (Extensions to the road system appear to have been made more recently.) In both States the rivers and the coastal seas supplement the road system—in former

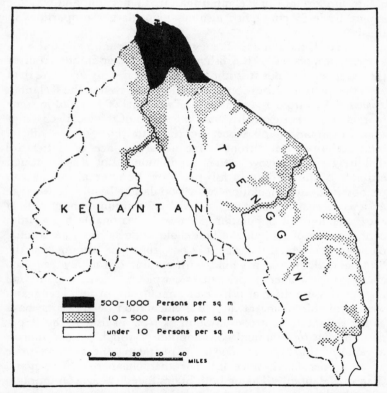

FIG. 4.—Approximate Density of Population in Kelantan and Trengganu.

times most of the traffic was water-borne. But transport by them is too slow and too apt to be affected by local variations of drought or flood to be efficient in the service of a large-scale fresh-fish market. An additional factor in promoting market facilities in Kelantan is the railway passing through from the Federated Malay States to Siam ; Trengganu has no rail system apart from the small local line serving the Japanese iron concession inland from Dungun. The effect upon the Trengganu fish trade with its limited hinterland market, is that the system of communications essentially connects one centre of fish production with another, and not with centres of large demand inland.

These factors make it clear why the fresh-fish market is of much greater importance in Kelantan than in Trengganu, and why the export trade in cured fish assumes much larger proportions in the latter State.

In both Kelantan and Trengganu the internal market absorbs a large proportion of the fish produced, either in the fresh or the cured state. But it is impossible to obtain any accurate data as to the quantities thus consumed. Figures given in the Kelantan *Annual Reports* from 1927 onwards of amounts of fish sold in some markets represent explicitly only a fraction of the local consumption. Comparison of export figures with those of the official total Kelantan fish output is inconclusive, since the latter are definitely much below actual production. But a very rough estimate of the amount of fish taken by the internal market can be obtained by calculation on a population basis. If we take as a conservative estimate an average consumption of fish of 1 oz. per head of population per day (a reasonable figure from consideration of a sample of household budgets) the total annual consumption would be about 4,000 tons per annum for the State of Kelantan as a whole, with a value of roughly $345,000 (in 1940), at an average price of $5 per picul. No data of internal market consumption of fish or of annual output for Trengganu were available to me at all. But calculation on the same basis as for Kelantan, of a consumption of 1 oz. of fish per head per day, gives a total annual consumption of roughly 2,000 tons at a value of roughly $150,000 in 1940, taking the average price at $4.50 per picul, since fish is rather cheaper in Trengganu than in Kelantan. Taking both States together, the consumption of fish in the internal market is probably in the region of 6,000 tons per annum, with an average total annual value of about half a million dollars. It is likely that these figures are underestimates.

From the export statistics it is obvious that fish plays a much more important part in the trading economy of Trengganu than in that of Kelantan.

From 1910, when the two States first came under British administration, till 1939, the total value of fish exported from Kelantan was approximately $4 million ; from Trengganu it

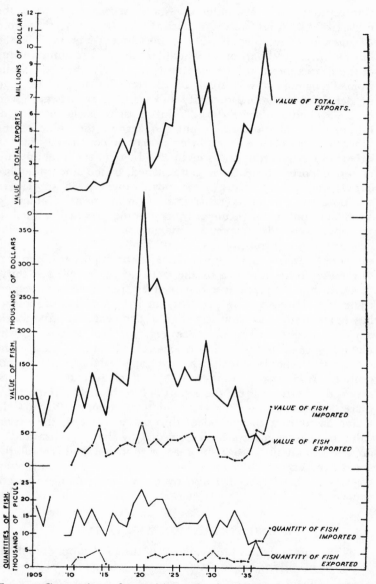

Fig. 5.—Comparison of quantities and values of fish annually exported from and imported into Kelantan, with annual value of total exports 1905–38.

35

was approximately $25 million. And in recent years, while the export of fish from Kelantan declined considerably, that from Trengganu maintained itself. In the three years 1936–8, the last for which comparison is available, the average annual value of the fish exported from Kelantan was just over $42,000, while from Trengganu it was about fifteen times as much. In 1939 no details of the actual quantity and value of exports are given in the Kelantan *Annual Report*. But as there is no mention of fish in a list of more important *variations* in the value of the different types of articles exported it may be concluded that the export of fish was roughly as in 1938. In 1939 the total amount of fish exported from Trengganu (dried, salted and otherwise cured) was 104,125 piculs (6,198 tons), valued at $694,151. In addition, 10,870 piculs of *bĕlachan* (shrimp paste) valued at $108,697, and 6,435 piculs of other marine produce, valued at $55,393, were also exported, making a total of all marine produce of 7,223 tons, of a value of $858,241.[1]

The difference in the importance of the export trade in fish to the two States is shown by the fact that during the last thirty years the highest place ever occupied by fish in the total export trade of Kelantan was 14 per cent. (in 1914) and the proportion has fallen until it is now less than 1 per cent. (approximately 0·9 per cent. for 1936–8.) (See Fig. 5). In Trengganu the highest place taken by fish in the export trade in the sixteen years since fairly reliable statistics have been available was 28½ per cent. (in A.H. 1349 ; 1930–1 A.D.). Even recently, when the export of iron-ore soared, the export of fish still held a respectable place, representing 6·3 per cent. of total exports in 1939. In Kelantan, during the period of thirty years' record, the export value of fish has always been exceeded by that of copra and of areca-nut, and since 1931, by that of rubber. And irregularly, it has been exceeded by the value of cattle exported, while since 1936 it has been exceeded also by the value of manganese ore exported.

FISHING POPULATION AND OUTPUT

No accurate figures are available of the total supply of labour engaged in fishing in Kelantan and Trengganu. The total

[1] I am indebted for these figures to the " Annual Report of the Customs, Excise Chandu, Fisheries and Marine Departments, Trengganu, for the year 1939 ", made available to me through the courtesy of Dato' Jaya Perkasa, M.B.E., and Mr. L. R. Birkett-Smith.

population directly dependent upon fishing production for a livelihood, and the number of persons engaged in the secondary processes of the industry, in fish-trading, the transport of fish and the cleaning and curing of it are also not accurately known. It is possible, however, to gain an approximate idea of the figures in the first two categories.

The Census of Malaya for 1931 gives the number of Malays engaged in fishing as 6,224 males and 377 females in Kelantan, and 9,838 males and 366 females in Trengganu. (The number of non-Malays engaged in this occupation is given as only 26 in Kelantan and 203 in Trengganu.) If we assume that since then there has been an increase in the Malay fishing population in proportion to the increase in the Malay population as a whole in the two States, the totals would be 6,750 for Kelantan (on a 1939 basis), and 11,570 for Trengganu (on a 1938 basis). These figures can be only approximate since the original Census could not rely upon any rigid distinction between those primarily fisherfolk and those primarily agriculturalists. But an estimate may also be made from another angle, based upon possibilities of using the available equipment. By taking the numbers of nets and boats in each State and multiplying the figures by the normal number of men required to operate each type, with allowance for the fact that full employment of the equipment is not always secured, a rough check upon the Census can be obtained. The detailed calculations, which involve estimates of the floating margin of labour moving from one type of equipment to another, estimates of the number of boats used primarily for other purposes than fishing, etc., need not be given here. But the results give an approximate labour supply of 6,500 men in Kelantan and 10,000 in Trengganu. These latter estimates, which I have calculated on a conservative basis, corroborate fairly closely the inferences from the Census. In the *Annual Report* for Trengganu for 1936 it was estimated that not less than 15,000 Malays in that State were fishermen. Unfortunately the basis of the estimate was not given, and the statement was not repeated in subsequent years. It would seem that this figure is rather high, since with a total of about 3,300 fishing-boats in the State this would mean that practically every boat was in full employment simultaneously, with an average crew of nearly 5 men. And while the greater concentration on certain types of fishing in Trengganu does need large crews for some boats, there are also a large number of small craft with a low average crew. A figure of 15,000 fishermen would imply either that there is in Trengganu an acute shortage of capital equipment (which observation does not suggest) or more probably, that in this figure were included a considerable number of men whose primary occupation was not fishing, but who occasionally participated in it.

To estimate the total population directly dependent upon the primary processes of fishing production involves still further approximation. But in the sample fishing area in which our census was taken the ratio of all persons counted to adult and adolescent males of effective working age was 3·44 to 1. The sample studied covered only about one-twentieth of the estimated total of Kelantan fishermen, but in default of other material it may be assumed that this ratio applies to the Kelantan and Trengganu fishing areas as a whole. On this basis the total Kelantan population of Malaya primarily engaged in fishing, their wives and their dependents, is about 22,000, and that of Trengganu is about 38,500 people, a total of approximately 60,000 for both States, or roughly about 10 per cent. of the total population.

Calculation of the physical volume of production in the two States is subject to a considerable margin of error. But there are several indirect methods of estimation. For Kelantan these are based on monthly returns sent in by some fishermen to the Superintendent of Marine and Customs, in Kota Bharu ; on estimated consumption together with amounts exported ; and on the results from the sample area studied intensively. My estimates indicate that the annual fish output of the State can be put between 6,000 and 9,000 tons, with a value of between $500,000 and $750,000 at 1940 wholesale prices. Since the estimates have been deliberately conservative, the higher figures are the more likely. For Trengganu the only material available is the export figures, but it is known that these represent a very large proportion of the total production. For the three years 1937–9 the average annual export of cured fish and shrimp paste was nearly exactly 6,000 tons, the average value being nearly $734,000. Converting this weight into an equivalent weight of wet fish, of about 12,000 tons, and adding the quantity estimated to be consumed on the internal market, on a minimum basis at least 2,000 tons, the annual output must total on the average about 14,000 tons, with a total value of a million dollars or so.

The possible error in these calculations is too great to allow

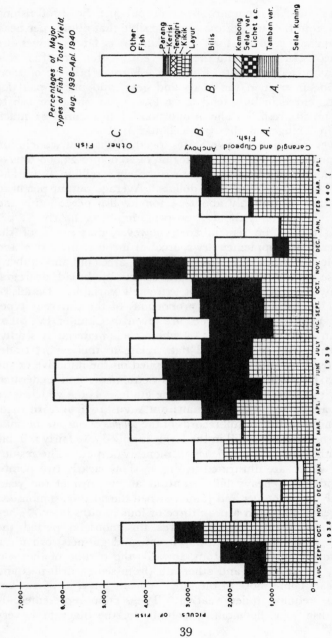

FIG. 6.—Output of Fishing in Kelantan. Monthly variation in quantities of fish obtained, August 1938 to April 1940. (Quantities in piculs; 1 picul equals 133⅓ lbs.) English equivalents of Malay fish names will be found in Appendix VIII.

Percentages of Major Types of Fish in Total Yield. Aug. 1938–Apl. 1940

C. Other Fish
    Parang
    Kerisi
    Tenggiri
    Kikik
    Layur
B. Bilis
    Kembong
    Selar var
    Lichet & c.
A. Tamban var.
    Selar kuning

of any significant comparison of output per head of fishing population in the two States.   It is possible that output per head in Trengganu is lower than in Kelantan as far as values are concerned.   In Trengganu the Chinese fish-curers are often able to buy at preferential rates from the fishermen because of their controlling interest in the boats and gear, and prices paid for fish to be cured in bulk tend to be lower than those for fish to be consumed fresh.   (The proportion of fish cured is much higher in Trengganu than in Kelantan.)

However, taking the two States together it is a fair conclusion that their total output is in the region of 20,000 to 23,000 tons of fish per annum, of a value of approximately a million and a half to a million and three-quarter dollars.   Average output per head of fishermen is probably about $1\frac{1}{2}$ tons of fish per annum and average cash income about $100 per annum, or nearly $2 per week.   (It must be remembered, however, that many of the fishermen have supplementary sources of income, from rice and vegetable cultivation, preparation of copra, areca nut and rubber.)

These estimates of output are essentially broad averages. Over any short period output is extremely variable.   Details of seasonal fluctuations and the proportions of the different types of fish caught at different times are given in Chapter IV.   For Kelantan and Trengganu as a whole such material is scanty, since nearly all the published data relate to the export trade. The statistics of monthly output compiled on the initiative of the Fisheries Department of the Straits Settlements and Federated Malay States do, however, give some guide.   Those which were made available to me for Kelantan from August 1938 to April 1940, though admittedly not full records of the *volume* of output, because many fishermen fail to fill in the forms supplied, are fairly reliable as an indication of the *variation* in monthly output.   The results, some of which are illustrated in Fig. 6, show clearly two points. They indicate a heavy fall in output at the turn of the year, especially in December and January when the north-east monsoon blocks sea fishing, and a peak three or four months later.   They indicate also that whereas towards the monsoon period the carangid and clupeid fish (from lift-nets and gill-nets) provide a high proportion of the output, towards the middle of the year anchovy (from seines) and other fish (from seines, drift-nets and other gear) come more to the fore.   For anchovy, in particular, the seine requires quiet weather.   The section represented by " other fish " in the monthly columns of the diagram is very large in some cases. This category includes over a score of kinds, the proportion of each being usually too small to be graphed separately ; in no case except that of *layur*, a scabbard fish taken in the seine, does any of them comprise more than 10 per cent. of the total production in any one month. Lift-nets and seines provide the great bulk of the output. It is unfortunate that the material available did not cover a longer period, to enable general trends to be more clearly freed, but the seasonal picture presented is fairly clear, and would appear to be typical.

EQUIPMENT—DETAILS OF TYPES AND COST

A considerable variety of equipment is used in the Kelantan and Trengganu fishing industry, and in the aggregate its capital value is quite large. This raises questions as to the source of the capital, and the level of investment by individuals, which will be taken up in the later analysis, particularly in that of the sample fishing community. Here it will be sufficient to consider the broad facts, with some description of the main types and their cost. A small amount of material on this subject has been published, and registers of boats and nets were kept locally, in somewhat unsystematic form. (See Appendix III.) But no estimates of the total value of the equipment had been made, nor was there any investigation of how its finance was arranged. In the last three sections of this chapter, therefore, I put forward some material on these topics, and indicate briefly some of the differences between Kelantan and Trengganu in these respects.

Boat-Types

Practically all fishing in Kelantan and Trengganu depends upon the use of boats, so that the aim of every fisherman is to possess at least one, and the lack of a boat is a fair indication that a man stands rather low in the economic scale. Nearly all the craft are individually owned. But the complex transactions involved in their sale, or in the raising of loans for their purchase, mean that the return from them is spread much more widely through the community than an investigator at first imagines (see Chapter V). Moreover, many boats are captained and kept up by men who are not the owners of them, and this is also reflected in the scheme of distribution of the returns from fishing. Again, the Malay of the east coast is a keen boat-trader, and this practice is significant in the general process of capital formation ; it would be an error to think, as one might do, that one of the main objects of a fisherman, having got a boat, is to keep that particular craft indefinitely.

The design of the boats, and the technique of handling them associated with this, are important in considering the general conditions of the industry. The craft are small. The largest are only about 50 ft. overall, including projecting bow-piece and stern-piece, and the majority are about 30 ft. overall or less. They are comparatively narrow, with shallow draught—a large craft has only about 7 ft. beam, with a draught of less than 3 ft., and smaller craft correspond. The size of the boats is dictated partly by the limited capital available, partly by the requirements of much inshore fishing but partly also by the need to work many of them from exposed beaches, necessitating haulage above the tide-line every evening. (Large heavy craft of the *pěrahu payang* type normally use only the harbours in the river mouths.) The craft have little freeboard—that of a boat 30 ft. overall, 5 ft. 9 in. beam and 2 ft. 6 in. deep in the centre was only 15 in. when light and 7 in. when loaded.

Despite their small size and low freeboard, the boats tend to carry a large press of sail, with the result that not infrequently they heel and fill in a gust of wind, or are pooped by a following sea. This is most apt to happen to a boat which has the special function of taking the catch of fish in to market, as it races in under insufficiently shortened sail. Such a craft does not normally sink altogether ; it is towed in to shore awash, but the catch is usually lost, together with floorboards and other gear. Again, in making the beach through a heavy surf, when sail and spars have been stowed and the boat is being steered in, a capsize sometimes occurs as the steersman attempts to ride a breaker. Much of the catch is often lost here also, and even when fish are recovered by enthusiastic helpers who rush out, by custom what they get is their perquisite to keep or sell as they wish ; the crew have lost all right to it Such accidents usually take place only in the rough weather just before and after the monsoon. In such conditions also it is not uncommon for a boat to overturn on the way out through the surf in the early morning. This usually prevents it from going out again that day, since loose gear such as floorboards, ropes and anchor stones (Fig. 9a) are liable to be lost and the boxes of the crew, containing their food and their smoking and betel materials, get sodden. Loss of food, etc., is not important ; the crew can get more from their homes

or beg from their friends. But loss of gear is more serious ; if
the boat is one of a net group this may immobilize the whole
group for the day. The point of these remarks is that the

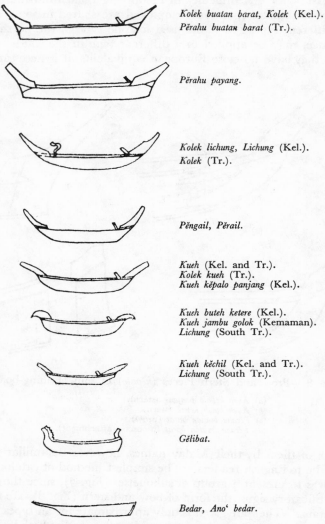

Kolek buatan barat, Kolek (Kel.).
Pĕrahu buatan barat (Tr.).

Pĕrahu payang.

Kolek lichung, Lichung (Kel.).
Kolek (Tr.).

Pĕngail, Pĕrail.

Kueh (Kel. and Tr.).
Kolek kueh (Tr.).
Kueh kĕpalo panjang (Kel.).

Kueh buteh ketere (Kel.).
Kueh jambu golok (Kemaman).
Lichung (South Tr.).

Kueh kĕchil (Kel. and Tr.).
Lichung (South Tr.).

Gĕlibat.

Bedar, Ano' bedar.

FIG. 7.—Boat types in Kelantan and Trengganu. The prow faces
left in each sketch ; each boat is approximately to the same scale,
of about 1 in. to 20 ft.

Malays, though expert seamen enough, are tempted by economic pressure to take risks with their craft and suffer loss in consequence.

At least a dozen different types of fishing-boat are in use on the east coast. A difficulty in identifying them from literature is that the names commonly employed are shared in some cases by different types ; a name used for one type in the north, for instance, may be applied to a different type in the south. But since they have no close European equivalents, it is necessary to

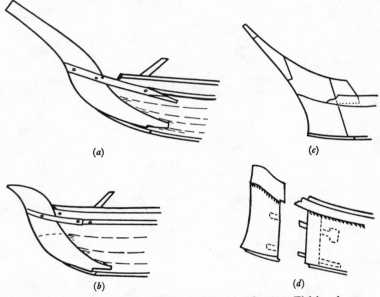

(a)   (c)

(b)   (d)

FIG. 8.—Prow and Stern Pieces (*kĕpalo*) of some Fishing-boats :

(a) *Kueh kĕpalo panjang* (stern).
(b) *Kueh buteh ketere* (stern).
(c) *Pĕrahu buatan barat* (prow).
(d) *Pĕrahu buatan barat* (monsoon attachment).

speak of them by their Malay names, however unfamiliar these may be to English readers. The simplest method of cataloguing them is to present a group of silhouettes (Fig. 7), since the chief identificatory signs, the form of bow and stern (Fig. 8), are fairly constant. The scale given is only approximate for all types since there are no definite standards of length ; individual builders work to the dimensions they fancy best. The craft are all carvel-built, the planks being secured by wooden pegs or less frequently by copper nails. What the silhouettes cannot indicate are the bright colours in which many of them are painted, and which make them a picturesque sight when drawn up on the long open beaches (Plates IIA, VIB) or as a fleet under sail.

On the whole the different types of boats have specialized functions. The *pĕrahu payang*, as its name suggests, is used primarily for work with the large purse seine known as *pukat payang*. The *kolek buatan barat* is the boat now most favoured, especially in Kelantan, for work with lift-net (*pukat takur*) and mackerel net (*pukat dalam*). The large *kueh* and *lichung* are used with most large nets; the small *kueh* concentrate mainly on hand-line fishing and work with small drift-net (*pukat tĕgĕlang*). But this specialization is not absolute. In the off season for work with the larger nets almost any type of craft will go out hand-line fishing, though it is recognized that the *kolek buatan barat* and the large *kueh* are less suitable for this. In fishing with hand-lines there is a general tendency for each boat, irrespective of size, to get much the same amount of catch, so those with a larger crew tend to obtain a smaller proportionate return per man for a day's fishing.

The distribution of the various types of boats is not at all even throughout the different fishing areas. (Table 17 shows this for Trengganu.). This is due partly to concentration on different types of production in the different areas, partly to inequalities in the possession of capital for boat investment, but partly also to definite local preferences. *Pĕrahu payang*, the craft of most specialized function, are found only where the *pukat payang* is used, mainly in the villages north of Tumpat, towards the Patani border; in Besut; and at Kuala Trengganu, Dungun and Kemaman. *Kueh* of various types, on the other hand, are in use almost everywhere in Kelantan and Trengganu, because of their general utility. But *bedar*, *ano' bedar*, and *sĕkochi* are types much more favoured in the south than in the north, while the *gĕlibat* rarely appears north of the Pahang border.

Of the larger craft the most marked preferences are shown in the case of the *kolek buatan barat*. At Tumpat near the Kelantan River mouth, nearly all the larger boats are of this type, built locally. At Kemerak, a little south of this, there are only about half a dozen of these craft, and nearly all large boats are *kolek lichung*. The reason given for this was that since the beach was rather steep they were easier to haul up than the *buatan barat*. Comparative cost may also have been a factor, since the community was not wealthy, and the *lichung*, many of them in older condition, were cheaper. In the Perupok area, still farther to the south, an interesting situation obtains. Formerly, it was stated, all the craft were *lichung* and large *kueh*; but by 1939 many of these had been replaced by *kolek buatan barat* for lift-net work. It was noticeable that this process had gone much further in Perupok, where the more wealthy and successful fishermen were, than in the poorer village of Pantai Damat. The relative advantages of the three types of craft were put thus : in good weather the *kolek buatan barat*, with its straight stem and longer keel, is much faster, both in paddling and in sailing. The *lichung* on the other hand, with its half-moon contour, lifts to every wave and loses way, while the *buatan barat* goes straight through. The *kueh* is " number two " in the list of preferences for fine weather— " It bears up a little as against the *lichung*." In bad weather, however, the position is reversed. Here the *lichung* lifts over the waves, while the stiffer *buatan barat* tends to take them inboard. And in surf off the beach the latter is more liable to capsize ; for this reason stubbier prow and stern pieces are substituted in the monsoon (Fig. 8), while the *lichung* needs no alteration. The *kueh* in bad weather stands intermediate between the other two types. During the monsoon, then, many *buatan barat* are laid up, while *lichung* and *kueh* can take more advantage of a comparative lull.

It was said in the Perupok area that the *kolek buatan barat*, as its name implies (*kolek* " of northern building "), was originally introduced from Patani, and came in in numbers during the last few years. Certainly the process of replacement by them was going on while I was there, and some men were using *buatan barat* they had bought in Patani. (War restrictions in 1940 made it almost impossible to import these craft.) But in the last few years Kelantan boat-builders, particularly at Tumpat, have been constructing them. Those of Tumpat are well-liked, but a criticism was passed on those built by a man of Paya Mengkuang. It was agreed that his boats were well-built and well ornamented, and of good timber, but as against a Patani-built boat they were high enough at bow and stern but too low on the quarter, so that the break of a wave was apt to come inboard.

At Pantai Bharu, to the south of Perupok, the large craft were mostly *lichung* ; this was explained by the comparative absence of capital. At Kuala Besut, over the border, in Trengganu, *buatan barat* were plentiful. It was stated, as at Perupok, that they originated in the north, but a man about

VIa MENDING THE MACKEREL NETS

The nets are spread out to dry on the upper beach, and mended in the afternoon. A pandanus-leaf cover protects a pile of dry net. In the background are the bamboo racks for drying lift-nets.

VIb CARRYING OFF A LIFT-NET FOR STORAGE

Towards night, the crew bear the heavy net from the drying rack to the house of the master fisherman and stow it away for protection.

VIIA   LIFTING OUT A FISH-TRAP
This trap, of heavy rattan, is of a type used to take snapper (ikan merah, Lutianus).

VIIB   THE STEERSMAN OF A SMALL SAILING CRAFT
In his right hand he holds the sheet and in his left hand the steering paddle, kept against the gunwale by the pressure of the water.

30 years old said that they had been in Besut for longer than he could remember. He argued cogently that their recent introduction into Perupok was not because they had been unknown before, but because the Perupok folk had no river mouth, and the *lichung* with its rounded stem was more easy to haul up on the open beach. Only recently had the Perupok men overcome their objection to the greater trouble of handling the *buatan barat*. At Ayer Tawar, a few miles south of Kuala Besut, *lichung* only were in use, and here the reason given was again the difficulty of hauling up *buatan barat* on the steep beach. Below Besut, along practically the whole length of the Trengganu coast as far as Kemaman, there are practically no *pĕrahu buatan barat*. The reason here cannot be steepness of beaches, for there are a number of river mouths offering easy facilities; it cannot be lack of capital either, for a number of the villages are comparatively wealthy. One reason is probably preference for the all-weather qualities of the *lichung*. At Merang it was definitely stated that " *kolek* ", i.e. *lichung*, were preferred to *buatan barat* because in sailing they were steadier. At Kemaman, however, *pĕrahu buatan barat* are found again. They are there called *lohor*, the term used for them in Pahang, and are used for fishing with the smaller *pukat payang* and with seine nets. They are not built locally, though there is much building of *pĕrahu payang*, *kolek* and *kueh* ; they are imported either from Beserah in Pahang, or from Kelantan.

This brief analysis of boat-types is of more than technical interest. It shows that a wide range of types is available to the east coast fisherman. It shows also that for a given end alternative types are usually possible and that, apart from the limitations of capital, choice is exercised partly on a basis of local usage, but largely on the grounds of relative functional advantages of each type.

The æsthetic criterion is not an important influence in choice of types. But in Besut the absence of *kueh buteh ketere* (with hooked " nose ") was explained by one man as being due to the bow shape not being " pretty ". And at Kuala Marang the much higher price of a *kolek* (*lichung*) than that of a *kueh* of nearly the same length was explained by saying that the former was much desired for the shape of its bow and stern pieces. The æsthetic factor enters definitely in the admiration given generally to these bow and stern pieces. It is seen also in the decoration of all types of craft in bright colours—as many as half a dozen being used on one boat ; in the occasional painting of birds or mythological animals on prow and stern; and particularly in the carving and painting of the accessories such as *bangar, sanggur* and *chaping*. The two former, crutches to hold masts and sails inboard when taken down, are often highly sculptured in floral or bird designs, and even occasionally in the form of a figure from the local shadow-play. Such carving is done by specialists, and may cost up to $10 extra, a pure concession to love of ornament. (Plate XIVB shows a well-carved *bangar*.)

Cost of Fishing-boats

I collected data on this subject along the Kelantan coast from Tumpat to Melawi, and at practically every fishing centre of importance on the Trengganu coast. Since there is a great deal of trading in boats all along the coast, prices tend to an equilibrium; the local differences in price for the same type of new craft are due primarily to preferences for different sizes and for quality of timber and workmanship. But a large number of boats are bought second-hand, especially by poorer communities. Details of the various types were as follows (in 1939–40):

Pĕrahu payang. This is the most expensive type, costing from $300 to $500 new, according to size; second-hand ones often sell at $150 to $200.

Kolek buatan barat. New, the price ranges normally between $200 and $300. The average new price of fifteen such craft bought between 1935 and 1940 was approximately $250. After 1939, when their importation from Patani to Kelantan was forbidden, their price rose slightly. Of twenty-three second-hand boats of various ages the average purchase price was $155 each, the range varying from $50 for a very old boat to $250 for one only a year old.

Kolek (lichung). In Trengganu large new craft of this type used for seine-net work cost from $250 (for a boat with an overall length of about 35 ft.) to $350 or even $400. Smaller craft cost proportionately less, and those used for lift-net work cost between $120 and $200 new. The poor village of Ayer Tawar bought boats of this last type second-hand for $50 to $100. In Kelantan the displacement of these craft by *buatan barat* in recent years has meant that only second-hand ones are acquired, and their rate of economic as against technical depreciation has been heavy. Their prices in the present condition range between $20 and $90 according to age. Small old *lichung* used for line fishing in Kelantan change hands at $10 to $20.

Pĕngail. This is a modification of the *kolek* (*lichung*), and costs from about $170 to $200 new.

Kueh. Since *kueh* can be of almost any size there is great variation in their price. A common large type, from 27 to 30 ft. long, costs from $100 to $150 new, but larger ones cost from $150 to $200. An exceptionally large one at Kemaman cost $245. Second-hand, the common type fetches from $50 to $100, but in Kelantan these boats have suffered somewhat from the competition of the *buatan barat.* Medium *kueh*, with the hooked short prow, cost in the region of $100 new, and small *kueh*, about 17 to 20 ft. long, cost new about $50. Such craft when old are bought cheaply, from about $10 upwards. The average price of ten second-hand small *kueh* was $22.50 apiece.

Gĕlibat (or *dĕlibat*). This, a modified form of small *kueh* common at Kuantan in Pahang, costs there about $40 new.

Bedar. Sea-going vessels of this name are very large, but the river-mouth and fishing-craft much used in Besut and Trengganu are of the size of medium and small *kueh.* *Bedar* proper, about 23 ft. long overall, cost about $120 new ; smaller *ano' bedar* used as passenger ferries cost $45 to $50.

Jalurer and *sĕkochi.* These medium-sized craft used in Pahang cost about $150 new. But smaller *sĕkochi* cost only about $50.

The implications of this material for capital formation and investment are discussed in Chapter V. But it is evident that there is a wide range of opportunities for the acquisition of capital equipment by these fishermen. Owing to specialization in building, some types may be acquired a little more cheaply on one part of the coast than on another, as, for instance, *kueh* at Kuala Trengganu, or *kolek buatan barat* in the north of Kelantan (from Siam). But from comparison of prices along the coast it seems that the difference is small.

Types and Cost of Nets

More than a dozen different types of nets, with several types of trap in addition, are in common use by the fishermen of Kelantan and Trengganu. These nets vary greatly in size, cost, labour required in operation, and the kinds of fish they catch. Though most belong to well-known and widespread types, their variations are such that it is not easy to differentiate them by simple English labels, so in this book they will frequently be referred to by their Malay names. At times, however, to avoid overburdening the account, I have used descriptive English terms, which will be understood to apply to the types specified below, though the terms could include other types as well.

The main types of nets, with some details of their cost and use are as follows [1]—all costs being given as in 1940 :

Pukat Tarek : This, the commonest type of net in both States, is a seine of about 1 in. mesh, up to 200 fathoms long and 4 fathoms deep, made of cotton thread. (When the term " seine " is used in the text of this book, this net is meant.) It is normally operated from a single large boat, and is commonly hauled up on to the beach to take the catch. It takes many types of fish, but a variety with a very small mesh is used particularly for anchovy. These nets vary considerably in cost, according to size. In most areas of Trengganu from $300 to $400 was a normal price in 1940, though small ones costing from $150 upwards are also used. In Kelantan these nets are apt to be larger, better and dearer than in Trengganu ; some may cost $1,000 or even more when new. In some areas of Trengganu it is a common practice to make up seines from old nets of other types— from lift-nets at Tanjong Kampong (Kuala Trengganu), or from *pukat payang* at Kuala Besut.

Another variant of this type of net is the *pukat tarek sungai*, a river seine, smaller and less costly than that used at sea.

Pukat Payang : This is a large net up to 200 fathoms long and 4 fathoms deep, used for taking pelagic fish of many types, and operated from a single large boat. The net is purse-shaped, the central part being made of cotton thread and having a mesh of about $\frac{1}{2}$ in. ; the mesh increases out to the wings, which are of ramie fibre, with about 1 ft. (or 1 cubit) mesh at the ends. The wings serve primarily as a guard to prevent the fish from escaping as they move towards the centre " belly ", but occasionally a shark or saw-fish is caught in them. *Pukat payang*, like ordinary seines, are very variable in size and cost. There are two major variants, a smaller one costing about $100 and a larger one costing from about $250 to $500 new. The smaller nets are worked from a *kolek* (*lichung*) type of boat, the larger ones from a *pĕrahu payang*. The latter craft has a crew of 15–18 men.

Pukat Pĕtaram : This is essentially of the same type as the *pukat payang*, and may even be a smaller version of that net, though used in somewhat different conditions. Some of them, of very small mesh, cost only about $30, since they are cut down from second-hand large sea-fishing nets. The fibre wings are then discarded and the ordinary large outside mesh of the original net is used for the wings of the re-made net. The *pukat pĕtaram* is used at sea, but in the monsoon season, when the bar across a river-mouth is high, it is used in the estuary, taking *kikek*, prawns and rays, the last being the most lucrative catch. Under the name of *pukat kikek* this net is not approved by the

---

[1] A description of six of the main types of the nets used in Trengganu, with a brief account of their use and the types of fish taken by them is given by C. C. Brown, *op. cit.* My own observations in Trengganu were supplemented by this and by a short memorandum (in Malay) kindly supplied to me by Dato' Jaya Perkasa.

Trengganu authorities, since it is thought to hold too many small fry, but as a *pukat pětaram* (possibly with a slightly larger central mesh) it is passed. According to the Kuala Trengganu fishermen the *pukat chang* used in southern waters, mainly by Chinese, is a variety of *pukat payang* akin to this, but with cords instead of meshed wings.

Pukat Tangkul : This net, termed *takur* (or *takul*) in Kelantan and Trengganu dialect, is a type of lift-net, the fish being held in the belly as the net is raised. (The term " lift-net " used in the text will denote this common type.) The net, about 150 feet square, and of graduated mesh (Fig. 16), is made of cotton thread. It takes five boats to work, and is operated from ten to twenty miles out to sea, always in connection with fixed coco-nut frond lures known as *unjang*. The fish taken are primarily types of horse-mackerel (*Carangidae*), though at Kuala Marang and Batu Rakit, in Trengganu, I was told that pomfret are also taken by it, which is unusual on this coast. The cost and working of this net are described in detail for Kelantan in later chapters. The price in Trengganu, approximately $200, is much the same as there. A smaller type known as *pukat takur kěchil*, costing about half the price and worked from smaller boats, is also occasionally used in Kelantan.

Pukat Dalam : This is a large gill-net, of 1 in. mesh (knot to knot), 100 to 200 fathoms or more in length and 7 to 9 fathoms deep. Because of its depth it may be conveniently referred to as a " deep gill-net " or from the predominant fish it takes as the " mackerel net ". It is made of cotton thread, and is composed of sections, termed in Kelantan *uta'*, and at Kuala Trengganu at least, *bidang*.[1] A number of sections joined together form a circuit (*likung*). In Kelantan about 20 sections form a circuit, each section costing about $10 new. In Trengganu a smaller number of sections is used, and their cost is rather higher, though a complete circuit of about 120 fathoms costs about the same figure of $200.

This net is used either by day or by night, the latter being most favoured from the chance of greater profit. By day, two boats are used in combination and the fish taken are commonly jewfish and pomfret in a mixed catch. By night, a single boat is used, with a triangular float known as *nong* (in Kelantan) to mark the end of the net for completion of the circle. Here the fish taken are primarily scomber (a type of mackerel) and a carangid known as *sělar gilek* ; both appear in large shoals intermittently in dark phases of the moon.

Pukat Hanyut : This is a drift-net of fairly large mesh, made of ramie fibre, and composed of six to eight sections. Each section is about 25 fathoms long and 4 fathoms deep and costs about $14 to $16. In Trengganu, however, some of these nets are made of cotton thread at the foot, and at Kuala Marang sometimes the entire net is made of thread, the high cost of ramie being given as a reason. This net is used from a single large boat at night, with a crew of about six men. It takes a variety of large and medium-sized fish, including pomfret, jewfish, shark, dog-fish, dorab, and also rays.

There are several types of drift-net on this coast, but when the

---

[1] *Bidang* is ordinarily used in Kelantan as the classifying word when numbers of nets are mentioned, as *dua bidang pukat*, two nets.

term is used without qualification in this book, it is *pukat hanyut* that is meant.

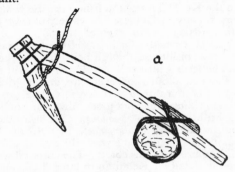

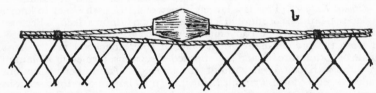

FIG. 9.—*a*. Anchor of Fishing-boat. *b*. Float attachment of small Drift-net (*Pukat těgělang*).

Pukat Talang : This is also a drift-net, akin to *pukat hanyut*, but with a larger mesh, to take primarily the *talang*, a fish of the horse-mackerel type about two to three feet in length.

FIG. 10.—Rattan Sachet for stone sinker of small Drift-net.

Pukat Těnggělam (*pukat těgělang* in the local dialect): This is a drift-net also, smaller in size and in mesh than the *pukat hanyut*, undyed, and made of cotton thread normally, though in Trengganu ramie fibre is sometimes used. This net has wooden floats and is weighted with small stones neatly enclosed in wicker sachets, which submerge it well below the surface ; in contrast to the *pukat hanyut* it may therefore be referred to as the " submerged drift-net ", or " small drift-net". It is up to 45 fathoms or so long, about 3 fathoms deep, with a mesh of the breadth of three fingers (Figs. 9*b*, 10).

It is frequently worked by a single man, but often two or three men lash their nets together as sections of a larger unit. The fishing is normally done a mile or so from shore, in a comparatively small boat. The net costs in Kelantan $10 to $12 new, but in Trengganu, where it may be much larger, it may cost up to $50. The *pukat těgělang* is the general utility net of the small fisherman or crew-man of the larger net groups. With it he takes many types of fish in the shore waters, varying according to season and including jewfish of several kinds, small dorab, cat-fish, shad, a herring known as *pěrupok*, and also prawns. If specialized for the last, it is known as *pukat udang*, prawn net. A drift-net of similar type, but specially adapted for the taking of cat-fish (*ikan duri*) is known as *pukat duri*.

Pukat Sudu : This is a purse-net of the V-shaped *pukat payang* type, but made of cotton thread. It is used in conjunction with a coco-nut frond lure and with three boats, somewhat after the fashion of the *pukat takur*. Its primary purpose is the taking of pomfret (*bawal*), and the *pukat bawal*, though listed as a separate type in the Trengganu net register, may be essentially the same. At Kuantan in Pahang, however, other fish are taken by it as well. The cost is usually between $100 and $150 new, for nets about 20 fathoms long, but larger ones may cost up to $250.

Pukat Tanggut : This net, of the purse type, is worked from a single boat with a crew of three or four men, and takes a variety of fish, but especially pomfret. Its cost is about $80 new.

Jaring : This is a gill-net, varying greatly in size and use, but commonly employed in Kelantan and Trengganu waters as a *jaring tamban*, a drift-net about 30 fathoms long for " sprats " and other small clupeid fish. In this form it is made of cotton thread, of fine construction and small mesh, and is not dyed—in contrast to most of the other nets. It is operated by day from one or two boats, each with a crew of about four men. The net costs about $40 new. A small type of this net is also used by dorab line fishers to obtain their bait, and then costs only a few dollars.

Pukat Todak : This is a net adapted for taking small garfish, and has a mesh of about one inch. Only very few of these nets are in operation on the Kelantan-Trengganu coast, and their cost is small.

Takur Baring : In Kelantan this is a small push-net used for taking shrimps (*udang baring*) in water just off the beach ; it is operated by two or four men (or even with a woman included) working waist-deep or shoulder-deep. Its cost is about $20 new. In Trengganu there is a larger variety called by the same name, but operated from a boat at sea in depths up to seven fathoms. Such a net costs about $40 at the present time, though at Kuala Trengganu it used to cost $50 or so before many men had learnt the art—there was formerly only one man on the Tanjong who made them.

In addition to the nets described, all of which with the exception of the *takur baring* of Kelantan are used from boats at sea, there are two types of hand-net which are very popular, especially in the monsoon season when the boats cannot put out. One is the *jala*, common in all parts of Malaya. This is a circular casting-net weighted by heavy rings ; it is flung in a spreading sweep with a beautiful motion to cover a fish or a shoal in shallow water. (I once counted 505 small fish taken in one cast.) The major expense in this net is the cost of the tin rings, which take about 9 lb. of metal, costing about $4, with 35 cents for casting them. The other type is the *saup* or *sodok*, a triangular net with a long handle, used out in the surf in the monsoon season to scoop up grey mullet. The cost of this net is small, being about a dollar. Apart from nets there are various types of fish-traps (*bělat*) of which the largest is the *kelong*, a very elaborate erection of stakes leading the fish in to the central enclosure. The coast of Kelantan and Trengganu, because of its exposure, is not favourable to the use of these *kelong*, but in recent years several have been erected off Tumpat in Kelantan and another was put up in 1940 off Kuala Besut. This last cost $1,000, but there is great variation in their cost, according to size. These traps take a great variety of fish. A portable trap of a simple widespread kind known as *bubu* (a generic name) is used in Kelantan for taking snapper (*ikan merah*). It is constructed of rattan, and costs about $2 or so apiece.

There is some specialization in the different types of equipment and fishing in the various coastal areas, and an indication of this in respect of the main types of net used is given in Fig. 3.

CAPITAL INVESTED IN FISHING-BOATS AND GEAR

In assessing the capital value of the equipment in an industry such as Malay fishing there are three methods which can in theory be followed. The first is that of basing the calculation on the *value of the equipment new*, that is, what it would cost at present prices to fit out the industry on the hypothesis of all the capital having been dissipated at a stroke. Given the total number of items of each type of equipment, and the present prices for new items, the calculation is simple. This was the method followed by the Fisheries Department in estimating the total capital value of all licensed fishing boats and gear in the Straits Settlements and Federated Malay States in 1931. But since the Malay practice is to buy a great number of boats and nets second-hand the figures have little practical relevance, except as setting a maximal limit.

A second method would be to take the *actual purchase price* of each item, thus obtaining—when correction had been made for any possible changes in value of money—the capital funds originally invested in the industry. Starting from this basis and considering the results in conjunction with returns from the equipment in the interim, conclusions could be formed about recoupment of capital, rate of profits, comparative advantages of different forms of investment, and the like. This method, always difficult for empirical investigation on a large scale, is impossible for all the fishermen of Kelantan and Trengganu, with practically no written records. For the community of the Perupok area I collected some data of this type, but even here it was not feasible to cover the whole field, and I had to be content with samples.

The third method, which is followed here, is to estimate the *current value* of the equipment, that is, what it would fetch at current market prices should it be transferred to other owners. The calculation is based upon knowledge of the new prices for the various types of equipment, and estimates of their average rates of depreciation, checked by sample data of prices actually paid for items of equipment of various age. It assumes that the various items of equipment now in use are spread evenly over the time range through which they are employable, from a few items just acquired brand-new to a few so old that they are on the margin of profitable employment and barely now saleable at any price. A completely even distribution is in fact unlikely, but my enquiries showed that fishing-boats and nets were fairly well spaced out at all stages of physical depreciation. But since in the case of boats depreciation of value appears to accelerate in the latter stages of a life of an approximate average of 15 to 20 years, I have taken as the current market value of the total equipment of any class a little more than half the present price of acquiring this equipment in a new state.

Details of the calculation have been omitted here, but comparison of the numbers of the various types of boats and nets (see Appendix III) with the prices given earlier will allow the results to be checked.

A review of the total capital in equipment of the Kelantan and Trengganu fishermen gives the estimate at current values (in 1939–40) shown in Table 1 on page 56.

In estimating the level of capital investment per head account must be taken of the fairly wide margin of error in the estimates of the numbers of fishermen. But if the number of Kelantan fishermen be taken at 6,500, a probable figure, the level of capital per head at current prices (1939–40) would be about $38. The

TABLE 1

VALUE OF FISHING EQUIPMENT IN KELANTAN AND TRENGGANU

			$
Kelantan :	Value of	fishing-boats . . .	150,000
	,, ,,	boat gear	17,500
	,, ,,	nets and traps . . .	80,000
	,, ,,	line-fishing gear . . .	1,200
			248,700
Trengganu :	Value of	fishing-boats . . .	238,750
	,, ,,	boat gear	27,750
	,, ,,	nets and traps . . .	100,000
	,, ,,	line-fishing gear . . .	1,600
			368,100
		Total value :	616,800

corresponding level for Trengganu, with a probable 10,000 fishermen, would be about $37. The difference is not significant, and all that one is entitled to infer is that for both States together the average value of capital equipment per head of fishermen is between $35 and $40.

FINANCE OF BOATS AND NETS

The variation in the command of free capital by fishermen in the different areas along the Kelantan-Trengganu coast is considerable. This account is based on a brief comparative survey in 1940. (A detailed analysis of the process of financing production in the Perupok area of Kelantan is given in Chapters V and VI.) The most significant differences occur in the purchase or use of the larger types of boats and nets. In essence the situation may be described in terms of contrast of principles : cash or credit ; build or buy ; own or borrow ; unitary or multiple control. In any given case, however, these principles may not be exclusively followed.

The methods of obtaining the necessary capital equipment for fishing, in more detail, are as follows :

i. *Full purchase, with unitary ownership.* Payment of the full price by a single buyer at the time of purchase is fairly common with boats, though rare with nets. The inducement is a substantial reduction of the price for cash down. A fisherman at Kuala Marang told me that he bought his lift-net for $200 cash, whereas if he had got it on time payment the price would have

been $250. In Kelantan a difference of $30 or $40 is usual.
Buying of nets for cash, when it does occur, is normally for
second-hand ones.

ii. *Putting up the capital for manufacture.* If a man has enough
capital to pay cash for a new net he usually prefers to have it
made for him, and so avoid the entrepreneur's profit. He is
said to " make it himself ", though he buys the yarn and pays
piece-rates to women home-workers to spin the thread and make
up the net in sections for him.   Men of substance usually manage
to have one net in use and another in process of manufacture,
ready to substitute for the first when this gets too old or torn, or
when a good offer is made for it. Sometimes a man manages to
have two nets completed and uses them alternately. With this
is linked the practice of making nets specifically for sale.

In Kelantan many nets are made locally. But the Bachok
district, especially the Perupok area, makes many nets for sale
elsewhere, some even going to Trengganu and to Patani. In
Trengganu the practice varies according to local skill and command of free capital. Purse-nets and seines are commonly made
in the area where they are used, but lift-nets are often imported
from other areas. The Besut area mostly makes its own nets,
though the poor community of Ayer Tawar, which uses lift-nets,
usually gets them from Kelantan. At Batu Rakit most nets are
made locally, but some are got from Batu Lipo, a little farther
down the coast. Only one man in Batu Rakit, I was told, makes
nets for sale—about two or three a year—whereas in Batu Lipo
about fifteen men make nets for sale. Kuala Trengganu supplies
its own nets, while Kuala Marang, the centre for much lift-net
fishing, makes some but buys most from Batu Lipo and other
centres to the north. I was told of only one man there who
made nets for sale. Chenering, a small village between Kuala
Trengganu and Kuala Marang, makes about ten *pukat sudu* a
year ; it does not use them but sells them to villages to the south.
At Dungun, seines, lift-nets and some other nets are made, but
some lift-nets are also bought from Kuala Trengganu. At
Kemaman many nets are made, but lift-nets were formerly
bought from Bachok and in recent years have been obtained
from Batu Lipo.

Boats are often made to order, cash being advanced by the
client to the builder for the purchase of materials, and the balance
paid when the craft is finished. Tumpat, Bachok, Kuala
Trengganu and Kuantan (in Pahang) are important building

centres which serve many of the other fishing communities.   But there is a great trade in second-hand boats.

Where capital is put into the manufacture of nets or boats it is normally supplied by single individuals, not by men in combination.

iii. *Purchase on time payment.*   This is the commonest method of acquiring capital equipment, both boats and nets, but especially the latter.   The high price of the equipment in such cases, as opposed to purchase for cash, means that interest is charged, though concealed.   The amount of cash initially put down varies from nil to about half the purchase price.   At Dungun, for instance, I was told that the normal practice is to pay half down, except where the parties are kinsmen, when no money is handed over at the beginning.   The advantage of this system to the buyer is that it allows payment to be made out of current income from the equipment, and the repayment is often proportioned to the takings.   A detailed analysis of the workings of this system in the Perupok area is given later.   The time payment system is known as *běli běrutang* (*běrhutang*), " buying with a debt ".

iv. *Purchase on borrowed capital.*   When the seller of equipment is unwilling to stand out of any substantial part of his capital for long, cash may be borrowed by the buyer from someone else. This is a common practice.   A percentage of the takings from the boat or net is then handed over to the lender of the money whenever the periodic distribution of returns is made.   This percentage is not repayment of principal, which is a separate affair ; it is interest, though the Malay prefers to conceal it under a name equivalent to " share " or " commission " (*bagian* ; in more sophisticated circles a form of the English word *commission* is sometimes used).   In Kelantan the " share " allotted thus to the creditor is normally half that which is allotted to the item of equipment in the distribution.   In Trengganu the " share " is often a smaller fraction (see under *Daganang* below).   In Dungun borrowed money—usually from Chinese—pays interest at a rate which may vary according to agreement but which is commonly 5 per cent. per month, or the equivalent of one crew member's share of the total takings.   In Paka the conditions were said to vary also according to agreement, but the lender often took as interest two shares out of the specific allotment to the net per $100 of capital he put in.   At Kemasik interest is taken by dividing the net's share into three parts, one-third going to the lender of the money and two-thirds to the user of the net.

v. *Acquisition in partnership.* Whereas boats are nearly always held by individual owners, nets are often held in some form of multiple ownership. In Kelantan the practice is known as *masok konsi* (entering a combine), and in Trengganu and Pahang as *bĕratang* (*bĕrantam*, to club together to pay). The most usual form is for one man to put up the initial capital for the purchase or manufacture of the net, and then to admit several others into partnership. They put up no cash, but agree to acquire an interest in the net at a figure somewhat higher than its cost. In consideration of the work they do on it in repairs, etc., they share in the proceeds after the agreed figure has been passed in the net's allotment from the takings. In a sense they buy the net on time payment but put in their capital in the form of labour. In some places, however, as at Paka and at Ayer Tawar in Trengganu, where the community is poor and no individual can muster the full initial capital or is willing to undertake the sole responsibility of borrowing or buying on time payment, all the partners assume a more equal rôle. Here the finance may be a combination of contribution of savings, borrowing, and time payment.

vi. *Borrowing of equipment without transfer of ownership.* Temporary borrowing of equipment, even when the borrower has command of liquid capital, is not uncommon, and at times a boat may be used for a whole season by a man other than its owner. When the owner is a widow, or a man unable from illness to go to sea, or a man who has surplus equipment, the use of the boat or net assumes the character of a regular hire. In Kelantan, a half share of the specific takings of the equipment is normally given to the owner by the borrower. This principle also operates in Trengganu, and I was given instances in Batu Rakit and Kuala Trengganu. But at Kuala Besut, if a *payang* boat and net are run by someone else than the owner, the practice is for the share of the equipment to be divided into three, one-third going to the user and two-thirds to the owner.

Towards the south a different practice obtains. At Kemaman and in north Pahang, in particular, the owner of boat or net or both is often a Chinese fish-dealer, who may take only a small percentage of the takings of the equipment, or even may forego it altogether, getting his interest and return of principal indirectly through his monopoly of the purchase of the fish.

This last point raises the important problem of the relation between Chinese capital and Malay fishing. It is of little significance in Kelantan, but is a serious one in south Trengganu, Pahang and farther to the south. As the Chinese in south Trengganu and north Pahang are not fishermen but fish-dealers, the problem may be approached from the angle of fish-buying.

FISH-BUYING AND MONEY-LENDING

In Trengganu the system of borrowing cash or equipment has often crystallized into a financial relationship between fishermen and fish-buyers, especially those who cure for export. The marked character of this relationship in Trengganu, as against its relatively small development in Kelantan, is to be equated with the much greater importance of the export of cured fish from the former State. It is important not only for an understanding of the fishing economy, but also because of its bearing on social relations between Chinese and Malays.

The term used for a fish-buyer who stands in this special relationship with a set of fishermen is commonly *daganang* (*dagangan*), but there are variant forms. Dato' Jaya Perkasa, whose opinion on Trengganu fishing matters is of great weight, holds that the correct form is *laganang*, and in this was corroborated by Tungku Wok, who added that other forms were the dialect (*pĕlat*) of the fisherfolk.[1] Specific inquiry from a number of fishermen produced *laganang* and *naganang* at Kemaman, and *daganang* at Kuala Trengganu, Dungun and elsewhere ; since the last-named seems to be the most widespread I use it here.

The *daganang* system varies in intensity. In its mildest form it corresponds to the Kelantan *tangkap* system (see Chapter VII). The buyer merely acts as a kind of insurance agent for the seller and takes a commission for his guarantee. But his association or contract with the particular net-group usually rests upon more than a simple agreement to look after their interests for a consideration. He has usually lent money to the fishermen for purchase of net or boats, and so has been able to attach them to himself on favourable terms, having a pre-emptive right over the catch and taking a commission as well to cover his interest charges. The system may go even further. The *daganang* may be the actual owner of the fishing equipment, and the fishermen may thus be virtually wage-earners at piece rates. Sometimes

[1] *Daganang*, correctly *dagangan*, is derived from *dagang*, meaning literally a foreigner, with a secondary meaning of merchant. *Laganang*, correctly *langganan*, means a regular customer. (R. J. Wilkinson, *A Malay-English Dictionary* (Mytilene, 1932).)

the *daganang* is not only the lender of capital for the purchase of equipment, or owner of that equipment, but also the lender of consumption goods, particularly rice, to maintain the fishermen during the monsoon season. He thus gets a double profit, and attaches his clients to him more firmly still by their obligation to repay him during the fishing season. It is mainly Chinese who have elaborated the system in this way.

Since the *daganang* system varies from one locality to another its economic results can be fully appreciated only by a regional analysis, which takes into consideration how boats and nets are owned and how earnings are divided. Details of this are given for the main fishing areas on the Kelantan-Trengganu coast in Appendices IV and V, and only some brief general conclusions are given here.

The *daganang* concept covers in practice three rather different elements. One is a commission for services rendered in ensuring the receipt of the sale price of the fish ; the guarantor has the right of buying the catch if he wishes, but if he does not exercise this right he has the obligation of seeing that the buyer does not evade payment or reduce it from the contract price. The second element is a commission which is really interest on the loan of capital, irrespective of whether the lender controls the sale of the fish or not. The third element is exclusive control of disposal of the fish because capital has been invested in the equipment ; here it may be immaterial whether any direct return for the investment is received since the controller gets sufficient profit by handling the catch—he may even *give* boats and nets to the fishermen when the price for their catch is absolutely at his discretion.

It will be clear from the details given in the Appendices that the essentials of the first element occur in Kelantan and in the north of Trengganu (under the name of *tangkap* or *tětap*), but that it is only as one moves southwards that the second and third elements become intensified. The third in particular is associated with the more developed Chinese control of fish-curing and fish export, where a larger command of free capital enables them to neglect more obvious and immediate returns on their investment in fishing equipment in favour of middlemen's profits.

How do the Malay fishermen view this relationship with the Chinese ? Opinions vary. At Kemaman the answer of a group of fishermen was that they liked it, " because they can get hold of cash easily ". In several other places, however, the fishermen complained at the low prices they received from the Chinese to whom they were bound to sell. A man at Beserah, asked how the fishermen liked the system, replied : " It's not that we don't like it, Tuan—we get enough to eat, but there's no chance of saving." There, in 1940, some difficulty arose through an application to the government by some Chinese dealers for permission to employ Chinese fishermen. This was a novel move in that region, but they justified it by the argument, which was not contradicted, that they could not get enough fish to handle since all the Malay fishermen were tied to other Chinese dealers. The Malay fishermen objected to the proposed competition. As one of them put it to me, they were in good relations with the Chinese ashore, as dealers, but if Chinese came out to sea it would be different. The only work open to Malays was fishing ; where else could they go if the Chinese displaced them there ? The upshot was that a licence was granted to the applicants for one year on condition that only half of the fishermen in the group should be Chinese and the other half should be Malays.

There is no doubt, however, that in south Trengganu and north Pahang, as in many areas on the west coast of Malaya, the heavy indebtedness of the Malay fishermen to Chinese dealers, coupled with the control of fish prices by the latter and the threat or reality of active Chinese competition at sea, are serious problems. The administration has realized this, and has made efforts to meet them, as by the formation of coöperative Malay associations for fish marketing. But so far the success of these measures has been only moderate.

VIIIA HAULING IN A SEINE
A large crowd of men, women and children gathers to assist and get a few fish.

VIIIB NET-MENDING ON THE BEACH
The worker has doffed his sun-hat in the late afternoon. In the background a boy is learning net-craft.

IXA LIFT-NET FISHING: " LISTENING " FOR FISH
*The boat of the master fisherman goes in above the net; at the stern a man is about to submerge
and " listen " for the shoal of fish.*

IXB LIFT-NET FISHING: HAULING THE NET
The boats are drawing together as the crews haul on the corners of the net.

CHAPTER III

STRUCTURE OF A SAMPLE FISHING COMMUNITY

This and succeeding chapters analyse in detail the economics of a sample fishing community, on the Kelantan coast of Malaya. As a background to the study of the workings of the economic system some explanation is needed of the nature and history of the community, the composition of its population and how the occupations of its people are distributed.

THE PERUPOK AREA, KELANTAN

The area for intensive study had to be as far as possible representative, with a flourishing fishing industry using a variety of methods, in fairly good touch with a range of markets, and preferably one where fishing could be observed in relation to other occupations. The choice was made of a community in the Bachok district, comprising mainly the villages of Perupok, Kubang Kawoh, Paya Mengkuang and Pantai Damat, which formed a major part of the two parishes (Mukim) of Perupok and Paya Mengkuang. The total population was about 2,000, and as a basis for quantitative work a social and economic census was taken for the coastal section of the area, as a solid block. This census, which covered 1,301 people, in 331 households, has formed the basis for the statistical material on occupations, property, etc., in this and later chapters, though much of the descriptive account refers to the area as a whole.

An outstanding feature of the area for the fishing industry, as in general on the east coast of Malaya, is the broad sandy beach with its gentle slope. This runs to the north-west as far as the mouth of the Kemassin River (Fig. 11, at end of book) and southeast as far as the mouth of the Melawi stream, a total distance of nearly 8 miles, with no substantial watercourse outlets to break it. It fronts the villages mentioned for nearly a mile. For them it serves as a landing-place for all the boats, an initial market-place for the fish and a focus for much net-mending and other activity. Above the beach for most of its length are rows of coco-nut palms and among them, a few yards inland, begin the houses and sheds of the villagers. Parallel to the beach, a hundred yards or so

behind, runs a sandy road which serves as one of its main functions
the traffic of the small motor-buses which carry most of the fish
to the inland centres.   Branch tracks at intervals give the buses
access to the beach.   More houses, thickly placed among palms
and orchards, lie beyond the road for a little distance till a
watercourse (*alur*) is reached.   Beyond this for about half a mile
are rice-fields, dotted with dwellings and clumps of trees, and
stretching back to the Kemassin River.   This river, which here
runs parallel to the coast, forms a boundary between the fishing
community and the wholly agricultural communities of the great
Kelantan plain.

No very definite boundary separates the villages from one
another and they form a geographical unit, marked off by open
land from the little town of Bachok to the south and from the
village of Kubang Golok to the north.   (For convenience the
unit may be referred to as the Perupok area.)   But socially they
have some individuality, and show rivalry particularly in fishing.
Their main marketing centre is Perupok, hardly more than a
village, but having a square of shops and a market-shed in the
middle for the sale of rice, vegetables, hardware, cloth and other
staple items.   (Fish, however, is rarely sold there ; the market
for it is the beach.)   A few shops are also concentrated at the
other end of the area, along a tiny street in Pantai Damat, and
some are scattered along the road, while occasional solitary stalls
offer snacks and vegetables.[1]   This diversity of market facilities
is surprising, since Bachok is little more than a mile away.   But
it is characteristic of much of the Kelantan coastal area and of
the dual fishing-agricultural peasant economy of the region.

ITS ECONOMIC HISTORY

A peasant society, whatever be its traditional roots, need not
be a changeless entity ; it may have undergone considerable
internal change.   This is so of the Perupok community, which
is in some ways of very recent growth, and it is advisable here
to trace briefly its development in the last seventy years or so.
There are, of course, no written records, but data for the earlier
periods were obtainable from the memories of a few old men
who could look back on the time of the Angin Běsar, the great

[1] See plan of the area in Fig. 12 (at end of book), and an analysis of the marketing
system for consumer's goods in Rosemary Firth, " Housekeeping Among Malay
Peasants ", *London School of Economics Monographs on Social Anthropology*, No. 7 (London,
1943).

cyclone of 1880, which is the most remote dating point used by these coastal folk for events within living memory.

Till the last half of the nineteenth century settlement on the Perupok coast was scanty ; most of the people lived inland near the Kemassin River, at Kampong Sungai and Kampong Panjang. There they cultivated rice but went to sea occasionally. The land down to the shore was covered in jungle, in which tigers roamed—as recently as about 1880 they came out from the watercourse at the back of the present road to take cattle. The unsettled conditions of the east coast in the middle of the century are shown by stories that raiders used to come up from Johore, in boats of twenty paddlers or more, seeking slaves. The grand-father of a local woman was seized by one of these bands when he was playing on the beach, but was lucky enough to escape later, and returned home. Even by about 1880 there were only half a dozen houses by the shore, mostly on the site of the present Perupok village.

The main types of fishing then were line-work for dorab and sea-bream, and netting *pělato* (a small carangid). A variant of the lift-net, known as *takur chokeh*, was used, from one small boat, in waters close to shore ; a simple form of lure, a small " parent *unjang* " (see p. 99) was employed. Fishing was largely for home consumption, but there was barter of fish and anchovies for rice and timber with the inland folk across the river ; the fishermen did not harvest enough rice for all their needs and the inland folk had little cash to spare for fish. One dorab then brought 3 or 4 gantangs of rice (about half its value in money in 1940). Barter with kinsfolk in the form of reciprocal gifts—a practice still current to some degree—was then very common. But money was in use. Rice, if bought, cost 25 cents for 7 gantangs. (The old headman who told me this, after a vivid description of the ravages of the Great Cyclone, said impressively that immediately afterwards the price of rice rose to a dollar for 8 gantangs—which is about half its price in 1940 !) About that time the fish from the Bachok and Perupok areas served a circumscribed local market and were not taken to Kota Bharu as at present ; that town was served by fish brought from Tumpat up the Kelantan River. The Perupok fish sold was taken by dealers in baskets on their carrying poles to the inland villages along footpaths, as there were no roads near.

About 1890 prices still were low. An old man from Kota Bharu stated that there mackerel cost 1 cent for 6 and *sělar kuning*

5 cents a hundred—prices in 1940 being 5 or 6 times this—while rice was 1½ cents per quart—about one-quarter of the 1940 price.

By about 1900 the fishermen had become more venturesome, with an improvement in their equipment. The present larger type of lift-net had been developed, using five boats, the present technique of fish lures adopted, and the lures themselves were set several miles farther out from the shore. According to local opinion, the change was of internal origin, in the Bachok district, not the result of outside influences. About 1900 there were three such nets in the area—one in Perupok village, one in Kubang Golok, and one in Pantai Damat. During the next twenty years their numbers grew to seven—including one in Kubang Golok ; the cost of a net was then about $500. The local population was still not very large, but people had begun to move into the present villages from others to the south along the beach, near Bachok. At this time mechanical transport was still lacking, and the fish were shared between the carrying-pole dealers, who bought for the fresh market, and Chinese and Malay dealers who bought for drying and export to Singapore. The former, who numbered a hundred or more, used to gather in groups of fifteen to twenty round a boat, and buy first, the dried-fish dealers waiting to take the remainder at a lower price. The expert fisherman sold the catch himself, and there was not the same elaboration of middlemen as at present. The dried-fish export was by no means simply in Chinese hands ; quite a few Malays were engaged, and often took their fish down to Singapore themselves. The women of the family also used to cook fish at night, as at present, and take it for sale to the inland villages, but in the absence of roads they went on foot and reached Melor, Peringat and Kubang Kriang, but did not go as far afield as Pasir Mas, which they do now.

About 1920, however, the road to Bachok was put through and altered the fishing economy considerably. Buses began to run, and opportunities of a more extensive fresh-fish trade developed, while the trade in dried fish tended to fall off. With the opening up of the fresh-fish market came also a change in the organization of the nets at sea. Formerly the fish were brought in by the boats actually engaged in working the net, or were sold to dealers who went out in their own craft to collect them, as is done still on many other parts of the coast. With the need to seize the fresh-fish market early in the day, however, came the development of the present system whereby the carrier of the catch is not a free middleman but is a member of the net-group organization.   This change led also to a change in the method of distribution of the fishing yield—perhaps about fifteen years ago.   It was responsible also for an increase in the wholesale price of fish, which fifteen or twenty years ago was only about $8 a boatload—for drying—whereas now a similar catch sold for the fresh market brings several times that figure.

About 1926 there were ten lift-nets in the Perupok area, among their owners being several expert fishermen still practising. From this time onwards the economy rapidly began to assume its present form.   Houses grew up along the road in numbers, bus transport became more regular, shops increased with the growing population, and about 1933 the market-place of Perupok was laid out in proper form, with permanent benches replacing the bamboo platforms.   As the buses tended to take more and more of the fresh fish the numbers of the carrying-pole dealers declined—though in 1940 there were still thirty or forty engaged— and they tended to take a subsidiary place in the marketing scheme on the beach.   Moreover, the life of the fishermen became more completely divorced from agriculture, and many of them came to live near the beach, with the sea as the almost sole source of their income, cultivating no rice, and in some cases owning no land.   During the last two decades, too, money transactions became the rule, and barter became rare.   In short, the life of the community finally merged much more completely into the general economy of the State.

COMPOSITION OF ITS POPULATION

The Perupok area, as a cross-section of the Kelantan coastal region, is not isolated, and there is considerable movement of population into and out of it.

The resident population consists of three elements.   Most important are the local Malays, describing themselves as " people from here " (*orang sini*), that is, born in the area roughly from the north of Bachok coastwise of the Kemassin River to its mouth. Of next importance are the non-local Malays, specified by their origin—Jelawat people, Tumpat people, " inland people " (*orang darat*), etc.   In this category are a few Malays from Trengganu and from Patani in southern Siam, which has a Malay population closely allied to that in Kelantan.   There were in all about 5 Malays from outside Kelantan and 20 from other areas in

Kelantan, as permanent residents. There were no Malays from the west coast living in the area. These non-local Malays are easily absorbed, but their immigration has certain economic results for them. One or more of their children may be adopted by their kinsfolk in their original home. Adjustments are necessary for them to maintain income from rice, coco-nut and rubber lands there ; such lands must be worked by their kin (and the proceeds shared) or leased, or visited periodically. They suffer when they hold a circumcision or marriage feast, since they cannot hope to assemble such a large body of kinsfolk contributors as a local person can, and they may be at a disadvantage for other types of economic coöperation as well. These points are mentioned to show that while Perupok is a comparatively new development as a coastal settlement, the shift of population has been quite local, and that there are solid reasons for this.

The third element in the area is non-Malay, mainly Indian and Chinese, and numbers about a score in all. (A rough count was made but the figures are not included in the census referred to above, which covers only local and non-local Malays.) There was one Siamese, an old woman who had married a Malay long ago and had become a Muslim. She was accepted as a full member of the community and was regarded as the head of the little group of houses where we lived. The Indians, known generically as *orang Këling*, are mostly from South India. They are Muslims, and as such they are socially more acceptable than Chinese. They all wear the Malay sarong—though retaining the Indian shirt—and speak tolerable Malay. They tend to congregate together, but of about ten in the census area two were married to Malay women and two others were living with Malay women, in separate dwellings. The children of the former were completely accepted into Malay society. Nearly all were shopkeepers or coffee-shop proprietors, though one, who used to sell nuts on the beach as a hawker, went out fishing for a time. The Chinese, also numbering about ten, remain separate in most social affairs, though some Kelantan Chinese of long standing in the State have assimilated a great deal of Malay speech and customs. These people wear Chinese dress and speak poor Malay. One old man, however (said locally to be possibly a Japanese), did embrace Islam, mainly, it appeared, in order to be able to marry a Malay woman ; it was easier to change his religion than his domestic life. The Chinese keep a couple of shops, selling general merchandise, and a coffee-shop and eating-house ; there is also a Chinese bus-owner.

In the Perupok area these foreign elements had no great economic importance, and they had Malay competitors in each occupation. This is the case in many of the coastal villages. Inland, however, especially in the towns and new villages at cross-roads, Indians and Chinese provide most of the shop-keepers. Elsewhere in Kelantan, too, there are some Siamese communities. One, a few miles to the north of Perupok, had a temple and a seminary staffed by Buddhist monks, with a settle-ment of Siamese rice-cultivators, who also made tiles and did plastering and other construction work.

Movements of population out of the area could not be accurately recorded, but they were not in great volume. There was no marked tendency for young men to emigrate to the towns. But there were some losses to the community. Young women married out into other areas ; young men went as religious students to Bachok, Kota Bharu and in one case even to Mecca. Several families had sons away at work—one was a policeman, another a rubber-tapper ; another was working in Trengganu ; another in Siam. As far as could be ascertained, these losses amounted to only about twenty cases.

EXTERNAL ECONOMIC RELATIONSHIPS

Economic and social relationships with other areas are fairly common, since many families have had kinship or marriage connections in the last few generations in Kota Bharu and other Kelantan centres, and exchange of visits takes place at intervals. And occasional visits are paid to neighbouring villages for ceremonies or recreation. But the economic contacts are of main interest for our analysis. These are of two kinds, relations up and down the coast, and relations inland through a rough quadrant of a circle with Kota Bharu and Kuala Kerai (the main towns) at the extremities of the arc (see Fig. 2).

Coastal relationships arise primarily through the movements and the requirements of the fishermen. During the monsoon season, when the open beaches (as that of Perupok) are practically blocked for work at sea, men from Perupok and the vicinity go up to Tumpat, which has a river-mouth, a partly sheltered harbour and a northerly range of fishing. Some take their boats and also their wives and families ; they rent rooms there and stay for a month or two, fishing mainly with drift-nets. Again, about April or May, when lift-net fishing falls off, boats and crews may go off to the south, to Trengganu or Pahang, where fishing is more constant. Conversely, at times fishermen from other areas put in at Perupok. When mackerel were very plentiful there in May 1940, Tumpat men came down to fish, and several boats landed on Perupok beach in the morning to sell their catches. In March and April also, seine and purse-net boats, totalling 16 in all, sold fish on the Perupok beach because of the distance from their homes. They included boats from Kandis, Tawang, Au, To' Bali, Semerak and even Besut (Trengganu) to the south, and Kemassin to the north. About $250 worth of " foreign " fish were thus bought by Perupok dealers. Again, some fishermen from Perupok and the neighbourhood have made a practice of going far afield for a long period each season, as to Beserah in Pahang or Menaro in Siam. After the outbreak of war in 1939 restrictions were put by the authorities on such movements from one State to another, particularly across the Siamese border. But sailing up and down the coast is not easy to control, and some movement still took place.

These coastal relationships provide an extra income to the community, and facilitate the acquisition of boats of different type or superior build and the spread of improved techniques of fishing. The fishermen themselves have no strong feelings about the entry of " outsiders " in this way ; the only real friction occurs when in lift-netting one party takes fish from the prepared lures of another (p. 121).

Seasonal importation creates another coastal relationship. After the monsoon is over large sailing vessels come down from Siam with rice and salt. Most of the rice is sold in Trengganu, but fish-curers at Perupok buy large quantities of salt. Smaller craft also come from the Perhentian and Redang islands off the Trengganu coast, bringing bananas, pumpkins, nuts and turtles' eggs. Like the rice and salt, these are sold for cash on the beach, but in smaller lots, or to middlemen who sell them again to consumers. Occasionally a Perupok fishermen takes his boat over to the islands to get turtles' eggs, but usually only if he has kinsfolk there, or has a debt to collect through having sold a net or boat to someone there. Visits to Singapore by steamer are made by dealers in fish or copra (and by travellers en route to Mecca). But these are costly, and rare.

The inland contacts are more complex. Apart from the many small economic relationships with kinsfolk in the agricultural villages, of which one feature is the exchange of fish for rice, fruit and vegetables, there are three main types of market relations.

One is the demand of the coastal people for rice, cloth, oil, iron tools, coffee, sugar and other goods. This is satisfied partly by imports from Kota Bharu by local shop-keepers, and partly by occasional visits by the people themselves to Bachok, Jelawat and Kota Bharu itself. A second contact is the daily movement of the fresh-fish dealers by bus, bicycle or on foot to a string of inland settlements, of which Melor, Ketereh, Kota Bharu and Kuala Kerai are nodal points. This is supplemented by the less frequent but longer trips of the dealers in dried and cooked fish, who may go further afield, even to Dabong, far up in the jungle along the main railway line The third contact is the entry of women from across the Kemassin River, bringing vegetables, betel leaf, areca nut and spices for sale in the Perupok market. Whereas neither type of fish dealer usually brings back wares from the inland centres for local sale, these women may take back fish for retail disposal in their villages. Another type of link with the interior is given by men who go off in search of work as rubber-tappers or other estate labour in a bad fishing season. This is of small importance ; though the season in Perupok in 1940 was poor, only two men went inland for work. More important—because more regular—are the visits of fishermen inland to get timber, bamboos and rattan for their work.

Such relationships are characteristic of these fishing communities. Their economic life is not static and self-contained ; market relationships with the outside world are frequent and complex, and the economy of the community depends on them.

DISTRIBUTION BY SEX AND " ECONOMIC STAGES "

A complete analysis of sex and age distribution of the entire population of the Perupok area was not possible. But the census, covering approximately two-thirds of the people, gave the results summarized in Table 2. The classification by " economic stage " in the table has been introduced to meet the difficulty of getting records of age in years in such a community. The people take little interest in absolute age ; what concerns them most is physical capacity and rôle in economic and social life, which are to a great extent a function of age. The economic investigator is in much the same position. Without being able to assign to individuals any exact age in years he can still place them in broad categories as regards the economic scheme—whether they are young dependents upon others, growing participators in work, full working members, or persons of declining capacity. These categories are rough but useful ; they correspond broadly to childhood, young manhood and womanhood, adult life, and old age ; and they are often matched by expressions in the native idiom. Estimation of age has been used to some extent in an initial classification of the Perupok population, since three local checks are available and are used by the people themselves. These are : the Great Wind (Angin Běsar), the cyclone of 1880 ; the Pasir Puteh " rebellion " of 1915, which arose partly through a misunderstanding over a change in the form of taxation and is memorable to the peasantry through the death of its leader, known by his nickname of To' Janggut, " Grandsire Bearded " ; and the Red Water (Ayer Merah), the great silt-laden flood of 1926 which caused widespread damage. Events and people's ages are commonly measured against one or other of these historic time-points.

TABLE 2

DISTRIBUTION OF POPULATION IN THE CENSUS AREA BY SEX
AND " ECONOMIC STAGE "

Economic Stage. Category.	Males.	Females.	Total.
a	241 *	224 *	465
b	63	17	80
c	315	349	664
d	33	59	92
Totals	652 *	649 *	1,301

* Includes 16 small children, sex not accurately ascertained, but believed to be 7 male and 9 female, and listed as such.

Category a comprises children roughly up to 15 years of age, i.e. still dependent on their parents or other kinsfolk for satisfying their major wants. The limits for boys were not hard to fix : one was the circumcision ceremony, the other was going out fishing. A lad not yet circumcised or " not yet at sea " fell in category a. Girls were more difficult to place since there are no such definite marks of social and economic transition, and

the borderline between *a* and *b* females is not at all precise.
Category *b* comprises " young people ", that is males roughly
between 15 and 20 years of age, and females rather younger.
These young people are at work, but have either not yet married
or are only in the first year of married life and so rarely have an
independent household or a separate budget. Completion of
about a year of married life, at which time the pair normally
set up house for themselves, has been taken as the upper limit
for this group.

Category *c* comprises the fully effective working group of the
population (an age group of roughly 20 to 60 years of age), and
begins after about the first year of married life, when full economic
responsibilities have usually arrived. Few persons in the community escape these responsibilities. By the Mohammedan
Offences Enactment of Kelantan the failure of a parent or
guardian to secure a person's marriage at the appropriate stage
may be made the basis of legal action in the religious Court.
Except in cases of insanity or severe permanent illness all
individuals marry, and only a few, once divorced, remain
unmarried thereafter. (In collecting the material an attempt
was made incidentally to compute the effective reproductive
group, but the basis for estimation was too vague and this
sub-classification has been discarded.)

Category *d* comprises men and women no longer taking part
fully in the economic processes ; it contains all men and women
roughly over 60 years of age, but possibly contains a few women
under that age.

The table shows that the sex distribution in the area is almost
even. (A more accurate record of the sex of the small children
mentioned in the footnote would make no significant difference.)
Again, the total number of persons in group *c*, the fully
effective working group, is approximately equal to that of the
other three groups. From the first result it is clear that the
population is well balanced for division of labour between the
sexes. From the second result it appears that the effective
working group should suffice for support of the whole community.
The situation here is still more favourable if account be taken
of the fact that groups *b* and *d* are not merely passive consumers ;
they are producers, though of less efficiency than those in group *c*.
If their individual contributions are estimated at one-half the
value of those of individuals in group *c*—probably a conservative
estimate—then a theoretical total of 750 man-power units is

obtained for production. If every individual in the community (including children and old people) is taken as one consuming unit, then this means that 57 per cent. of the community's strength is available to meet its consumption demands. Such an estimate, rough as it is, invites comparison with others from peasant areas where, for instance, a large proportion of the men have been drawn off to meet demands for estate labour.[1]

Given the approximate equality in the numbers of the sexes as a whole, the most striking feature of the table is the disproportion between the numbers within each " economic stage ". In group b the proportion of males to females is more than three to one, whereas in groups c and d the women outnumber the men considerably. The significance of this lies in the fact that women tend to marry at an earlier age than men do. The marriage of lads fairly soon after puberty is common, but less so than that of girls, who in conformity with Muslim practice generally, and Malay attitudes in particular, are provided with a spouse as soon as is feasible. Metaphorically, the pressure of the marriage conventions forces women in at one end of the economic machine and out at the other at an earlier age than men. Analysis of marriage conditions in groups b, c and d [2] reinforces the same point. There are 46 more young men not yet married than there are young women, while there are 53 more women without husbands (divorced or widowed) than there are men without wives. There is a disequilibrium here, since the young unmarried men do not have family responsibilities as a rule, whereas the older divorced women or widows often have to care for children.

The average size of the household in the Perupok area is small. In the 331 cases investigated the most frequent number of persons per household was three, and the range from one to thirteen. Fig. 13 gives a diagrammatic representation of this distribution, and also shows the proportion of persons of each " economic stage " in households of different size. The household is by no means always a single economic unit. In fishing, the men of it often join different boat or net groups, and if it contains more than one married couple there is a tendency for each to

---

[1] A similar calculation for a fishing-agricultural community in Polynesia, of almost the same size, gave a figure of 52 per cent. (See my *Primitive Polynesian Economy*, 1939, 41–2). On a rough estimate from population data from sample areas in Nyasaland, the comparable figure would seem to be less than 45 per cent. (see Margaret Read, " Migrant Labour in Africa . . .", *International Labour Review*, XLV, pp. 605–31 (Montreal, 1942).

[2] A detailed analysis of the material is given by Rosemary Firth, *op. cit.* 8, 23–35.

have a separate budget, for the purchase and consumption of rice at all events.[1]

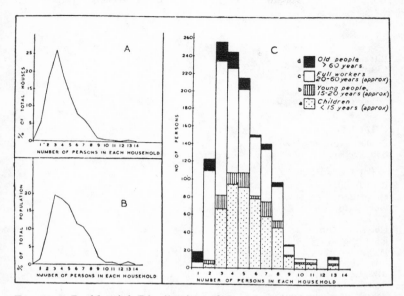

FIG. 13.—Residential Distribution of Population of Perupok census area, 1940.

OCCUPATIONAL DISTRIBUTION

Though the community is dependent upon fishing, not every man is employed as a fisherman. The total number of males above the age of about 15 years in the census area was 411, as will be seen from Table 2. The primary occupations of 401 of these were recorded, and the results are summarized in Table 3.

This shows the dominant bias of the community towards primary production. Rather more than 75 per cent. of the adult males are fishermen, 10 per cent. are middlemen catering mainly for fish marketing and the supply of consumer's goods to the fishing community, another 10 per cent. are nearly all catering for the material or social wants of the fishing community and only about 3 per cent. are on the fringe of the economic effort.

[1] For further details see Rosemary Firth, *op. cit.* 9–13.

TABLE 3

PRIMARY OCCUPATIONS OF ADULT MALES

Economic Age Category.

Occupation.	b	c	d	Total.	Group Total.
Lift-net experts	—	19	—	19	
Lift-net crew	49	169	8	226	
Other net-users	—	8	1	9	
Line fishermen	2	37	5	44	
Other fishermen (occasional)	1	4	2	7	
Ill, long period	—	2	—	2	
Total fishermen	52	239	16		307
Dealers (fish, etc.)	2	27	3	32	
Shopkeepers	—	8	—	8	
Total middlemen	2	35	3		40
Boat-builder	—	2	—	2	
Carpenter	—	3	—	3	
Goldsmith	—	2	—	2	
Tailor	—	1	—	1	
Coco-nut climber	—	1	—	1	
Rice planter	—	2	—	2	
Wage earner	—	2	—	2	
Bus driver	1	8	—	9	
Mosque official	—	2	1	3	
Religious teachers and students	4	5	—	9	
Bomor (medicine-man)	—	3	1	4	
To' Dalang (shadow player)	—	1	—	1	
Penghulu (headman)	—	1	1	2	
Total craftsmen and officials	5	33	3		41
Miscellaneous jobs and charity	—	3	3	6	
Retired (past work)	—	—	7	7	
Total miscellaneous	—	3	10		13
Total occupations	59	310	32		401
No data	4	5	1		10
Total adult males	63	315	33		411

Of the men of the older age group (*d*), approximately half are still engaged in some form of fishing ; they have tended to fall back from being *juru sělam* (lift-net experts) to ordinary crew-men, or from working with the lift-net to line fishing and other types which demand less effort and less regularity of work. A few of them have turned to fish dealing on shore. Nearly all the young unmarried men, or young men recently married (category *b*) are fishermen, most of them working with the lift-net. This last is to some extent due to their lack of capital, but it is in contrast to the situation which exists in some other parts of Malaya, of the young men abandoning their fathers' occupation for wage-earning or idleness in the town. Even Kota Bharu, the chief town of Kelantan, has its " corner boys ", who loaf about and live by odd jobs or by sponging on their relatives. This position on the Perupok coast is undoubtedly due partly to the lack of a systematic and prolonged school education, and partly to the lack of other immediate economic opportunities. For a correct appreciation of the whole position account should be taken of several young men not resident in the area at the time of our enquiry, who had been sent away by their parents to receive more extensive education than the local Koran teacher provided, and who would certainly not take up fishing if and when they returned. The religious vocation always draws away a few young men. Some, however, return to ordinary economic life, as had one of the most successful fish dealers of Pantai Damat.

Apart from the classification given above, a large number of the men have secondary sources of employment. Among the fishermen, netting with the lift-net in groups of 20 to 30 men is the major employment. But most of these fishermen go out with the gill-net when mackerel are in season, and many of them go line-fishing or with drift-net at the appropriate periods of the year. The classification in the table is based upon the fact that at times, when two or more forms of fishing are possible, certain men keep to one rather than to another, from differences in their skill or in their command of capital. Those listed as primarily line fishermen, for instance, are men who carry on this type of work when the majority have gone over to lift-net fishing ; they do this partly because of their superior skill at line work, and partly because they own or can go out in the small boats appropriate to this work. Those listed as primarily occupied with nets other than lift-nets are men who have an interest in drift-nets or seines ; most of them rarely take part in lift-net fishing, and then only to fill some temporary gap in a group. The expert lift-net fishermen, who have had special training for the delicate work of locating and identifying fish of the required type, are in most cases the owners of the nets they use, and except for work with the mackerel net they rarely engage in other types of fishing. All these factors of comparative specialization find reflection in the field of income, and will be treated later.

But primary work as a fisherman does not mean that a man is debarred from any other type of occupation. A number of

TABLE 4
SECONDARY OCCUPATIONS OF ADULT MALES

(The list at the top of the table shows primary occupations ; that at the side, secondary occupations of men in the primary categories)

	Fisherman.	Fish dealer.	Shop-keeper.	Boat-builder.	Carpenter.	Coco-nut climber.	Rice planter.	Bus owner.	Mosque official.	To' Dalang.	Peng-hulu.	Total in each Secondary Occupation.
Lift-nets made for sale	6	2	1	—	—	—	—	1	—	—	—	10
Drift-nets made for sale	1	—	—	—	—	—	—	—	—	—	—	1
Fish-traps made	1	—	—	—	—	—	—	—	—	—	—	1
Paddles, etc., made	2	—	—	—	—	—	—	—	—	—	—	2
Boat-builder	1	1	—	—	—	—	—	—	—	—	—	2
House-builder	1	—	—	1	1	1	1	—	—	—	—	2
Fisherman	—	3	—	—	1	—	—	—	1'	1'	—	9
Sinnet rope-maker	2	—	—	—	—	—	—	—	—	—	—	2
Turtle egg importer	2	—	—	—	—	—	—	—	—	—	—	2
Vegetable grower	2	—	—	—	—	—	—	—	—	—	—	3
Fruit-tree grower	1	—	—	—	—	—	—	—	—	—	—	1
Rubber planter	1	—	—	—	—	—	—	—	—	—	—	1
Rice planter	20	—	1	—	—	—	—	1'	—	—	1	22
Coco-nut climber	2	—	—	—	—	—	—	—	—	—	—	2
Coco-nut monkey owner	2	—	—	—	1'	—	—	—	—	—	—	3
Copra maker	—	6	—	—	—	—	—	—	—	—	—	7
Bomor	2	1	—	—	—	—	—	—	—	—	—	3
Shadow-play assistant	2	—	1	—	—	—	—	—	—	—	—	2
Fish dealer	10	1	—	—	—	—	—	—	2	—	—	13
Rattan dealer	—	1	—	—	—	—	—	—	—	—	—	1
Goat, sheep breeder	2	—	—	—	—	—	—	—	—	—	—	2
Shop landlord	2	—	—	—	—	—	—	—	—	—	—	2
Well-head maker	—	—	—	—	1'	1'	—	—	—	—	—	1
Total men employed out of each primary occupation	62	14	2	1	2	1	1	1	2	1	1	88 men

Note.—A dash, thus : 1', indicates that a man follows this occupation as subsidiary to another secondary one. Allowance has been made for such combinations in giving the total number of men at the end of the table.

these men have secondary employments outside the fishing field. The situation of craftsmen is similar. It was not possible in the time available for our work to make a complete record of the secondary employments of all men in the community, but the more obvious cases were noted, and the results are given in Table 4.

At least 20 per cent. of the fishermen and 22 per cent. of the total men of the community thus have some subsidiary source of income (apart from any income from interest on loans of capital in boats, net or cash, and from any returns on leased rice lands, coco-nut palms, etc.). Among the fishermen, the making of nets for sale, fish dealing, and rice planting are the chief subsidiary occupations. Those listed as fish dealers are men who consistently make a practice of doing this work when not engaged in fishing ; in addition almost every fisherman from time to time takes part in a little buying and selling of fish on the beach, just as his womenfolk do. The making of lift-nets for sale is an important source of income. The lift-net experts are the main entrepreneurs in this activity. The main secondary occupation of the fish dealers is the preparation and marketing of copra, much of it from coco-nuts sold by the more wealthy fishermen. A few fish dealers, like some men from other occupations, turn to the sea for a secondary employment when business on shore is bad, or when there is a keen demand for extra fishermen. A complete record would probably have enlarged considerably the number under this head. It will be noted that two of the Mosque officials engage in some fish dealing ; their income is small, and they both operate in a very small way on the beach.

No detailed account of reasons for choice of employment can be given here. But it is clear that personal preferences are involved. In this area as a whole, comprised between the beach and the right bank of the river, of which our census area is but a part, the two primary employments are fishing and rice cultivating. There is some tendency to a zoning of these, but many men pursue them both as the seasons allow. On the whole, no great preference seems to be shown for one employment over the other by the men who undertake them both ; the one gives fish, the other rice, both staples of diet. But from some people I received an expression of opinion on their relative merits.

The most definite was that of a young man who was normally a lift-net fisherman, but who, through his wife, had a couple of plots of rice to cultivate. He said that he preferred planting rice to going to sea—there one got wet, one had to paddle, and with several boats concerned, the benefit might be small. Here, in the rice-field, one got a living : " It's number one work ! " His father-in-law, a padi planter and occasional medicine-man, agreed. He stressed that his work lay in the fields ; he hadn't been to sea since he had begotten children ; he raised goats and cattle, and he hadn't travelled either to the west or the east (local expression for Siam and the southern States of the East coast). Another point of view was expressed by an old woman who was harvesting with her daughter. The girl's husband was an energetic cultivator of rice and vegetables, but couldn't go to sea—he got sea-sick. He had tried a couple of times, but he couldn't rouse himself and work for illness.[1] The old woman commented bitterly : " Other people eat well ; we eat ugly stuff," meaning that they were kept short of fish as a result. There is no doubt that it is the desire for plentiful fish in addition to a wish for extra cash that sends many of these rice planters to sea. One woman, harvesting alone, explained her husband's absence by saying that he was out with a lift-net ; people asked him, and he had gone because he was " hungry for sea-food ".

One of the notable features of the Kelantan peasant life is the freedom of women, especially in economic matters. Not only do they exercise an important influence on the control of the family finances, commonly acting as bankers for their husbands, but they also engage in independent enterprises, which increase the family supply of cash. Petty trading in fish and vegetables, the preparation and sale of various forms of snacks and cooked fish, mat-making, spinning and net-making, harvesting rice, tile-making, the preparation of coco-nut oil, the selling of small groceries in shops are some of the occupations followed by women. From the material of our census it was clear that at least 25 per cent. of the adult women of this community have some definite occupation which yields a regular income. And if casual or intermittent work be also taken into consideration—such as selling husband's fish, fish-gutting, etc., probably some 50 per cent. of the adult women are gainfully employed from time to time.[2]

---

[1] Sea-sickness is an unexpected frailty in a Malay. But I met another case of it, also in a padi planter. He, too, said that he was ill whenever he went out. But he had made a secondary employment for himself in the preparation of coco-nut fibre and the twining and plaiting of ropes used by fishermen.

[2] A detailed analysis of women's occupations in relation to the economy as a whole is given by Rosemary Firth, *op. cit.*, 19–23.

Our general survey of fishing communities along the Kelantan-Trengganu coast showed that this occupational analysis of the sample community was fairly typical. As one moves southwards, however, into Pahang and Johore, conditions change : Malays play less part in middlemen's activities, and the rôle of women in trading is less important. But more regional investigation in detail would be necessary for precise comparison.

CHAPTER IV

PLANNING AND ORGANIZATION OF FISHING ACTIVITIES

In this and the following five chapters I give a detailed analysis of the fishing economy in the Perupok area, as an example of the type of relationships which characterize Malay fishing communities. In the first place, one must dispose of the idea, still popularly held, that the Malay fisherman leads an easy, even lazy life ; that all he has to do is to launch his boat, catch enough fish for the family supper and return to idle away his time on the beach or in the coffee-shops. European visitors to the villages in the morning or early afternoon may see men sitting on the platforms of their houses or lounging about under the coco-nut palms. They tend to form a false impression ; they do not realize that these men are often fishermen who have been out all night at sea and are now resting before another night's work, or fish dealers whose work begins in the afternoon when the day-fishermen come in, and who will probably be busy till late in the evening curing the catch or going far inland to sell it. Another erroneous idea is that the fishermen do not plan their activity but take things as they come, and are essentially unorganized. It is true that the scale of their organization is not very large, and that there are men here, as everywhere, who are content to work in a rut. But within the scope of their organization their planning is often careful, even anxious, and may be looking months ahead.

Within the limits of the natural conditions, the capacity of their equipment and the labour supply available there are many choices to be made. It is necessary then to indicate what kind of opportunities are open to the fishermen, how they select and plan their activities, and on what basis they make their preferences. This may be done by showing how they use their labour and capital in different types of fishing.

As a preliminary, however, it should be pointed out that a basic factor affecting the organization of the fishing industry is the relative preference of the consumer—whether fisherman or not—for different kinds of fish. In poor seasons or bad weather the Malay will take what fish he can get, and pay high prices for it. If he cannot get fresh fish, he will take cured fish or *budu* (anchovies pickled in brine), though he regards them as inferior. Some fairly wealthy consumers buy imported tinned fish (*ikan sardine*) occasionally, though this is a delicacy for the peasant rather than a direct substitute for local fish.   But given a chance of selection, definite preferences are exercised, and illustrated by price differences.   There are some individual variations here, but, broadly speaking, all Malays share the same range of preferences.   Among the commoner fish, those which are esteemed most highly are Spanish mackerel, pomfret, snapper, bonito, and gizzard shad.   The first three (almost the only table fish eaten by Europeans in Malaya) are apt to be taken mainly to the urban markets, where the effective demand in terms of price is usually greater than in the villages.   In keen demand also are dorab, moonfish, large horse-mackerel, prawns, squid (? really a small cuttlefish), mackerel and grey mullet, according to season.   A little lower in the scale of preferences are the small horse-mackerel and small herring, etc., which form the bulk of the lift-net catches along the Kelantan coast.   In close competition with them are sea-bream, silver-bellies, scabbard-fish, various kinds of jewfish (some more liked than others), garfish, pike, crabs, and various other types which constitute the bulk catches from seines, small drift-nets and some kinds of hand-lining.   Among the coarser fish which tend to be a drug on the fresh fish market when other kinds are plentiful, but for which demand may be fairly keen in times of general scarcity, are sharks, rays, soles, flatheads, *Coryphaena*, *pĕrupok* (a large herring), and small fish such as *mudin* and *kirun* (see Appendix VIII).   From the point of view of a fisherman, of course, this rating expressed in relative prices has to be balanced against the probable bulk of his catch in each case.

PREFERENCES IN THE USE OF LABOUR AND CAPITAL

There are eight major types of fishing practised in the Perupok area.   In six of them, nets are used from boats at sea ; in one, hooks and lines are used from boats ; and in one, hand-nets are used in the shallow water off the beach.   Several of these types may be sub-divided according to variation in the kind of equipment or in the method of using it for different kinds of fish, so that at least a dozen ways of fishing are open to men with different skill or interests or different amounts of capital at command.

Preferences for different types of fishing are governed in the first place by some physical factors—changes in wind and weather and the seasonal run of the various kinds of fish.

All the forms of fishing in the Perupok area are subject to seasonal variation, and some of them are possible for only short periods of the year. But the most marked seasonal rhythm is that of the north-east monsoon, which usually breaks towards the end of November, and with frequent high winds and torrential rain, lasts approximately till the end of January, or even later. This is succeeded by a period of light, variable winds, often from the south or south-east, and this in turn gradually merges into a long period of steady alternate land and sea breezes. These changes in the period from February to November, however, are not at all regular, and even the monsoon itself is apt to vary from year to year in its onset, its duration, and its intensity. The monsoon of 1939–40, for instance, was characterized by absence of gales, though it was longer than usual before the north-easterly winds finally died away and regular sea fishing became possible and easy. The fishermen complained, too, that subsequent fishing was poor, that the lack of really heavy gales had not given the water its usual post-monsoon turbidity, that consequently food for the fish was lacking, and in the clear water they saw the nets more easily and tended to " bolt ".

The economic effects of the monsoon are most marked. It puts a stop to all sea fishing, except on the rare days when the surf is moderate enough to allow boats to be launched. Consequently it cuts off the major source of income. This necessitates at least an attempt at saving some portion of the income of the preceding few months. Where savings have been insufficient, it leads to borrowing in cash or in kind, or to the accumulation of debts in the shops. On the other hand, more time is available to the men for repair of boats and nets, and for craft-work as a subsidiary source of income. And, again, some income can be gained by fishing for grey mullet, which come in to the heavy surf off the beach and may be taken with a long-handled scoop-net or a casting-net. The capital needed for this is small, so the work is open to almost any fisherman. The results are sufficient to provide fish for the household meals, and often enough petty cash for coffee, cigarettes and small household wants.

The major productive rhythm is well shown by Fig. 14, which gives a synoptic impression of fishing activities in the Perupok area for a year (1939–40). This illustrates the marked break in all forms of sea fishing in December—the height of the monsoon that year—and the filling of that break by fishing with the scoop-net. It shows also how fishing with drift-nets (*pukat hanyut* and

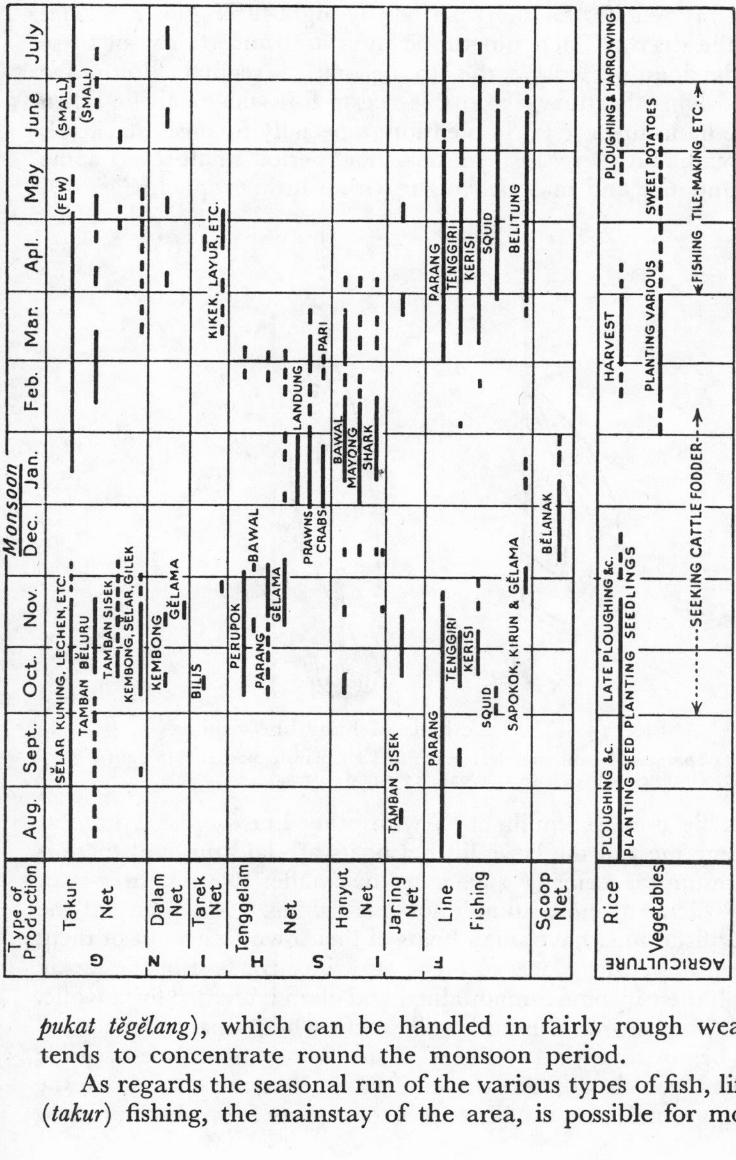

FIG. 14.—Cycle of Production, Perupok Area, 1939–40.

English equivalents of the Malay fish names are given in Appendix VIII. Solid lines indicate regular fishing; broken lines, irregular fishing. (The comparative diagram of other employments similarly indicates their main periods.)

pukat tĕgĕlang), which can be handled in fairly rough weather, tends to concentrate round the monsoon period.

As regards the seasonal run of the various types of fish, lift-net (*takur*) fishing, the mainstay of the area, is possible for most of the year. But it is interrupted not only during the monsoon, but for a couple of months, usually May and June, when the fish are too few and too small to be of commercial value. This gap is possibly relieved for short periods by night-fishing for mackerel with the deep gill-net, but the fewness and uncertainty of these periods does not allow this to become a regular alternative occupation. Fortunately, at this season fish which can be taken on the hand-line are fairly plentiful, especially Spanish mackerel, sea-bream and *Coryphaena*. For a short period small squid come in abundance and may be caught with a little grapnel (Fig. 15),

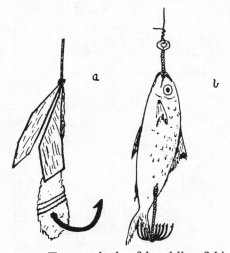

FIG. 15.—Two methods of hand-line fishing.
a. Spinner of pandanus leaf and fish for Spanish mackerel (*těnggiri*).
b. Grapnel with whole fish bait for squid (*sutung*).

this being done in conjunction with other line work. Hence at this time most of the large lift-net boats are laid up, and there is a premium on securing a place in the smaller two- or three-man craft which are most suitable for line fishing. Not many of the lift-net fishermen have small boats of their own, but most of them succeed in getting places as crew, and since the cost of the tackle is small their income is maintained and often increased for a while. Even while lift-net fishing is still worth while, Spanish mackerel have begun to appear. Every lift-net boat carries one or more spinners, which are thrown out on the way to and from the fishing grounds and usually take two or three fish, which are sold for the benefit of the crew.   When the lift-net yields are low, but places in small craft are not available, or it does not yet seem time to break up the net organization completely, some of the crew go line fishing for sea-bream in their own large craft, with four to six men.   As the line fishing in its turn begins to give a decreasing yield, the lift-net fish tend to appear once again in quantity, and so gradually a change of employment takes place once more.

Concentration on fishing with the small drift-net (*pukat tĕgĕlang*) is determined partly by the monsoon interruption in lift-net work, and partly by the presence of quantities of jewfish and large herring at the beginning of the monsoon and of *landung* (? a sea-bream) and prawns immediately afterwards.   The season for the latter, in particular, tends to overlap that of the lift-net for a couple of months, so that to some extent there is direct competition between them as alternative employments.   For any man, the decision is usually determined by whether he has a small boat or not.   Much the same competition exists between lift-net fishing and that with the large drift-net (*pukat hanyut*) in the early part of the year.   The latter cannot be run in conjunction with the lift-net or the small drift-net, since though it is carried on by night, and they are worked by day, its boats normally arrive back in the morning after the others have gone out.   In fact, the large drift-net fishermen are usually specialists, combining this work either with mackerel netting, or with line fishing for large dorab.   These fish, sometimes called the wolf-herring, are taken when small in the drift-nets, but when large are caught throughout most of the year on hooks operated from set lines with a bait of sprats.   The dorab fisher works from a one-man craft which is useless in rough weather, so this occupation dovetails fairly well with drift-net fishing conducted from a large boat which can stand heavy seas.

The other three main types of fishing, with deep gill-net, seine and *jaring*, are all directly competitive with lift-net work. The seine, in particular, depending on bulk catches either of anchovy or of other fish used primarily for drying, is in some areas preferred to the lift-net and so absorbs the major part of the labour and capital.   This is not so, however, in the Perupok area.   There this type of fishing is practised for short periods only, and very few nets are employed, as an occasional alternative to lift-net work in calm weather.   The deep gill-net (*pukat dalam*), depending primarily on scomber (a mackerel), and the *jaring*,

depending on sprats, pilchards, etc., are held available to meet
the appearance of shoals of the appropriate fish, which come
irregularly, for a short time, and often only at long intervals.
The deep gill-net, for reasons explained below, is a favourite
with the Perupok fishermen, and is a kind of " second string " to
the lift-net with them.   (Around Kuala Trengganu it serves the
same function for the seine fishermen.)   The *jaring* is somewhat
of a " third string " ; it is not as common, but is resorted to
when shoals of sprats are reported or appear in quantity in the
lift-nets.

The second set of factors governing preferences is of an
economic order.   Here comparative yield on the one hand, and
comparative capital costs on the other, are calculated.   In the
Perupok area the yield from lift-net fishing is the most regular,
and at times is high.   And though its capital costs are large, they
tend to fall primarily on a small group of men, with the major
cost of the net on one individual alone.   Hence this occupation
tends to attract a large number of men who have little or no
capital to invest, but who seek a market for their labour, and are
satisfied with a fairly constant income.   A considerable proportion
of the Perupok lift-net crews, especially the marginal crews
brought in when the number of nets increases, are agriculturalists
who welcome several months' fishing as a source of added cash
income, but who have no free capital to invest in the fishing
industry.   Fishing with the small drift-net tends to compete with
that by lift-net after the monsoon, when the yield per man from
fish and especially from prawns is, on the whole, distinctly higher
than that from the lift-net.   Hence there is a tendency for men
to carry on with this net for a month or so after the lift-net season
begins, and then drop back into the latter as prawns decrease.
But two things tend to check this competition.   One is that a
small drift-net costs about $10 and not all lift-net crew men can
raise this sum ; the other is that the regularity of lift-net work
imposes a certain allegiance on the crew.   If a man should drop
out for a few days every now and again when prawns or other
drift-net fish are plentiful, the leader of his group will probably
resent it, and " throw him away ".   Only if the group as a whole
is not working is he really free to choose each day the kind of
fishing that gives the better immediate yield.   Choices thus have
to be long-period ones.   This applies also to most other forms of
fishing, and is one reason why large drift-net (*pukat hanyut*) fishermen are not usually associated with a lift-net group, though they

may occasionally enter to fill a temporary gap. With deep gill-net and *jaring* the case is rather different, though the latter tends to find its alignment with drift-net rather than with lift-net work if it is pursued at all consistently.

The deep gill-net offers an irregular yield, but a high one.[1] Another advantage of the deep gill-net is that while competitive with the lift-net (a night out with the former almost necessarily means the sacrifice of the following day with the latter) it can be closely linked with it. The same type of large boat is used, and the crew is also large. Moreover, the composition of the net in sections, each made and sold as a separate component, at about $10 apiece, facilitates small-scale investment by crew men. Hence, when shoals of mackerel appear, it is customary for all lift-net fishing to be dropped except by those groups where few sections of deep gill-net are held, largely because of poverty. The lift-net owner, his boat captains and their crews re-form into another type of organization, using probably two of the boats of the lift-net group, with most of the crews concentrated into them, and dividing the proceeds on different principles. The mackerel gone, they re-form again. The deep gill-net thus tends to provide a useful bonus to the lift-net fishermen in return for a brief interruption of their normal occupation. To a lesser extent this is also the case with the *jaring*.

From an economic point of view line fishing is in an inter-mediate category. Though the yield from certain types is fairly high to men of skill at certain periods of the year, it is often low to others. But the cost of the tackle is small, and so line fishing forms a convenient stop-gap occupation for the lift-net fishermen.

Other factors influencing choice of occupations are technical, and include comparative skill and man-power required. Here the contrast is between lift-net and deep gill-net on the one hand, and drift-nets and line fishing on the other. (Seines are in the former class also, but are not important for the Perupok area.) With lift-net and deep gill-net the crews are comparatively large, and the expert skill required in locating the shoals of fish and making a successful cast are primarily concentrated in the person of a single man, termed the *juru sělam*. The other crew members,

---

[1] The mackerel which are taken by it appear only for about six nights or less in a month, and then by no means every month. In the nine months from October 1939 to June 1940 shoals were taken by night in only four periods, totalling 17 nights in all. Appendix VII shows the large sums obtained by a single night's work, giving the crew in some cases as much per man per night as would be gained in a week with the lift-net.

though they may be also skilled, are there as labour power.
Consequently a man with no particular qualifications as a fisherman, but who can wield a paddle competently and haul on
a rope, gravitates to these occupations. In working with drift-
nets, however, the crew is small—two or three with *pukat tĕgĕlang*
and five or six with *pukat hanyut*. So also with line fishing, where
the common practice of pooling the catch makes an unskilled
man not welcome. Most of all is individual skill needed in
fishing for dorab, where the single man in the boat stakes all on
his own craftsmanship. Hence despite the high yield of this
occupation to experts, their numbers remain few.

A further set of factors governing choice may be loosely
comprised in a " social " category. Here come such influences
as a man's desire for work of an individual kind as against
participation in group activity (the most marked contrast being
between line fishing for dorab and fishing in a lift-net group) ;
attachments to kinsfolk or friends, leading them to follow the
same employment ; disagreements leading to the break-up of
fishing associations ; or liking for travel to other areas. I have
known all these factors to operate in the Perupok area.

A few examples, particularly on the relation of lift-net fishing
to other occupations, will show how individuals are moved to
apply their capital and labour in one direction rather than in
another.

First consider two dorab fishers. One is the acknowledged
expert of the neighbourhood. " If he can't get fish, who else
can ? He is really clever ! " By his own statement he likes
sea-work, boats and trading. He has fished for dorab for about
ten years, and has a small boat for this work. In addition, he
also engages in drift-net work by night, and in fishing with the
jaring, for which he employs a large boat, gained by saving the
capital. He also imports drift-nets from the west coast and
re-sells them for a small profit, and finances the manufacture of
three or four lift-nets a year, for sale only. He has never operated
consistently as a lift-net man. Discussing his line fishing with
me one day, he said that he had just earned $1.30, and $1.50 the
day before. He added that the lift-net group of a neighbour
had sold its catch that day for $25, and asked rhetorically :
" How much per man ? Sixty cents ! Isn't it better that I
should go dorab fishing ? "

The other dorab fisher has wider interests. He prepares a
little copra, buying coco-nuts on the palm and using a monkey

to climb for them ; he cures fish for sale, and his wife makes a fair income by selling cooked fish in inland markets. But his main work is at sea. He has one section of a heavy drift-net (which he joins to that of his friend just cited above), a submerged drift-net, and a section of *jaring*. The last is used, when opportunity offers, in a medium-sized boat which he owns but which is normally attached to the lift-net group of a neighbour. He himself runs the boat for sprat fishing but not when it is in the lift-net group. His main occupation is fishing for dorab, for which he has a small boat. In the 1940 season he said that he probably would not be able to use his sprat net, since his larger boat could " rest " only when there were no fish for the lift-net. Asked why he himself did not go lift-net fishing with this boat he answered : " Too many people—there are rows ; I don't like rows." But he agreed that there was " a little more profit outside the lift-net group ". " Why don't others do the same as you do ? " I asked. He replied that as far as dorab fishing was concerned they were afraid lest they overturn at sea, with no one to help them ; with the lift-net there were always plenty of their comrades to come to their aid should the boat upset. And he advanced the same feeling of insecurity as a deterrent to drift-net work, saying succinctly : " Night ; darkness ; heavy seas ! "

Comparison of lift-net work with that with the small drift-net brought several points of view. Asked why do not more crew men of the lift-net go out with this smaller net when the larger cannot be used, one man said that they were poor and couldn't buy the nets. " They drink coffee and eat borrowed rice, how can they buy a net ? " This was the general reply. But one man, himself a user of the small net in question, added that these lift-net crews were not shrewd—they could get second-hand equipment cheap and build up their capital, as he himself had done. But poverty and lack of business sense are not the only hindrances. Hardly any expert fishermen with the lift-net are users of the small drift-net. Some are too elderly, others are well enough off to live comfortably otherwise. One such man whom I knew well had no drift-net—he said at first, jokingly, that he was too lazy to beat the net on the sand (an essential cleansing operation after each day's use). Then he added more seriously that a lift-net expert has other work to do, in looking after his net, arranging for boats and crew, and so on. " If I went out with a drift-net it would be awkward." When it is a question of continuing drift-net work after the lift-nets have begun operations again, opinions varied. One man, who spent much of his time in line fishing and drift-net work, said that the latter was all right before the lift-nets got into full swing, but that then the fish of the drift-net lost their value. (This was correct, as I saw from the beach market.) He said that he himself preferred lift-net work, and had done twenty years of it, going out with the net of his brother's wife's brother. When his brother took a second wife, however, the association had been severed. Another man, who spent most of his time with a lift-net group, stressed another point of view. After the lift-net season had started some time and he was still fishing with his drift-net, I asked him why he was not with his group. He replied : " There are prawns here ! The boat is mine, the net is mine, why should I go ? I get cash this way." As the prawn season finished I saw him back with his old group. But later on, when the lift-net fishing declined again, he invested in a small variety of this net (*pukat takur kĕchil*) needing fewer boats and men, and organized a group of his own.

The rôle of deep gill-nets as a " second string " to lift-nets is illustrated by the complaint of an ordinary crew member when the former were getting only a few hundred fish apiece. " If there are no lift-net fish, and no deep gill-net fish, how is a man to eat ? How can one stand it ? . . ." and more to the same effect. But the linkage between these two forms of investment is not automatic ; several experts with the lift-net, and some other prominent fishermen, have no deep gill-nets. Occasionally, when the mackerel fishing is very tempting, lift-net fishermen may buy sections of deep gill-net ; if the demand is great, drift-nets may have to be bought instead and converted. The use of a man's own sections of deep gill-net does not depend simply on his own preference. Unless he owns a complete circuit, which is rare, he is dependent on his associates ; and crew also may be a difficulty if the area is temporarily over-capitalized in these nets. At the height of the 1940 season some line fishermen were being " borrowed " by deep gill-net groups to make up their numbers. Normally there is no clash between the claims of lift-net and of deep gill-net, the general consensus of opinion going one way or the other. But I once saw friction over this. A lift-net expert who had no sections of deep gill-net (and told me he didn't want them), objected to one of his boat captains dropping the lift-net work (even though prospects were poor) and going out with a deep gill-net in which he and his brother had

sections.   The captain went, and the result was a " divorce " ;
the expert took on a new captain and boat.

I give these instances to show how choice in different types of
fishing is not a simple resultant of relative skill and command of
capital ;   personal   preferences   have   to   be   balanced   against
participation in organized group activity.

Apart from the problem of how choices are made between
types of employment, there are other important questions of
fishing organization : how time is utilized in the employment,
who directs the work, how effective this direction is, the nature
of intra-group and inter-group relations, and the extent to which
extra-economic factors, such as ritual, affect the organization.
These problems cannot be discussed in detail for every type of
work.   Since fishing with the lift-net (*pukat takur*) is the major
occupation in the area, and its organization is the most complex,
the remainder of this chapter will be devoted to it.   But the
principles involved are essentially the same for other types of
coöperative fishing.[1]

THE USE OF TIME IN LIFT-NET FISHING

One wishes to know concretely, how far efficient use is made
of time in this kind of fishing ; how much is wasted by inefficient
management, lack of judgement or unwillingness to take risks ;
how much is lost through purely physical limitations on the
employment of equipment ;   how much through conformity to
social conventions.   Some idea of the relative influence of these
factors can be gained from seeing the number of days on which
nets did or did not put to sea in a given period, and the reasons
given by the fishermen.   The results of analysis of my six-months'
record of lift-net fishing from this point of view are given in
Table 5.   (The number of lift-nets at sea each day is shown in
Fig. 22, p. 264.)

For convenience in comparison the material is presented in
the form of " net-days ", counting each day spent by each net
as one unit ; this is necessary since the number of nets in use
during the period fluctuated between 17 and 21.   The figures
are only approximate, since some of the reasons are not exclusive.
For instance, if a net does not go out because of insufficient crew,

---

[1] A description of the organization of deep gill-net (*pukat dalam*) fishing for mackerel
has been given in " The Coastal People of Kelantan and Trengganu ", *Geographical
Journal*, CI, 198–201 (London, 1943).

this may be due to previous poor results, due in turn either to actual lack of fish on the grounds, or to the expert's inefficiency ; or to bad judgement by some of the crew as to the possibilities of the weather. But the general picture is substantially accurate for the period, which was from the middle of November 1939 to the middle of May 1940.

TABLE 5

UTILIZATION OF TIME IN LIFT-NET FISHING

Reason for Time Lost.	Number of Net-days lost.	
A. Rough weather, impossible to put to sea . . .	817	
Rough weather, difficult to put to sea	112	
Light, contrary or no wind—difficult to reach fishing grounds	129	
Strong current, difficult to use nets	48	
" No fish "—scarcity at fishing grounds	375	
Other fishing preferred	120	
Illness of expert	4	
		1,605
B. Net not ready for fishing	41	
Mending badly-torn net	6	
Boat overturned on going out, fishing abandoned . .	1	
Insufficient crew (specific reason given)	26	
Previous poor yield	20	
Break-up of net-group on disagreement	30	
Court case over fishing	1	
Painting boat	2	
Buying boat	1	
		128
C. Muslim sabbath (Friday)	499	
Muslim holiday	57	
Preoccupation with funeral	5	
Preoccupation with marriage	1	
		562
D. No specific reason given or observed		167
Total net-days lost .		2,462
Total number of net-days out fishing. . . .		1,034
Total number of net-days theoretically available . .		3,496

The total available time (in theory) for lift-net fishing in the period of record was 3,496 net-days. In this period a total of 1,034 net-days represented the time in which fishing was actually carried out ; there was no fishing for a total of 2,462 net-days. Thus approximately only about 30 per cent. of the theoretically available time was utilized. In the table I have arranged the reasons for the lost time in four categories to allow the weight of the major factors to be clearly seen.

XA HAULING A MOTOR-BOAT OFF THE BEACH, 1963
This job needs about twenty men, as compared with eight to ten men for a sailing craft (compare
Plate VA).

XB BEFORE THE BOATS COME IN
Women, children and a few carrying-pole dealers patiently await the arrival of fish (Perupok,
1963).

XIA   FISH DEALERS GATHERING TO BARGAIN

A lift-net boat at Kuala Marang, Trengganu.   When fish are scarce the dealers rush out into the water before even the sail is lowered.

XIB   UNLOADING THE CATCH

A boat from a stake-trap (kelong) *has arrived at Besut.   The catch, mostly* ikan chĕrmin *(Caranx sp.) is unloaded into baskets and dealers cluster round to buy.*

Section A comprises the more purely physical factors, which either prevented fishing altogether, or appeared to limit severely the possibilities of obtaining a catch. During and immediately after the monsoon the strong north-easterly winds often made the launching of boats on the exposed beach quite impossible. About one-third of the total time lost was due to this. Nearly another one-third of the total time lost was also due to conditions distinctly unfavourable to fishing ; although a few boats—often those of one net only—put out, the great majority decided that wind or current would prevent them from reaching the fishing-ground, or casting their nets ; or that if they cast their nets they would get no return for their labour. The illness of the expert organizing fisherman (the *juru sĕlam*) intervened seldom. Where an expert was unable to go out himself he " borrowed " another to perform his work, and the fishing did not suffer.

Section B comprises broadly the technical reasons which caused a loss of time, either through accident to equipment, failure to bring the organization of equipment to bear at the right time, failure of the organizer to perform his expert function of securing fish, or breach in the organization of the net-group through disagreements among personnel. The loss of time in this specific category is only approximately 5 per cent. of the total. The comparatively poor results secured by some nets, and distrust by the crew of the ability of the particular expert probably account for some of the time lost in category D, and with fuller information might possibly raise the time lost in category B to about 10 per cent.

Section C comprises the social and religious conventions which made a drain on the time theoretically available for fishing. In all, they took about 23 per cent. of the total time lost. But it must be noted that while religious law prescribes attendance at mosque at midday on Fridays—and thus prevents lift-net fishing—it does not prohibit work as such. Thus a great part of Friday is spent by the members of each net group in mending their net and dyeing it, and some part of the other holidays as well is often so occupied. Though the time is lost to fishing, it is a necessary expenditure for the maintenance of equipment.

Section D gives the time lost for which I obtained no specific reason, and includes such factors as apathy of the expert or crew, his failure to get together a sufficient labour force for the net, or sufficient boats to work it. But in the main it covers those cases where, although the majority of nets went out, some did not, either because the expert thought the labour was not worth while, or because he and the crew had lost heart after a series of poor catches, and preferred to try other methods of fishing till the situation improved.[1]   When the lift-net fishing was poor some of the experts proceeded to sea intermittently on a kind of " day on, day off " principle, while others went out much more consistently.   In the former case it was not possible for me to say in many instances how far the apathy of the expert, his considered judgement, his preference for other types of fishing, or his failure to secure a crew were responsible.

In general the time lost under category D represents failure to use equipment to the full in cases of normal risk.   Considering the general principle that output tended to vary directly with the number of days at work, it seems that a considerable increase could have been obtained by a fuller use of opportunities and skill.   At the same time, it cannot be expected that a full use of the total of 167 net-days lost would have yielded a proportionately greater output, since many of the catches made by the nets that went out were obtained from the lures of the nets that remained at home.

A broad conclusion emerges from the table.   While a high proportion of the theoretical time available for employment was lost—the abnormally unfavourable season undoubtedly being responsible for some of this—about two-thirds of the loss was due to causes beyond the effective control of the fishermen.   Less than one-third was due to their own inefficiency in management or to conformity to social conventions.   This generalization, it is true, derives from a study of only a six months' fishing period. But since it is based upon a close empirical study it is of much greater value than the " impressions " of Malay indolence and inefficiency which are commonly put forward with assurance, but on no quantitative basis.

Turning to the individual differences in the use of resources in this field, considerable variation is seen.   It is not necessary to analyse the " time-sheets " of each net-group in detail (see Appendix VI), but the variation may be shown here by a table of frequencies.   Since five of the net-groups were formed only during the period of record, and one split up, the number of days on which each net went to sea has been expressed as a

---

[1] I observed no case in which a good catch one day caused the crew to lie off the next—the result was to increase their keenness to go out, not diminish it.   Among the reasons for this are the range of opportunities for expenditure on consumer's goods in the local shops, and the avenues for local investment of cash in land, boats and nets.

percentage of the total number of days on which, in theory, it could have been taken out.

TABLE 6

VARIATION IN USE OF TIME BY DIFFERENT NET-GROUPS

Percentage of Available
Days Utilized.

%						Number of Nets
Under 15	.	.	.	.	.	nil
15–20	.	.	.	.	.	1
20–25	.	.	.	.	.	6
25–30	.	.	.	.	.	5
30–35	.	.	.	.	.	5
35–40	.	.	.	.	.	nil
40–45	.	.	.	.	.	3
45–50	.	.	.	.	.	1
Over 50	.	.	.	.	.	nil
		Total nets concerned				21

Thus more than half of the net-groups utilized less than 30 per cent. of the time available (in theory) to them, and none of them as much as 50 per cent. of the available time. The general reasons for the low proportion of time used have been explained. The individual differences are due to several causes : the energy and power of organization of the expert is an important factor, as naturally also is his skill, which by its success keeps his crew attached to him. But another factor is the command of free capital by the expert at a prior stage, allowing him to manu-facture a new net in time to be able to substitute it at once for his old net when he sells this. The time lost by failure to have a new net ready was considerable with some net-groups in the period under review. This question of net-capital will be considered later.

ORGANIZATION OF LIFT-NET FISHING

An outline of the technical processes of lift-net fishing is necessary to make clear the economic organization.

The *pukat takur* is a large net, roughly square in shape, with a slight sag in the centre. Its outside measurements vary between 25 and 27 fathoms by the tape, or rather more by Malay measure-ment, which uses the double arm-stretch. The net is composed of seven sections of different mesh (Fig. 16) with the smallest in the centre, and to the outer section are lashed the edge-ropes which take the strain when the net is being hauled. At each corner is a strong rattan loop to which a rope is attached when the net is in use.

Five boats are necessary to work the net. One, the *pĕrahu pukat*, carries the net to and from the fishing grounds, and holds one corner when the net is being cast and hauled. Three other boats hold the other corners. Two of them, upstream as regards the current which runs continually up or down the coast, are

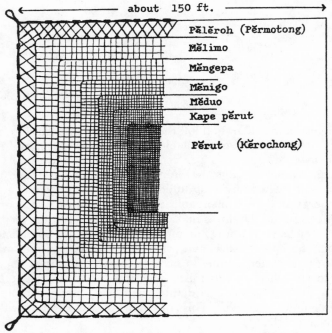

FIG. 16.—Lift-net (*Pukat takur*) : diagrammatic plan.

The mesh of the *pĕrut* will just take a thumb-tip ; that of the *mĕlimo* will take two fingers ; that of the *pĕlĕroh* is about 3 in.

known as *pĕrahu atas haruh* (upstream boats) or *pĕrahu bĕlaboh* ; the third is the *pĕrahu bawah haruh* (downstream boat). The fifth boat, that of the expert who is the leader and organizer of the group, is the *pĕrahu sampan* ; its work is manipulation of the fish lures and watching the shoals of fish.

Takur fishing is distinguished from most of the other types of net-fishing by the use of lures or artificial shelters known as *unjang*, or descriptively as *rumoh ikan*, " fish-houses ". These are constructed normally of coco-nut fronds tied at intervals on to a long rope [1] and are of two complementary types. The one is a fixed structure known as the *unjang ibu*, the " parent *unjang* ", and is set down on the fishing ground at the beginning of the fishing season, after the monsoon, and renewed when required, as after a storm. It is weighted at the bottom by a heavy stone or bag of sand, and sustained at the top by a long bamboo float. Each net-group has several of these *unjang*, often distributed over more than one fishing ground. Those of each group are often known by a distinguishing mark on the bamboo float, which projects above the water for several feet and can be seen from a distance. One net-group leaves a cluster of twigs at the top node of the bamboo ; another ties a branch at the top ; another splits and binds it in the funnel shape known as *sako*. Some *unjang*, however, bear no particular sign, and are known simply by their position on the fishing ground relative to others or to shore bearings. The other type is known as the *unjang ano'*, " child *unjang* ", which is carried out daily on the boat of the master fisherman, and is used to transfer the shoal of fish from the " parent *unjang* " to the interior of the net.

For a day's fishing the boats leave before dawn, and shape their course to the fishing ground indicated by their leader. Arrived there, he proceeds to investigate his *unjang*, looking for fish first from the bow of his boat, and then submerging himself below the surface of the water—hence his title of *juru sělam* (*sělam* meaning " to dive "). Under the water he relies mainly not on his eyes but his ears ; he hears the fish if they are there, and distinguishes by the volume of noise whether the shoal is large or small, and by the quality what kind of fish it is. This " hearing " of fish, surprising as it may seem, is well authenticated. It demands a considerable degree of training to interpret the sounds correctly, and the skill thus acquired is the principal factor entitling a man to be called *juru sělam*.

When a shoal of sufficient size has been located the work begins. The net is cast, and sinks. The " parent *unjang* " is

---

[1] In the Perupok area coco-nut fronds are the rule. Branches of other trees will also be used, but they are difficult to get. *Chělagi* is one kind used ; another is *tejor*, which the fish are said to prefer to coco-nut fronds ; another is *kěsinar*, which has the advantage of lasting a year as against the couple of months of coco-nut fronds. In Trengganu, branches are more common for *unjang ibu* (known there as *unjang běsar*). *Tejor, gělam* (paper-bark) *papi*, and *ru* (casuarina) are among the types used, either alone or mixed with coco-nut fronds. The *unjang* used by the line fishermen may be made from *daun paku piar*. The complementary *unjang ano'*, known in Trengganu as *unjang mělor*, are made from coco-nut fronds alone.

then released from its float, and allowed to sink also, attached to a long cord. The boat of the expert, the *pĕrahu sampan*, then paddles cautiously into the centre of the net, paying out the " child *unjang* " as it goes. This serves as a substitute shelter for the fish, which, if all goes well, rise to it from the sinking " parent *unjang* ". By this time the *pĕrahu sampan* has returned outside the net, and all wait for the fish to rise. If they have done so and have not " run " in another direction, the " child *unjang* " is slowly drawn up into the boat again, and the boats at the corners of the net begin to haul in. As the fronds of the *unjang* begin to come in the fish retreat, and when the loops at the corners of the net begin to appear above the water, the *unjang* is pulled in strongly, the sides of the net are hauled up, and the fish are taken.

The carriage of the fish to shore brings another element into the organization. Since five boats are needed to work the net, if one of them goes off with the catch, or is heavily laden by it, no further cast is possible that day. It is common, then, in lift-net fishing for a sixth boat to participate in the group, doing nothing in the work of the net, but acting simply as carrier of the fish. This is the *pĕrahu pĕraih*, the " dealer's boat ".

The principle on which this boat is associated with the group can vary considerably. The variations are discussed on pp. 111–12. It is necessary to say here only that in the Perupok area the *pĕraih* is usually an integral part of the net group, his function being that of seller and carrier. Should, however, there be only five boats available, then the " downstream boat " acts as fish carrier.

THE RÔLE OF THE FISHING EXPERT

The key of the lift-net organization is the expert fisherman. His functions are diverse, being economic as well as technical.

In the first place he is normally the contributor of the major capital of the group. The net is usually his (the precise structure of the system of ownership is analysed in Chapter V) ; he normally owns at least one of the boats and may have a financial interest in others of the group. Some of the *unjang ibu* are made and put down by him ; the *unjang ano'* is his property ; and on his shoulders falls most of the burden of organizing the technical side of the undertaking, keeping boats and crews together, and dividing most of the cash returns. The one phase of the activity with which, as a rule, he has nothing whatever to do is, curiously enough, the actual selling of the fish. This is the job of the *pĕraih*.

The success of the expert as an organizer rests fundamentally on his technical skill ; his energy and his command over his men are important, but if he cannot locate his fish and manipulate his gear to secure a good catch, his crews melt away and his organization breaks down. His technical skill in turn depends partly on his flair, which varies among Malay fishermen as among any set of craftsmen, but also to a large degree on his training.

The term *juru sĕlam* is a generic one applied to all experts in locating fish by " diving " ; there are *juru sĕlam takur, juru sĕlam pukat tarek* and *juru sĕlam payang*, each particularly adept in dealing with the requirements of the type of net in which he specializes. But their training is broadly the same ; it consists essentially in going out as pupil to an expert, who acts as teacher, and learning the craft by imitation and practice. One of the most important elements of the craft is the technique of submerging oneself in the water and " listening " for fish. The expert holds on to the boat with one hand and keeps his head and body below the surface, listening for the noises which the fish make. His first task is to locate the fish in the vicinity, identify what kind they are, and form an opinion as to their quantity. All these facts are important, because his net is adapted to take only certain kinds of fish, and if these are not plentiful his crew will waste time and labour in casting the net. No large net is ever cast at random ; the expert always explores the prospects beforehand. Hence his ability to interpret the sounds made by the fish is one of his essential functions. These sounds are said to be due to pelagic fish swimming through the water in shoals or to demersal fish feeding at the bottom ; some are due to fin and tail movements of large fish. Experts, when asked to describe them, used various noises and similes. The commonest fish taken in the lift-net, *sĕlar kuning* (a small horse-mackerel with a yellow stripe along its body) was said to make a sound " like the wind ", " $\bar{o} \ldots \bar{o} \ldots \bar{o} \ldots \bar{o}$ ", or " *ro - o, ro - o* ". *Lechen*, another small horse-mackerel taken in the same net, was said to make a noise " like parched rice " " *to ta to ta to ta* ", while the shark which comes after the fish and is a pest to the fishermen was said to make a noise with its tail, heard as " *peyup, peyup ; sĩu-up* ". The scabbard-fish (*layur*), taken by the seine, was said to have " a good voice—like a crow ", making the sounds " *ok, ak ; ok, ok oʻ aʻ* ".[1]   The experts are apt to describe these sounds in different ways, partly because each man has his own mental associations for them, but partly because, being usually a specialist on one type of net, he is most familiar with the kinds of fish which that net is best adapted to take.

The training of an expert takes time, and involves him in some cost.   One of the acknowledged older experts of the Perupok area took a year to learn, when he was a young man.   At the end of the time he gave his teacher the customary present of a jacket, with a cloth and $5 ; he also made a practice of inviting him to any feast which he gave.   The convention that a gift is the proper return for expert knowledge is common to all forms of transmission of the arts—including those of the manipulation of shadow-play puppets, and of magic and the art of healing.   In Malay ideas it is a breach of the ritual code for a teacher to sell his knowledge in a direct commercial way.   It may be asked how the ritual element enters into what is a technical process. The answer is that in theory at least the expert relies not simply on his own ability but also upon the bounty of God (and less explicitly, upon the assistance of spirits) to obtain his results. The old expert just mentioned said that his teacher taught him to ask for fish from Tuhan Allah before " diving " and also to make appeal to Nabi Kidir.   Before descending into the water the expert should wash his face over and " remember " God in some such words as—

> " Peace be on you !
> I ask the favour of Allah,
> The bounty of Allah."

And as he goes down he should touch the water first with his foot, saying the creed : " There is but one God . . ." The old man added : " He thinks of Tuhan Allah ; if he does not so remember, he can't do anything."

Another younger expert, who had learnt net-dyeing and technical and ritual matters of boat-handling from this old expert who has just been quoted, had been instructed in fish noises and other fish-craft by his own father.   He was a pupil for two years in all, and it took him over three months to " hear " the fish noises and to separate them.   At first he could not distinguish them from those of the sea and the waves.   " About other people I don't know ;   I myself took three months."

[1] Descriptions by experts of sounds made by other types of fish have been given in " The Coastal People of Kelantan and Trengganu ", *loc. cit.*, 198–9.

Incidentally, I remarked to him that Europeans were ignorant of these fish noises. He replied in surprise : " How can they know, when they don't go to sea ! " This attitude was part of a general Malay belief that no European knew anything about the sea or fishing—a reasonable inference when one considers the life of white people as seen by the peasants of Malaya. These people were only half-convinced that Europeans handled sailing craft and nets by being shown photographs by us. A common form of expression was, " Do people really go to sea in your country ? Then, of course, they must be Malays ! "

Much of the technical information which an expert accumulates is shared by the more skilled fishermen—knowledge of winds and currents, fishing banks, the ways of different kinds of fish and the depths at which they are mostly to be found, identification of position by shore bearings or by the nature of the bottom, the efficient handling of boats and gear, etc. But the identification of fish noises is acknowledged to be acquired only by training, and to demand a " flair ".

A question of interest is the degree to which all this knowledge, including that of " hearing " fish, is kept as a secret, giving its possessors special privileges and possibly forming the basis of a class organization, even of a hereditary character. The answer is that this is not so. A *juru sělam* is not born, but made. The experts do form an amorphous group with earnings higher than the average because of their specialist knowledge and skill, but this group is not a closed one. Any young man can be taken on as a pupil, and there are always a number of trained men who for various reasons are not acting as net-group leaders but as ordinary fishermen, stepping in and out of the higher earning category as demand for their services varies. Moreover, though a lad whose father is an expert obviously starts with an advantage if he wishes to take up the craft, there are many men who have attained their position from the ranks of the ordinary fishermen. Of 15 lift-net experts in practice in 1940, the fathers of 8 had been experts also, while those of 7 had not been, including the father of Japar, acknowledged to be the most efficient and successful of all experts of the Perupok area. Out of the 7, the fathers of 3 had been boat captains, which gave them an advantage since, as one of them said, a boat captain may at times take the place of a fishing expert. But the fathers of the other four had been ordinary crew members, not even owning boats. It is thus not necessary for the occupation to be inherited. There is, however, an opinion that an expert should have a like ancestry behind him. A man whose own father had been a fishing expert said of another man, now a boat builder, that he used to be a *juru sĕlam* but retired because he didn't get fish. He had no " ancestry ". And a noted magical practitioner said of his own craft : " If a magician has no ' origins ' he can't ever attain anything—with fishing experts, it's the same." In practice, however, this assertion (based on mystical, not rational grounds) was not borne out ; success was fairly evenly distributed between those who had " origins " and those who had not.

The fishing experts of the Perupok area are nevertheless closely related by kinship. Of the 15 men mentioned above, there are four pairs of brothers and one group of three brothers ; a number of these are kinsfolk of one another and of the remaining experts. Some, moreover, have kinsfolk among the experts of other areas. But they have ties just as close with other fishermen who are not experts, and there is nothing in the nature of a closed marriage group.

THE FISHING EXPERT AND HIS CREW

The relation between the expert and his crew is one of free association, either party being at liberty to break the bond at any time. He is the leader and commander of the crew—he is sometimes described as the " head " (*kĕpalo*)—but they are not simply wage-earners and they are not bound to obey him. They are all workers on a share basis, in a coöperative enterprise to which they contribute their labour as a complement to his skill and capital. While he leads them, and sometimes drives them, in matters of technique and organization, he shows what to the outsider is often a surprising readiness to consult them on matters of policy. Their mutual relations are governed not by any set formula of rights and duties, but by a number of practical assumptions about what is reasonable in the circumstances of their work.

These assumptions include views about punctuality, energy and steadiness of work, and the blameworthiness of slacking and cheating. They include also broad agreement on the responsibility of the expert for the safety of the equipment, on his setting aside a large proportion of the proceeds of the catch to meet the cost and depreciation of his capital, on shares being given for special work, on a " fair " rate of return being secured weekly to the ordinary crew, and on a provision for them each day of some fish for home consumption and for sale for petty cash. Moreover, it is assumed that the fishing expert will assist the crew-members financially by small loans in times of stress, especially during the monsoon periods.

The source of labour for the lift-net crews in the Perupok area is comparatively local, with a radius of about a mile from each net. Since the boats may go out well before dawn, the crews cannot live very far away. Some of the labour is drawn from the agriculturalists of the villages near the right bank of the Kemassin River ; these men are available especially in the season between rice planting and harvest. But the backbone of the labour force for each net is the group of people living near the beach and devoting themselves almost wholly to fishing. There is a distinct tendency for each net to find a fair proportion of its crew in its own immediate neighbourhood. But where villages are close together, as at Perupok, men come from one to work with a net of another ; there are no barriers against this.

The labour is that of men and youths alone ; no women go to sea. Women and girls, however, as also small boys, lay and take up the skids for the boats when they are hauled down and up the beach morning and evening (Plate VA). There are no rigid age limits for going to sea. Boys begin to go out about the age of fourteen, and the phrases " not yet at sea ", and " at sea ", roughly denote two separate age grades. But younger boys sometimes act as crew, while at the other end of the scale men go to sea until very late in life. The fact that shares in the proceeds of lift-net work, as in most other kinds of fishing, are allotted on a *per capita* basis and not by assessment of the amount of work done by each crew member, is an incentive to old men to carry on while they can, and to lads to try and earn money for cigarettes, coffee and sweets. But a high proportion of old men and lads in a crew is the sign of either a newly-formed net group or an inefficient *juru sĕlam*. The better experts like and get a crew composed mainly of young mature men, of maximum working powers.

In contrast to the economic organization of many more primitive communities the Malay fishing unit is not primarily dependent upon kinship as the tie of association. But kinship often enters. When I asked men why they were working with a particular group, I often received such replies as " Younger and elder brother ", or " He's my uncle and I've got to help him ". One fishing expert in Pantai Damat said of another that the latter could get a crew easily—he had plenty of kinsfolk, who did not leave him in a period of poor fishing but hung on, whereas non-kin would have gone off elsewhere. He himself, he said, being short of crew, had just been to one of the villages near the river to ask three of his kinsmen to join him for a few days ; he had put it to them as a trial proposition, and if the net was unsuccessful they could drop out. They had agreed, largely because they were his relatives.

To test the extent to which kinship ties entered into the composition of a net group I examined the crew of this same expert some time later. He had five boats, and the prospect of getting a crew for a sixth. In his own boat he had three of his nephews and two men who were not kinsfolk of his. The carrier boat was his younger brother's, and this man had with him his step-son and a brother-in-law ; the other members of the crew were non-kin. The captains of the other four boats were all unrelated to the fishing expert, and with the exception of two who had a son each, their crews were not kin to them ; most, in fact, were not regular members of the group but men taken on as opportunity offered. None of them had any particular kinship tie with the expert. Thus out of a total crew of rather more than 25 men, only 6 were kin of the expert. From a more cursory inspection of other crews it seemed that this ratio of about one-quarter kinsfolk in the group was fairly representative. The relationship between fishing expert and carrier agent will be examined later from this point of view.

I have already mentioned the way in which the fishing expert in his rôle of organizer combines the issue of orders to the crew and the taking of advice from them. This must be now analysed more closely, from the point of view of the technique employed.

The expert in this rôle works partly directly and partly through the medium of his boat captains (*juragan pĕrahu*). Most important of these are the captain of the net-boat (*juragan pukat*) and the captain of the carrier boat (*juragan pĕraih*). Each is " number two " in the organization, but they have different functions. The captain of the net-boat is responsible for the net, its ropes and other appurtenances. The boats ordinarily are launched about the same time in the early morning, but some are quicker than others in getting away, and their speed is not exactly the same ; so also in returning from the fishing grounds in the evening. So while it is desirable that they should move as a fleet, and often they do so, each boat acts as an independent unit when sailing. It is the job of each captain to see that his boat arrives at the destination given by the expert or follows the course set by him, with the minimum of delay. The boat with the net aboard is likely to be more sluggish than the others, but the captain must see to it that his crew paddle if necessary in order not to keep the others waiting to begin work. The net-captain is often responsible for storing the net when it is not in use, though this depends upon the size of his house, and the absence of a special net-shed. And he, in particular, is expected to be on hand for all operations of net-repairing, dyeing, fixing of ropes, etc. The carrier agent has an even more responsible job in getting the catch safely to shore, and in selling it at the best bargain he can make. He must, therefore, be not only a seaman but also good at chaffering. Further, he has the job of collecting the receipts in time for the weekly division of the takings. Again, he is expected to obtain a good proportion of the crew, and to put down some of the " parent *unjang* ". As an attached selling agent he has the obligation to take part in the weekly work of repairing the net. This last point is important, because other types of fish carriers are exempt from this obligation.

The captain of each other craft has to look after the boat, sails, ropes and other gear, and has some responsibility for his crew. But he has no other functions apart from the work of the net and participation with all the group in carrying the net to and from the drying-stand on which it is spread after each day's work. (Incidentally, it is a picturesque sight in the late afternoon, when a score or more men in line, six to ten feet or so apart, bear off the net at their backs or on their shoulders—Plate VIb.) The captain of each boat is usually, though not always, its owner. In some cases a wealthy expert owns two boats. He runs one himself, and gives the other to be run by someone else, with whom he shares the boat's portion of the total takings. Sometimes a boat owner has a job ashore, or he may even go as an ordinary crew-man in his own craft, not wishing the responsibility. In one such case the reason given by one of the crew was " People are not frightened of him ! " Strength of personality is necessary to handle these tough Malay fishermen, and the success of a fishing expert depends to no small degree on this. When I asked Awang Lung why it was that his brother was always going out when no one else did, he replied that it was because he was forceful, and his men were afraid of him. He added that Japar was another of the same type—" When he speaks, all his men do as they are told ; when I speak, half do, and half don't." The power of leadership is important not only in the immediate technical work of the net, but also in braving threatening weather, and particularly in getting an early start in the morning. This last is important in the season of variable winds ; if a wind springs up off the sea early in the day the crews which start at dawn may find themselves unable to reach the *unjang*. Energetic experts who succeed in the thankless task of getting their crews off at 2 a.m. or 3 a.m. are therefore apt to do well. In theory, the boat captains are responsible for waking their crews and getting them out, but a keen expert keeps them up to the mark.

An expert who did not conform to these standards was one who, when I knew him, was an ordinary crew-man, going out with drift-nets. He had no house of his own, but lived with wife and child in a shed of another fisherman—the hallmark of poverty— and was a gambler to boot. He was a lift-net expert, but had been careless and did not work properly. His net was borrowed, but he was late in getting his boats down and got few fish ; as a result his crews were always small, which reacted still further on his efficiency. Finally, his prospects were so poor that the owner of the net took it back, and his group automatically broke up.

An expert is preoccupied with keeping his crew together, and for this reason he studies their interests. If the season is poor, or is coming to an end, he often consults not only the boat captains and the *pĕraih*, but also the crew at large, as to whether they should go out. The work is hard, and he would rather they should stop for a time and find more profitable occupations, then return to him when prospects improve, instead of leaving him in disgust at small or no returns for their labour. Just before the beginning of the 1939 monsoon a number of lift-nets did not go out because their crews did not think it worth while ; they had to paddle back nearly all the way. In the middle of March 1940, when fish were getting scarce, one expert asked his crew when they went out if they wanted to cast the net to get fish for home consumption. They replied that if they could get $10 worth of fish it would be worth while, but otherwise they didn't want the trouble. The upshot was that they did not cast the net at all, and fished for sea-bream with lines instead. One boat, getting 40 fish among 4 men, also trolled a Spanish mackerel. When asked, the crew advised that it should not be sold but divided for food. This matter of fish to eat is important. When a catch is small, say less than about \$4 or \$5 worth, it is customary for it not to be sold in bulk but to be divided among the crew, the expert getting his share like any ordinary crew-man. On this basis he gets a much lower proportion of the takings than if the catch were sold. But if he insisted on a sale he would soon breed discontent among his crew. Again, fish which are taken by trolling are the property of the crew of the boat which gets them. Should the expert happen to be the captain of the boat he has his voice in their disposal, but he usually leaves the final decision to the rest. Thus one boat got a Spanish mackerel. The expert suggested that it be cut up for food, but the crew objected ; they wanted to sell it. They were mostly young men without wives, and cash was more to them than a slice of the fish. " How shall we get on ? What shall we drink coffee on ? " they said. So the fish was sold, fetching 26 cents per man.

The subject of loans by the expert to the crew is discussed in Chapter VI. It is sufficient to say here that this is an important feature in the lift-net organization, and one by which wealthy experts are at an advantage.

FLUCTUATIONS IN LIFT-NET CREWS

A marked feature of the lift-net organization is the rapidity with which the crews fluctuate in numbers. For a net to keep an identical crew for a whole season's fishing is quite anomalous. The situation may be best understood from a rough diary of the position of several nets during the first three months of the 1940 season ; this may be compared with the average daily yields in Fig. 25, and the weekly records in Appendix VI, the nets being designated by the same letters in both cases.

January 1940. Fishing began in force in the middle of the month. There were 20 nets ready to fish or nearly ready, as against 17 at the end of the 1939 season, and the extra crews needed were a worry to the *juru sĕlam*. Net *M* had not a full crew ; *E*, who had given a lot of money to his crew during the monsoon, was finding that they were not turning out in full force—perhaps some having gone elsewhere ; *J* said that he had only 14 or 15 men instead of 25, and complained that the crew wanted money, but that if he gave it to them he couldn't even then be sure they wouldn't go out with someone else. A few days later *E* and *J* had managed to collect enough men to fish, but *O* and

R were held up for a week for lack of crews. By the end of the month *J*, having been out twice without result, had been immobilized again by loss of some of his men. At this time *A*, *B*, *C* and *D* had crews of 30 men or more, because of their success. *J* explained that the wind was strong and his crews would not have been able to reach the fishing grounds. But someone else commented that his net had been actually seen in the boat ready to start, and it was therefore really want of men— probably because his son, a new expert, was not inspiring confidence.

February. For the early part of the month every net was out except when blocked by high winds. At the end, however, crews began to fall off, owing partly to having turned to drift-net work or cultivation during the heavy weather. *J* and *R* lost a number of days, at a time when *D* had 45 men in his six boats. Towards the end of the month *L*, *M*, *R* and some others went to enlist men from the villages near the river. *R* was in special difficulty. He had borrowed a number of men who were also drift-net fishermen and they now wanted to leave him since the secondary season for this work was now starting. He told me one evening that he was determined to go to sea the next day— if the waves were light he would go with only three men to a boat ! He did go out for two days, got only $6, lost some of his crew and was then held up again.

March. At first most of the nets were out fairly regularly. *M* had 28 men in his five boats, while *D* had 35. *J* had 20 or less, and one of his boat captains said that it was doubtful whether he could continue to go out, with only four men to a boat. I asked how then could *L* manage, who recently had had the same. But *L* had since had two spectacular catches, and the answer was : " He has six or seven men to a boat—he gives them food." This did not imply a direct gift, but the flocking of men to his success : he now had over 30 as a crew. A fortnight later, however, after a series of poor catches, the crew of *L* was down to 24 men, and then to 21 men. On this last day *J* had 20 still, but *D* had 36, and some of the other Perupok nets had about 30 apiece. Towards the end of the month few nets tried to go out, since the lift-net fish were few, and line fishing was in full swing.

April. Lift-net fishing slowly developed again, as a few nets went out and got catches. But *R* was not out once. He made several attempts to get a crew, but failed. His son-in-law, who

XIIA PREPARING TO BARGAIN
A lift-net boat has arrived; the floor boards are removed to let the dealers see the catch. The boat captain takes out a few fish for his household.

XIIB WHOLESALE BUYING
Dealers are taking out the catch after agreeing on the price with the boat captain.

XIIIA HAGGLING AMONG DEALERS

Dealers are arguing about the relative prices of the baskets of fish which each is going to take separately to market. The crew are cleaning the boat.

XIIIB RETAIL SELLING ON THE BEACH

The fish are set out in heaps on the sand and buyers move around them. The retail trade is carried on by both men and women.

rather unwillingly had joined the net-group (he really preferred independent fishing), said to me once that he would have gone out had there been only 3 men to a boat—but there weren't. Winds were light, and a small crew could handle boats and gear at this time. All crews were now small. *B* and *N*, two successful nets, had only 21 and 20 men respectively instead of 30, and the carrier boat of *B*, which earlier had 7 to 9 of a crew, regularly now had 4 only.

The situation may be summed up by saying that in each lift-net crew there is a fairly stable nucleus, consisting mainly of the boat captains, their kinsfolk and close friends or neighbours. When the number of nets is small there is a reserve of labour, in which the semi-independent fishermen are one element. But when the number of nets increases this reserve disappears, part-time agriculturalists and independent fishermen are enlisted where possible, and the chance of successful experts attracting the floating margin from other less skilled men is greatly increased. And in the comparatively fluid structure of the crew organization lies the opportunity for a new expert to establish himself if he has confidence and ability.

RELATIONS BETWEEN FISHING EXPERT AND CARRIER AGENT

In the relation between the *pĕraih laut*, the carrier of a catch of fish to the shore, and the net-group from which he takes it, there are several possible variant elements. The carrier may be the regular and agreed sole client of the net-group, or a casual picker-up of what fish he can get ; he may be an outright buyer of the catch, taking what profit he can ; or he may be an agent, getting his returns by sharing with the rest of the group. He may participate in the actual fishing and net-repairs of the group, or he may have no part in them. Various combinations of these elements occur at different places on the Kelantan and Trengganu coast, and the same vernacular terms do not always indicate exactly the same type of relationship.

In the Perupok area a distinction is made between three kinds of middlemen : *pĕraih ratar* ; *pĕraih ibu* ; and *pĕraih sĕlendar*.

The *pĕraih ratar* (or *pĕraih mĕratar*), meaning " wandering dealer ", from the standard Malay *rantau* (*mĕrantau*), is most loosely attached. He is essentially a boat captain who has gone out line-fishing or setting down *unjang*, and wants to take back a load of fish and get some profit on the day's outing. If he meets a net from another area he will probably have to pay cash for the fish and his profit will depend on what he can sell it for on the beach. If the net is from his own area he is more likely to be enlisted or " borrowed " by the expert fisherman to act as carrier for a secondary catch in return for an agreed proportion of the takings from it. This proportion varies ; I have known cases where it was ½, but have been told of others where it was ⅓. (At some places on the Kelantan and Trengganu coast this scheme does not apply, and the *pĕraih laut* is a dealer purely and simply.)

The *pĕraih ibu*, the " parent dealer " or " parent middleman ", is the regular carrier of the catch and is a member of the net organization, working in each case for that net alone. His functions as selling agent, will be described later.

The pĕraih sĕlendar is in an intermediate position between the " wandering dealer " and the " parent middleman ". The " wandering dealer " goes where he wishes, where the fish are ; the " parent middleman " is bound to the net and shares fully in its organization. The *pĕraih sĕlendar* is on the outskirts of the net. He has already agreed on shore with the expert fisherman that he will be associated with that net. His function is to take any second catch after the " parent middleman " has gone off to shore with the first catch. His services are required only for a net which has only five boats, and part of his job is to take the place at the net vacated by the " parent middleman ". For this he gets half the proceeds of the catch he takes to shore. If there is no second catch, he and his crew get nothing. Hence the *pĕraih sĕlendar* is not a fixed adjunct to the organization. He is a boat captain who changes over to this rôle from other work when there is prospect of heavy catches being obtained. To the Perupok fishermen the great distinction between " wandering dealer " and *pĕraih sĕlendar* is that the latter is bound to help with the work of the net at sea, and the former is not. Should the " wandering dealer " be called by the expert fisherman to haul on the net, however, he receives an allowance for this, out of the expert's portion of the takings at the end of the week (see Chapter VIII).

In the Perupok fishing organization the " parent middleman " is the main carrier of fish. The relations between him and the expert fisherman are interesting. Early in my work on the beach I noticed that when I happened to ask an expert what his catch sold for he usually replied : " Ask the *pĕraih*." This may have been due at times to his ignorance, but the reason was mainly a feeling that it was not appropriate for him to meddle with what was not his immediate business. I took up this point with several people. One expert explained that the *juru sĕlam* does not ask the *pĕraih* what the fish fetched until Friday—the day for settling accounts. This rather surprising statement was confirmed by other men. When I asked why, the answer was given that expert and *pĕraih* (as also expert and boat captain) are not " in good relations "—they coöperate best at a distance. They do not become very friendly in ordinary daily life because the expert has to make the other two afraid of him—if they are not afraid they will not work well ! They do not normally drink coffee together in the shops—if they meet and one invites the other to a cup, good ; but not otherwise. With members of the crew it is different ; the expert maintains friendly relations with them since if he does not they will " run " elsewhere.

The " social distance " thus maintained between the leader of the net-group and his responsible subordinates tends to a more effective economic coördination. It rests upon the fact that they are bound by ties of mutual profit, which can endure a certain degree of coolness between them. Reproof can be administered by expert to *pĕraih* without breaking friendships, but, on the other hand, confidence must be shown in the honesty and capability of the *pĕraih*. The crew, however, whose financial stake is not so large, and who can be absorbed without effort into any other group should they leave, must be courted rather than shunned.

But while it is intriguing to find that the Malay fisherman is capable of visualizing the pattern of social relations in this situation of exercise of authority, the position just described must not be taken as universal.

Though the *juru sĕlam* may not ask the *pĕraih* what the fish fetched till the end of the week, I found that most of them seemed to know the answer by the evening—the crew or the dealers tell them, and they can easily check up with the version given them by their *pĕraih* later. Moreover, relations in a number of cases between *juru sĕlam* and *pĕraih* seemed to be quite amicable, especially when they were kinsfolk. Of 23 cases which I examined, in 15 *pĕraih* and *juru sĕlam* were not kinsfolk in any significant sense— they were " different ". In the other 8 cases they were thus : 3 of the *pĕraih* were sons-in-law to their *juru sĕlam*, 1 was a brother,

1 was a son, 2 were brothers-in-law and 1 was a cousin's husband Two of the *pĕraih* shared houses with their *juru sĕlam* fathers-in-law.

STRAIN AND RE-ALIGNMENT IN NET-GROUPS

To the outsider one of the most striking features of the organization of the lift-net groups is their fluidity of personnel, or, put another way, their brittleness as structural units. When disagreement occurs there is little attempt to smooth things over and make adjustments : the aggrieved party tends to throw up his place at once or get rid of the other party, whichever is the more feasible. Other social ties, as of kinship, do not operate to any great degree as shock-absorbers tending to take the strain of economic difficulties. This structural brittleness has the advantage that it does away with the need for people to work together for long periods while nursing grudges against each other, and so tends to prevent the accumulation of bad relationships. On the other hand, it does render long-term coöperation and planning difficult.

It is true that some associations, especially kinship ones, are of long standing. But the highly brittle character of the netgroups is shown by the fact that of the score of groups of any duration in 1939–40, only half finished the season with the same *pĕraih* with which they began, and some of them had made several changes. And two in which the net-owners were not the expert fishermen, had lost the experts, who had gone to start other groups. The cause was not a disagreement in every case—in one the *pĕraih* had bad health and wished to take a shore job—but it was mostly so. The local expression—" there was trouble " (*ada balo*) is the usual reply to any question as to the reasons for change. It may be noted that in most of the groups which retained the *pĕraih* throughout the season these were kinsfolk of the *juru sĕlam*.

The departure of a *pĕraih* from a net-group or of an expert from a net-owner can take place in one of several ways. He can go of his own initiative, being dissatisfied with his treatment ; he can be dismissed (*buang*, meaning literally " thrown away ") ; or the net-group may break up completely. A certain dignity is involved in taking the initiative, and a man is often careful to point out that he has not been " thrown away " for poor work, but has left because the catches were so bad. A brief account of the more outstanding cases that occurred under my own observation shows the usual kind of setting for such affairs. Early in

January 1940 Daud, who was expert for Selemen of Paya Mengkuang, parted from him. Selemen had set aside for distribution what Daud said was $25 and Selemen later said was $20. Hence Selemen said that he did not wish Daud to operate his net, and Daud said that Selemen was not " straight ". Another fisherman, commenting on this, said that Daud was the nephew of Selemen's wife, and it wasn't good for kinsfolk to part. When Selemen's net was going out later under another expert there were complaints too. The new man said that he was getting only one share instead of the three to which he was entitled, that Selemen did not " work straight ", and he wasn't satisfied. There was no immediate break-up of the net-group at the time, but there was a later re-organization in which the expert took over a financial interest in the net.

In the net-group of Awa'Loh, of Perupok, his son-in-law acted as his *pĕraih*, and the latter's elder brother as his *juru sĕlam*, running one of the old man's boats as well as the net. One evening this boat was washing about in the surf at the edge of the beach ; the old man saw it and told the *juru sĕlam* in emphatic terms that he was not fit to handle the boat. " His mouth was a little quick." As a result the expert left, and was absorbed into a group which another brother of his was just forming. As usual, opinions differed about the justice of the accusation ; the brother held that the boat was endangered because other people did not go down and help to pull it up.

When the net of Sa'Ma' got a series of poor catches, the *pĕraih*, Mat Taiyeh, lost his temper at sea one day and called it " pig-net ", " dog-net " . . . and other names. The offence here was twofold : first, in applying the terms for unclean animals to the net ; and second, in doing this at sea, where animals should not be mentioned by name at all. Several members of the crew, including the expert, spoke strongly to the *pĕraih*. Awang Lung's comment was " Mat Taiyeh's mouth is not right ; if I heard this in my net at sea, I would throw up everything for the day and return to shore." The group broke up because other men too were not satisfied with the net. Mat Taiyeh managed to get a new group going, which lasted with poor success for about three months.

These three cases epitomize most of the reasons for breach of a net-group—suspicions about money ; accusations of negligence ; and failure of the net to get fish.

So far I have considered the matter from the side of personnel.

But an important feature in re-alignment of net-groups is the changes in capital which they bring in their train. This can be studied by examining the boat-grouping.

CHANGES IN BOAT-GROUPING OF NETS

A notable feature of the organization of lift-net groups is the frequency with which boats change from one net to another. There are three main reasons for this : desire of a boat owner to change his craft for a more efficient one or to take advantage of a favourable offer to sell ; disagreement of a boat owner with his companions of the net-group ; and break-up of a net-group because of consistent lack of success. A fourth reason, occasionally, is the failure of a net owner to have his new net ready when he sells off his old one. In any particular case, a combination of these reasons may provide the motive for change. This tendency to re-shuffle is clearly facilitated by the system of distribution of takings from a net, which by giving individual shares to boats and men for each day's work, allows of a break-away at any time. No conception of a contract for a given period such as a season exists, though a net owner naturally is apt to feel aggrieved at the loss of equipment, with or without the loss of personnel.

Effects of this tendency to a continual re-shuffle of boats are a constant re-arrangement of the investment of capital as between different net-groups, and to some extent a loss of efficiency in production owing to the loss of time in changing boats.

I recorded all the changes in the boat-composition of the various net-groups in the Perupok area from October 1939 to June 1940, but the full complications cannot be presented here. To demonstrate the degree of re-arrangement it is sufficient to compare the position of the various nets at the beginning of the 1940 fishing season in January with that in the middle of the season, five months later. The results are given in diagrammatic form in Table 7. This shows that of the 20 groups beginning the season only two remained quite unchanged. Of the 102 boats in the 20 net-groups which began the season, only 68, or $\frac{2}{3}$ of the whole, remained in those groups by mid-season ; a re-shuffle to the extent of $\frac{1}{3}$ had taken place. (Actually this re-shuffle was slightly larger, since a few boats had left and returned again by this time.) By mid-season there were 19 net-groups left, with 101 boats in all, including one borrowed from outside the area. In the interim 19 boats had been bought, 7 of them from outside

TABLE 7

CHANGES IN BOAT-GROUPING OF LIFT-NETS IN PERUPOK AREA,
IN SEASON 1940

Grouping at :

Opening of Season	a a a a a	b b b b b	c c c c c	d d d d d d	e e e e e	f f f f f
Mid-season	a a a B R (x″)	b b U X″ X″ X″	c c b X″ (X″)	d d d d d I	e e e e e u	f f f u X″
Opening of Season	g g g g g	h h h h h (x″)	i i i i i	j j j j j	k k k k k	l l l l l
Mid-season	g g f s X″	h h h h k Q	i i J X″ (b) (U)	j j i x′ x′	k k k s x′ A	l l l l X′
Opening of Season	m m m m m	n n n n n	o o o o o	p p p p p	q q q q q	r r r r r
Mid-season	m m m m m j	n n n n n	o o o o o	p p p p ( )	q q q q ( )	r r r r X′
Opening of Season	s s s s s	t t t t t			U Formed in March, thus :	g G d b x′ X′
Mid-season	— Broke up March	t t t t X′ g				— Broke up in May

Explanation of Table : Each square represents a net, of which there were 21 (A to U) ; each letter of the alphabet represents a boat. Boats belonging to the same net at the opening of the season are denoted by the same letter. Small letters in mid-season indicate boats of unchanged ownership ; capital letters indicate boats which have changed hands. X′ represents boats bought within the Perupok area, not having been in a net-group previously that season ; X″ represents boats bought from outside the area ; x′ and x″ represent boats of similar origin co-opted or borrowed. Letters in brackets indicate boats borrowed.

the area, 4 from inside the area but formerly outside the lift-net fishing, and 8 from other net-groups. The total capital turnover thus involved was approximately $2,600, or roughly about 18 per cent. of the total capital invested in lift-net fishing boats on current values. Of this, about $1,200 represented payments or obligations to people outside the area, and about $1,400 to people inside the area.

The changing of boats from one group to another usually, though not always, involves also a change in the personnel of the group.

The importance of the change of boats in the present context is that it frequently demands fresh capital from some members of the group. Thus Group *A*, with a total boat-capital of about $550 at the beginning of the season, had increased this to about $750 by mid-season (excluding the value of a boat borrowed from Tumpat). The two new boats purchased cost $310, but against this could be set off $50 received from the sale of one of the boats, a medium size *kueh*, to Group *K*. In this case the majority of the new capital was probably found from profits during the season, which for Group *A* had been a successful one. Group *B*, with an initial boat-capital of about $900, had increased this by mid-season to about $1,050. But actually the fresh capital which had to be used was about $490. Four boats had been bought, at a total of about $700, and though three had gone from the group, only one of them had been sold, for $210, while the other two were taken by their owners. In this case the fresh capital was probably found partly from the profits of a successful season, but partly also from savings, since the fishing expert and principal boat owner was a comparatively wealthy man.

The purchase of a boat for a net-group is sometimes done to increase the working number to six, the ideal in this area. More often, however, it is done to replace a boat lost by sale or defection of the owner. The purchaser may be a man who previously has been an ordinary man of the crew, but now wishes to enlarge his economic opportunities by becoming a boat owner. In such cases he is often financed to some extent by the net owner, who thus tends to make a firmer tie between them. If the new boat owner wishes to leave the net-group, the net owner demands his contribution back. The boat owner then has to reconsider his decision, and may stay. In some cases, he raises a loan outside, repays his creditor, and departs. This is one factor in making for the complicated system of boat finance discussed in Chapter VI.

It often happens, however, that the boat is bought by the net owner himself, thus ensuring its participation with him. Of the 19 boats bought in the first half of the 1940 season, 6 were bought by net owners, 5 of whom were already possessors of other boats ; one was bought by a man who, I think, was a partner in the net and who already had another boat, and one by the son-in-law of a net owner. Of the remainder, 3 were bought by *pĕraih* to replace boats they had sold, i.e. to continue their work with the same net, 1 was bought by a *pĕraih* to enable him to begin work with a net, 1 was bought by a fishing expert to begin work in a new group, and the other 6 were bought by ordinary crew-members. It is clear, then, that the incidence of purchase of craft tended to fall most on those men most closely associated with the nets. In other words, men buy boats to serve particular nets, not merely to be able to join any net-group.

At the present time in the Perupok area there is no large " floating margin " of boats suitable for lift-net work waiting for entry into production. This may be due in part to a fairly rapid increase in the number of nets in the area in recent years, and may be responsible for a higher rate of purchase as against co-option.

But one result of this is that the sale or defection of a boat from one net-group is apt to cause repercussions in others. An example may be given from Group *H*. In the early part of 1940 the *pĕraih* was working with a boat borrowed from a relative in Telong, along the coast. Towards the middle of the season this boat had to be returned ; he thereupon bought from another boat owner in the group for $150 the boat formerly used for carrying the net. This boat owner had another craft, but to keep up the numbers he then bought a fairly old boat from a neighbour—when I left the price had not yet been fixed. This boat had formerly been lent to the expert of Group *Q*, who used it as his *pĕrahu sampan* ; he then had to look for another, and for some time had to borrow again.

The formation of a new net-group is often accompanied by all the forms of acquisition of boats—purchase, attraction from existing groups, co-option from outside the lift-net circle, and borrowing. The first three are seen in the formation of Group *U*, which originated after a dispute between the expert of Group *G* and his *pĕraih*. The group " divorced ", the *pĕraih* taking with him his own boat and one of the members of the group, who took on the responsibility for purchase of a net, and also put up the capital for two other boats. One was bought from the expert of C, the other from a man who had not previously used it in lift-net work. Another boat was induced to leave Group B and join, another was co-opted with its owner from outside the lift-net field, and a sixth boat came from Group F. This had been set free from D since one of the crew-members of D had bought a boat from the *pěraih* of I (who had previously worked with it in D) and this brought D up to full strength of six. Meanwhile, the expert of G had been left idle for several weeks without a full complement of boats ; he had filled in part of his time with line fishing and deep gill-net work. He finally got his group working again by getting a *pěraih* from S (which had broken up owing to lack of results), using two of his own boats, enlisting a neighbour who had previously allowed his boat to be used in F—though he himself had not run it—and getting the neighbour to buy a fifth boat from outside.

RELATIONS OF FISHING EXPERTS WITH ONE ANOTHER

The more successful *juru sělam*, as might be expected, show a competitive spirit, stimulated in part by the daily standard of comparison afforded by the prices of the catch on the beach. After my habit of recording sale prices became generally known I was often asked for details of their rivals' takings by experts and others, who regarded the written word as more reliable than beach rumour. In Perupok, where the more successful experts were, the rivalry was keenest.

On the whole, the social and economic relations between the experts are amicable. This is shown specifically by the common practice whereby an expert who is ill or who is unable for any other reason to go out " borrows " another if there is one to spare. Borrowing of boats and crew is also often done.

When an expert is ill and borrows another, it is usual to give the latter two shares in the distribution of the day's takings plus a gift of \$1 or so, if he gets a good catch. The man who is ill, if he is the owner of the net, gets the net's share as usual, but if he is not, then he gets no cash return but only a portion of the fish for his domestic use as a kind of charity present, like any regular member of the crew. Borrowing of crew and especially of boats (with crew) is essentially a matter between the experts. It takes place not only between those who are brothers, but between others also. The consent of the boat owner and crew is assumed, since they are only too ready, as a rule, to have the chance of earning something. Should the boat owner be asked to come by the other expert, he still has to get the assent of his own leader before going.

The relation of experts to one another, however, comes out most clearly in their behaviour over their *unjang*. These lures, set out from five to fifteen miles from the shore over a stretch of twenty miles or so up and down the coast, are the property of the individual fishermen who put them down, and are an important part of the fixed capital of the net-group of the owner. Soon after I came to live in the area I was interested in a Court case in which one expert accused another of stealing fish from his lure to a value of $40 or so, and of releasing the lure from its float, so that it sank and was lost. The evidence was conflicting, and the case was finally dismissed.

The facts appeared to be that the man had fished from the particular lure, and that being an old one its rope had apparently parted when it was being handled, and so it was accidentally destroyed. But what emerged was that fishing from the *unjang* of another *juru sělam*, though verbally called " stealing ", was a common practice, and was by the fishermen's rules a legitimate proceeding, with two provisos. One was that the customary share in the division of the takings should be paid over to the owner of the lure ; the other, less explicitly formulated but well understood, was that the two parties should be on good terms, that is in effect, that the owner was willing. It must be understood that when the nets go out they make for points on the fishing ground many miles apart, and an expert, not finding fish at his own lures, will often see them at a neighbouring lure which is not his own. If the owner is near it is courtesy to ask his permission to fish there ; if he is not, one is entitled to go ahead on the understanding that he would not object if he were present. Why should he ? He may be wanting to do precisely the same some miles away. This practice is fairly general along the whole of the Kelantan-Trengganu coast. Between the middle of January and the middle of March 1940, I noted 13 cases of fishing from the *unjang* of other experts in the Perupok area, and 8 cases of fishing from the *unjang* of men of other areas. And this was not a complete record. It leads to trouble where the parties are already at loggerheads, or where an expert going too far afield takes fish from the *unjang* of another area. The reciprocal use of *unjang* with or without previous permission works as between experts of a single area, since there are enough checks to enforce payment of compensation. Here a law which took a strict view of individual ownership and enforced a fine for supposed breach would go far beyond the conceptions of the fishermen themselves. It would actually hamper efficient use of capital, if rigidly applied. But when it is a matter of using *unjang* of men from other areas the owner has not much chance of getting compensation, and the system only works partly through implicit reciprocity and partly because of comparative ignorance. But while the law here has more reason it is correspondingly more difficult to enforce. Moreover, it must be remembered that the use of an *unjang* for a day's fishing when the owner is not using it does no real damage, since the fish there vary from day to day. A system of " live and let live ", or tacit reciprocity working through the good relations of experts, seems to be the most effective solution. This is reinforced by the fact that the direction of the wind often forces an expert to make far to the north or south of his own *unjang* and to trespass on other grounds if he is to make a cast at all that day.

THE RITUAL FACTOR IN ORGANIZATION

There is a further element in the organization of fishing, especially with the lift-nets, which so far has been mentioned only incidentally. This is the body of ritual acts and beliefs which accompany the various stages of the fishing process, and condition the organization of man-power and use of capital in several ways. They can be dealt with here only very briefly.

This ritual rests upon the assumption, common among men engaged in the pursuit of living things, that the fish are not unconscious of the activities and intentions of the fishermen even at a distance, and evade them if not appropriately treated ; moreover, that the actions of the fish are governed to some extent by the spirits of the sea (*hantu laut*) who have to be placated. In line with these general assumptions is a body of specific beliefs and ritual acts which are integrated into a system. This has two nodal points in terms of personnel—the *juru sĕlam*, in control of the net and the actual fishing, and the *bomor*, sometimes also a fishing expert but often not, charged with more general acts of placation of the sea-spirits. The ritual acts include avoidance of animal terms while at sea and substitution for them of other more neutral and more honorific terms ; avoidance of certain days deemed to be "unlucky" when carrying out the more crucial activities such as taking out a new net for the first time ; the bedecking of boats with garlands of flowers to please the fish, the sea-spirits and the boat itself ; careful treatment of the boat as an object more than timber, endowed to some degree with spiritual guardianship ; avoidance (in some areas) of wearing shoes or carrying umbrellas aboard it ; and, most important, the performance of ritual over both boat and net, and the offering of food and other substances to the sea-spirits to secure their coöperation.

In popular terms the general object of these ritual acts is to smooth the way for the fisherman by allowing him to "meet with fish", and when he does so, to stop them from "running". The idea as commonly formulated is not that the fish are conjured up by these rites (they are already there out in the sea), but that proper conduct and some sort of mediation are required in order that the fisherman may get into contact with them, find them at the *unjang* he tries, and get them into his net. People are successful not simply because they are better fishermen than others, but also because they "meet with fish" when others do not. Our concepts of mathematical chance or luck are thus translated by these fishermen into a concept of a more active personal kind, in which the fish as well as men exercise will or initiative. Expressions which are closely equated by the Kelantan fishermen are thus : "lucky" (*mujor* or *nasib baik*) ; "meeting with fish" (*běrjumpo samo ikan*) ; and "fish liking a man" (*ikan suko orang itu*). Skill in fishing is not denied, but skill alone is not finally effective. The point is put another way by the *bomor*, whose commonly expressed object in performing his rites is to "help" the fishing expert by asking for bounty, for large catches instead of small ones. The 1940 fishing season, being a poor one, called forth a great deal of activity of the fishermen in seeking the aid of ritual. Rice and rice flour blessed by holy men, talismans, and especially offerings by various *bomor* were all called into service. The unfavourable conditions were thus a boon to me in bringing out a great deal of this belief and practice which had lain unsuspected during the more normal conditions of 1939. The *bomor* who was the subject of most discussion was a religious man named Nik Rung (Haroun), and the spectacular results which followed in each case from the performance of his rites for a series of hitherto unsuccessful fishermen was almost in the nature of a "miracle of magic". When I asked Nik Rung what was the precise object of his performance he stated that it was to

help, simply.   He said that if a *juru sĕlam* was not clever he could
not make him clever, but what he did was to give him a little
more fish.   Modestly he said : " If a man can get $15 worth of
fish by himself, then I can get him $20 ; and if he can get $50
then I can get him a little more."   He also said that he himself
wasn't a man of the sea.   (He had, in fact, never been to sea in
his life, and this was one of the earlier criticisms of his ritual by
older practitioners whose attitude was " how can a landsman
pretend to know rites which will deal with the sea-spirits ! ")
He therefore did not know about fish, and could not give any
skill or knowledge to an expert.   But what he could do was to
increase the efficiency of the relation of the sea-spirits to the expert
—so that if the fish were there when the fishing began, they would
not " run ".

   This point of view was the essence of the attitude of the
fishermen and the *bomor* as a whole.   Our primary concern here,
however, is the consideration of the economic effects of the
existence of these beliefs and the associated ritual behaviour.[1]

   On the one side, the beliefs and ritual involve the fishing
expert in an expenditure of time and money.   Journeys are
made to secure blessed rice, or a *bomor*.   Several hours are spent
in arranging the materials for the offerings and attending the
ritual.   Gifts of cloth or money or fish are made to the holy
man or teacher who blesses the rice.   The offerings must be
bought and a present made to the *bomor* who performs the ritual,
or he must receive a substantial share of the takings of a successful
catch.   The religious men who pray over the boat must be paid
small sums.   If talismans are purchased they are expensive.   It
is difficult to make up a budget of an expert's expenditure annually
on ritual affairs, and it is probably very variable.   But the following figures are some guide.   In getting rice blessed by a holy
man one fishing expert took a quart of rice in a 10 cent cloth,
and gave some money for the service, 10 cents or 20 cents being
the common fee, but as much as a dollar being sometimes given.
Another man took a Spanish mackerel costing $1.50 ;   others
send Spanish mackerel two or three times a year, at a total cost
of about $2 or $3.   The religious men who perform the prayers
over a boat get usually 5 cents apiece, but wealthier men give
them 10 cents—there are usually from about seven to ten of

   [1] Comparison may be made here with my analysis of the economic effects of
magical beliefs and rites in somewhat similar conditions in an agricultural and fishing
community of the Solomon islands.   Cf. *Primitive Polynesian Economy*, Ch. V.

them—the total thus being between about 35 cents and $1. The offerings cost little, the main item being glutinous rice, of which anything from one quart, costing 8 cents, to several gallons costing a dollar may be used. The fee to the *bomor* if he is successful amounts to a dollar or so in cash or fish. The annual total for each fishing expert varies probably between about $2 and $10 on such items. Special items such as talismans cannot be reckoned on an annual basis, and only a few experts have them. Their cost varies greatly. One fisherman brought back from Mecca a piece of alleged ambergris—believed to be a powerful talisman for fish—which cost him $7.50 or $8. But it is a common idea among Perupok fishermen that pieces of " true " ambergris can cost up to $1,000. I cannot conceive of any Kelantan fisherman investing such an amount, however. A talisman of a somewhat amusing kind was the rice flour sold (so I was told) by a high official in Kelantan in March 1940, for the benefit of the Malayan Patriotic Fund. Portions were issued at a dollar each, and one expert fisherman attached great virtue to it. He ascribed a large catch made by another man to the efficacy of this " medicated " flour, and being a fairly wealthy man had bought $6 worth of it himself. It is sprinkled on the boats before they go to sea.

From the point of view of the technique and organization, the ritual and beliefs impose precautions in the treatment of equipment beyond ordinary care. Boats must be properly washed out after each day's fishing ; if left with the blood and scales of fish in them they would frighten off other fish on the morrow. When a net is being dyed the last steeping must be done by the expert, who thus has to give it his personal attention while engaged in reciting the ritual formulae over it. As in other spheres, the performance of the ritual would seem to give an extra measure of confidence to the workers and a fillip to their energies which, though difficult to measure, is undoubtedly one of the elements making for success.

The part played by magic in association with productive effort is a well-worn topic in anthropological discussion. It is only necessary to say here that as far as the Kelantan fishermen are concerned their ritual and beliefs are not so much productive of economic results as ancillary to those results, presenting a fairly simple and coherent scheme of ideas by which success and failure can be explained.

CHAPTER V

OWNERSHIP OF EQUIPMENT AND MANAGEMENT OF CAPITAL

One of the subjects that has received least attention in any discussion of Oriental peasant economic systems is the amount of capital they have, and how it is accumulated and managed. As far as the Malays are concerned the reason probably is the common European idea that Malay peasants, including fishermen, have very little capital and very little notion of how to handle economically what they do possess. It is true that by European standards the amounts are usually small, and in general are well below the level to meet any modern demands for technological improvement, particularly when it involves mechanization. Yet, as far as the fishing economy is concerned, it should be obvious that even at the level at which these people work, the depreciation of equipment (especially of nets) is rapid, and the outlay upon running repairs and upon investment in new equipment must be heavy. For the economy to maintain itself as it does, often with additions to investment, either there must be a continual transfer of capital from outside sources—of which there is as a rule little sign—or there must be a considerable capital flow from within the industry itself.

THE PLACE OF CAPITAL IN THE PEASANT ECONOMY

In this connection a number of problems arise. Some are more purely factual—what kind of capital goods are there in the fishing economy ; what is their value in money terms ; who possesses them ; how and by whom are they applied to the processes of production ? Other problems may be rather more abstract—how is capital conceived of by these people ; do they associate it closely with saving ; do they look upon it as something requiring a special kind of return when used ; what kind of choices have they before them for its employment ; do they continually seek new uses for it, or are they content to employ it along traditional fixed channels ; by what processes is it accumulated and dispersed ?

The answers to some of these questions, and to others of the same order, will be found in this chapter, though material is also given in other contexts in this book. From such data it should be clear that as indicated earlier (p. 5) it is incorrect to apply the term " pre-capitalistic " to such an economy except in a special sense—that is, only in regard to the small magnitude and different form of its capital operations and to the almost complete absence of a class of " capitalists ". As far as function is concerned, the system uses capital in ways that are frequently strictly parallel to those in a modern business economy.

In Chapters II and IV some account has been given of the types of boats and fishing gear, and their cost, as items of equipment in the technological processes of production. We have now to consider them as items of fixed capital in the economic processes of production. But in addition to this there is also the element of liquid capital, in the form of the money which flows through the economic system as the result of many different types of transactions. At any one time a fisherman with an amount of money in his possession has before him a range of choice as to how he will dispose of it. He may hoard it in his home as a stock of liquid capital, against general contingencies, or to be used for some specific act of consumption—as to get his son a bride, or to be spent on his own funeral. By keeping it in this way he runs some risk of loss by theft. Again, he may spend it on some object such as a new house, which gives him greater comfort and prestige. Alternatively, he may invest it directly in some standard item of fixed capital, as a boat or a net, or rice land or a coco-nut orchard, from which he can draw a fairly regular income, either from his own enterprise or by allowing someone else to use it in production. Again, he may decide to increase his capital by acting as entrepreneur in the manufacture and sale of a net, in which he will have to disburse his funds bit by bit in purchase of materials and payment to workers, and wait some months before he has an opportunity of gaining a return. If he is a fishing expert, and the season has just ended, he may have to decide how much of his free capital he ought or can afford to invest in future labour for his net-group—by advances to his crew to tide them over the monsoon with food. This involves some risk, since when the fishing season begins again they may still prefer to go to another expert with whom prospects are better. Further, in some circumstances he may consider the prospect of giving a feast. This is in part " conspicuous consumption ", but also has the function, while creating for him a set of long-term debts, of mobilizing for him an immediate considerable capital sum (pp. 177–182). Faced by situations of this kind he does exercise a great deal of judgement in disposing of his capital. Moreover, the decision is not usually his alone ; in Kelantan, at least, his wife has a very definite voice and often her opinion carries the most weight.

The concept of capital as a stock of money used to finance production is quite a clear one to the Malay peasant. It is described by the general term *modal*, and there is a definite idea that the *modal* laid out on equipment such as nets or boats, or in middleman's goods for re-sale, demands a specific return.[1] In any coöperative undertaking the return on the capital employed is given in the form of a share (*bagian*) of the total receipts. In the scheme of distribution from fishing the return to boats and nets is allotted as a separate item from the return to labour. It is specified as *bagian pĕrahu* and *bagian pukat* as against *bagian tuboh* (literally, " body-share "). In fish-dealing and other enterprises, however, the return to capital is often merged with that to labour. In the return to capital, a general distinction is also drawn between that element which is a partial recoupment of capital expenditure (*balek modal*, literally, " capital coming back ") and any profit (*untong*) made. But the concept of *untong* normally covers interest on the capital as well as true profits on the undertaking. This tendency not to force a distinction between interest and profits is probably to be correlated partly with the fact that so much production among these Malays is coöperative, with the provider of capital contributing also his labour and some of the management skill ; it is probably also attributable partly to the religious restrictions on the taking of interest, which do not encourage the recognition of it as a separate economic category.

VOLUME OF INVESTMENT IN FISHING EQUIPMENT

Before discussing the detailed processes of manipulation of capital resources some idea may be given of the general amount of investment in production equipment in the fishing industry of this community. For the reasons discussed in Chapter II the most useful method of expression of this seems to be in terms of

---

[1] As I was once photographing a boat the owner asked the cost of my camera. I replied : " About $150—about the same as that of this boat." He answered : " But *this* " (putting his hand on the boat) " can get a living ; *that* can only take pictures."

its current value rather than of its initial cost, or of its replacement value if wholly new equipment were to be substituted for it.

In calculating the capital value of the boats in the area, of which there were 206 in all in the middle of 1940, I have taken as a basis the data collected for a sample of 81, for which I was given details of the purchase. For those acquired in 1940 the purchase price can be taken as the current value; for those acquired previously the current value has been estimated by taking depreciation at $10 per annum in the case of *kolek* and large *kueh* less than 10 years old, and at $5 per annum in the case of medium and small *kueh* and other small craft, as also for large craft more than 10 years old. This method of estimation is supported by data I obtained about the original cost and recent sale of some craft, estimates of the present value of others by the owners or other fishermen, and offers made by would-be purchasers. The average current value of boats in each class in the sample has then been converted into total figures for the area as a whole on the assumption that the sample is a representative one. It will be noted that in the case of the *kolek buatan barat*, the most important element in the boat capital of the area, the sample is almost one-half of the total number of cases. For comparison, the original capital outlay incurred on the boats in the sample cases is given first; it will be realized that this original outlay represents a number of craft bought at various times, in various conditions of use.

The results of the calculation are given in Table 8.

TABLE 8

Estimate of Capital Value of Boats in the Perupok Area (in 1940)

Type of Boat.	Number in Sample.	Original Outlay.	Current Value.	Depreci- ation.	Average Current Value.	Total Boats in Area.	Total Current Value.
		$	$	%	$		$
Kolek b. . .	40	7,030	6,320	10	158	82	12,956
Kolek l. . .	8	395	287	27	36	13	468
Kueh (a) . .	7	745	570	23	81	24	1,944
Kueh (b) . .	8	345	274	20	34	17	578
Kueh (c) . .	18	534	415	22	23	70	1,610
Totals . .	81	—	—	—	—	206	17,556

Note : b., *buatan barat*; l., *lichung*; (a) large; (b) medium; (c) small. The total number of boats under *kueh* (c) is approximate.

This table shows the heavy concentration of investment in *kolek buatan barat*, the large craft used mainly for lift-net and deep gill-net work, the coöperative forms of fishing which are predominant in output. A number of the other large craft are also used in lift-net work. The 20 lift-nets operating in the 1940 season had altogether a total of 102 boats, comprising 69 *kolek buatan barat*, 18 large *kueh* and 11 *lichung*. In all, these represented a capital at then values of about $12,900, or 73 per cent. of the total value of all the boats in the area.

Taking the number of fishermen in the Perupok area as approximately 550, the average investment in boats, at current values, is in the region of $30 a head, or a little more, a level possibly rather higher than that for the Kelantan fishermen as a whole.

To the capital invested in actual boats must be added an item for sails, anchor-stones and ropes, floor boards, props, and skids, also for the decorative spar supports known as *bangar*, etc. Some of these are relatively expensive—a large sail of the brown variety favoured by these fishermen cost about $15, and a well-carved *bangar* for a large boat cost from $5 to $10. But when boats are bought second-hand, as many of them are, these items are usually included in the price. Then, again, there is the gear which each individual fisherman must have—a paddle, an oblong box for his betel and smoking materials and fish-hooks, a round box on a pedestal for his food, and normally a pandanus sun-hat painted in segments of bright colour. Sun-hat and paddle are cheap, but a betel box costs from $1 to $5, and a food box (which is not made locally) $2 or $2.50. It is impossible to give any close estimate of the total capital represented by all these items, but at a figure of about $20 for a large craft down to $5 for a small one the total involved may well be in the region of $2,500. It is probably safe to say that the total capital represented by the boats, their equipment and the fishermen's personal gear in the area is about $20,000.

An estimate of the total capital value represented by the nets and other fishing equipment of the community is rather more subject to error than an estimate of the boat capital. The short life of most of the nets—about two years for a drift-net, three years for a lift-net, and five or six years for a deep gill-net—means that their depreciation in value is much more rapid than that of boats, and any attempt to fix their value as a class at a given time must leave a fairly wide margin of error. Moreover, net-making is an important occupation of the community, and lift-nets especially are made for sale to buyers from outside. Such nets, though they are part of the general capital of the community, are not to be reckoned as part of its capital for fishing ; they belong to a different sphere of production, and yield a separate income. But the difficulty in classification is that some at least of these nets are equally available to fishermen inside the community, and are bought by them when a new group is started, or occasionally when a net of their own making is not yet ready for sea. And some expert fishermen try to have a second net always well advanced, to replace the used one should they get a good offer for it, or to sell in its turn, or simply to keep as an alternative to the one they are using should it get damaged. One can distinguish in theory between nets in use, nets in reserve, and nets for sale, but in practice this is not always easy. Moreover, in the actual technique of field recording it is not so easy to enumerate and check the number of nets in an area as it is with boats. After some time I knew most of the latter individually, could note their absence or significant change of position on the beach in the evening (indicating that they had moved to another net-group), and find out the reasons. But the lack of individuality in nets of the same type, the irregularity of their use, and the habit of storing them in the dwelling-houses made it more difficult to study them and keep an objective check upon their movements. The practice of mending the larger nets in the open, however, facilitated observation, and it is only with small nets that really approximate estimates of numbers have had to be used. For these latter types, however, the economic census taken of 331 households in the area provided a useful guide and check.

In calculating the current value of the nets and other fishing equipment it is not possible to use current sales figures to the same extent as in the case of boats ; the latter change hands freely at all stages of their life, while there is more tendency for the larger nets to be bought new and sold when used for some time to buyers outside the area. This is particularly so with lift-nets, while *pukat hanyut*, apparently a fairly recent introduction, are all bought new from one or two importers. A more accurate figure is therefore obtained by taking the average cost price and allowing for depreciation in accordance with the average age of the nets and their general life. But the average figures for each class have in all cases been related to my records of net transactions, covering about 70 cases in all, or about 20 per cent. of the total of the more important types. The values set for scoop-nets, casting nets, and line-fishing equipment, however, are only rough estimates.

The following is the total of the current value of this equipment, calculated on this basis.

TABLE 9

ESTIMATE OF CAPITAL VALUE OF NETS IN THE PERUPOK AREA
(IN 1940)

Type of Net, etc.	Number of Units in use.	Average Value per Unit.	Total Value.
		$	$
Pukat Takur . . .	20 nets	150	3,000
Pukat Dalam . . .	520 sections	6.50	3,380
Pukat Hanyut . . .	42 sections	8	336
Pukat Tarek (fish) . .	1 net	550	550
Pukat Tarek (anchovy) . .	1 net	250	250
Pukat Tĕgĕlang . . .	170 nets (e)	6	1,020
Jaring	26 nets	25	650
Pukat Takur Kĕchil . .	2 nets	40	80
Jaring Tamban . . .	10 nets (e)	5	50
Takur Baring . . .	6 nets (e)	10	60
Saup	100 nets (e)	1	100
Jala	20 nets (e)	5	100
Fish-traps (Bubu) . . .	7 traps	—	10
Fish-lines, hooks, traces, etc. .	500 sets (e)	1	500
Total value . . .			$10,086

Note : (e), figure estimated, not counted.

This total can be only approximate, but an investment of the order of $10,000 at current values is quite possibly rather under than over the mark. If calculated at original cost, allowance being made for what was bought in used condition then, the figure would be in the region of $13,000.

To sum up, the total capital invested in all boats, nets and gear in the area is about $30,000, or an average of about $55 per head of fishermen.

OWNERSHIP OF EQUIPMENT

The next question that arises is : how is this mass of equipment owned and controlled among the body of fishermen ? It will already be evident that we are not dealing with a community of an equalitarian kind, where each individual has a fairly similar control over the means of production. Ownership is not of a communal or group kind, but is highly individualized, though the number of individuals who own equipment in a fishing unit differs considerably according to the type of unit.

For the investigator the situation is complicated by certain linguistic and social conventions which, while familiar in principle to every anthropological field-worker, must be analysed and understood in their specific practice in the community studied. Possessive pronouns are used of goods with which the speaker is associated, but over which he or she has not in the last resort the right of disposal. To the simple question " Have you a boat ? " or " Have you a net ? " a man often replies in the affirmative, when he is merely managing the boat for another owner, or is financially interested in the net to the extent of having " entered the combine " as a profit-sharer, but without having put any capital into it. Or again, a man who is working even temporarily as an ordinary crew member may speak of the net or boat as " mine "—" My net got only 200 fish last night . . ." etc. Again, it is common for a wife to speak of a boat or net as " mine ", though in practice she does not handle it. There are three possible situations here. One is that the equipment is actually hers in law, having been inherited, say, from a deceased husband or being held in trust for heirs who are minors. Another is that the boat or net is part of the *pĕncharian laki-bini*, the property which has been accumulated by the joint efforts of husband and wife during their association, and in which she therefore has an equal share with him, should they divorce. In these cases the respect which is accorded to women in financial affairs in this community may well mean that should the question of its sale arise, it will be she who has the deciding voice. But a third possibility is simply that the boat or net is the husband's by law, having been his property before their marriage and that, as his wife, she assumes a proprietary interest in it. This identification of a person with the things with which he is associated is a common enough phenomenon, but it may give rise to inaccuracies if data on economic matters are being collected simply by the question and answer method. With a knowledge of the background, there is less difficulty.

These are merely some of the complications in the whole structure of ownership of fishing equipment not only in this area but elsewhere on the east coast. In the time at my disposal it was not possible to conduct an extensive inquiry into such problems as the exact distribution of property between husbands and wives over the area as a whole, so that in the following analysis, when the individual property of fishermen is distinguished, the question of the precise position of the wife is left undetermined.[1]

In theory, for boats, ownership is shown on the register kept at the Customs Station for each District. But in practice many names entered in the register do not correspond with the actual individuals who control the boats. A rough practical test of ownership both of boats and nets is, who receives the *bagian*, the share of that item of equipment in distribution of the yield of production. And the share of a boat often goes to someone whose name does not appear as owner on the official list. There are several reasons for this. One is that money may have been lent for the purchase of a boat, or on a boat as security for some other purpose. The lender may have his name inscribed on the register to strengthen his claim to repayment, but he gets only half the share, the other half going to the man who runs the boat and in practice is the owner of it as far as use and even sale are concerned. Or again, the actual owner may have the name of some other person inscribed on the register, for private reasons—in the south of Trengganu boats are said to be registered in Malay names, though in fact they are Chinese-owned. Or again, a boat may be sold, and even pass through several hands, the name of the original owner still remaining on the register when he no longer has any interest in it. Here the reason seems primarily to be apathy and a certain timidity in the face of officialdom ; the real owner appears annually to pay the licence fee, but says nothing about the change of ownership in the interval. Another reason is unwillingness to pay the small fee for registration of the change.

One case may be cited to show the complications that exist. A fish dealer sold a boat to a carrier agent about 1938, and the price was paid in full. The agent, Mamat-Klesong,[2] then sold the boat again to Mbong, a woman whose husband was Berheng ; she handed the boat over to her son Yah to run. In the middle of 1940 one of the neighbours came to me with the licence receipt to have it identified—few of these fishermen can read, and they

---

[1] The distribution of ownership of property in a polygynous union, to which we did devote some attention, is discussed by my wife in *Housekeeping among Malay Peasants*, pp. 35–41, 139–46.

[2] The number of personal names in Kelantan is very limited, and the formal *bin* (son of) is not often used in conversation. Hence people are distinguished by varying abbreviations, with familiar titles, nicknames, etc. One method of describing a man is to conjoin his name with his wife's. Thus, Mamat-Klesong is the husband of Klesong ; Mamat-Petimo, the husband of Petimo (Fatimah). Conversely, the women can be described as Klesong-Mamat, Petimo-Mamat, in distinction from Klesong-Awang, Klesong-Hamid, Petimo-Awang, etc.

were uncertain if the paper referred to the right boat or the right owner. It did cite the right boat by number, but the owner was cited as " Awang Mat bin Berahin of Kampong Qubang Golok ". This was clearly wrong, and was probably due either to muddled information given by Mbong in paying the fee, or to an office error. What was wanted, they said, was the name of Mamat bin Awang (i.e. Mamat-Klesong) as owner. " Why not in the name of Yah, or of Mbong ? " I asked. " Because she hasn't the cash to pay for changing the name." Seller and buyer, they said, had to share the expense, and she was unwilling. Moreover, they added, one had now to get a note from the headman—for a fee also—certifying the ownership before the clerk at the office would make the change. (I did not verify this, but if so it was probably in order to prevent wrongful claiming of boats, especially on decease of a registered owner.) The additional trouble and cost, and the suspicion of documents natural to a largely illiterate people made them wish to avoid this extra certification.

Such deliberate registration or re-registration of boats in the name of former owners means that official records cannot be relied upon as an index to actual ownership. (A similar situation, though probably not to the same extent, seems to exist in the case of land registration.) In discussing quantitative distribution of property, therefore, I have relied on empirical observation of the use of individual items, and a mass of data about rights, ownership, shares of yield, etc., obtained from many sources.

The distribution of boat capital in the community can be roughly gauged by reference to Table 8. For a total of about 550 fishermen there are altogether 206 boats ; that is, only about 37 per cent. of the fishermen can be boat owners. Actually, this proportion is rather less, when the boats owned by people who do not go to sea, and by fishermen who have more than one craft, are taken into consideration. In all there are 25 boats of this type, giving a total of 181 fishermen who own the boats that they run, or approximately 33 per cent. of the whole.

These 25 cases are comprised as follows : 9 boats are owned by people who do not go to sea—a widow, an old man, a boat-builder and shadow-play expert, and four fish dealers, a total of 7 owners. Eight boats are owned by lift-net experts, and 7 by other fishermen, the owner in each case having another craft which he runs himself. And in one case the owner, a dorab fisher, operates the two craft he owns alternately according to the season.

This last case is unusual; it is rare for boat capital to be left lying idle; if the owner is not using it himself he normally gives it out to be run by someone else and earn a dividend, if conditions allow. The person who runs a boat may be a kinsman, but this is by no means the rule—out of the 25 cases, only in 7 was the boat captain a kinsman of the owner. Two were sons, two sons-in-law, one brother, one the brother of a son-in-law and one a step-son.

As regards multiple-boat ownership, there are 15 men owning 2 boats each, and 2 men owning 3 boats each. Of the former, 6 men are lift-net experts, and of the latter, one is a lift-net expert. The expert has perhaps most incentive to invest in more than one boat, since by so doing he fortifies the position of his net-group. A common practice is for the expert to buy an additional boat. Then, if the net is successful, the man to whom he has handed it over to run will buy it from him by paying over each week not only the half of the boat's share to which the owner is entitled, but also the half which the captain himself has earned. Another method is for the expert to assist a man to buy a boat by a loan of the whole or a part of the purchase price; this is discussed in the next chapter. Occasionally, though rarely, an expert does not own a boat; one case of this is Q in the records, where the expert, being a poor man, bought his net on a low deposit, and used one of the boats of the man from whom he had purchased the net.

The distribution of net capital is more complex than that of boat capital. Excluding hand-nets, used only during the monsoon, there are about 270 nets in use in the community. But two types, *pukat dalam* and *pukat hanyut*, are made up of separate components, each of which is treated as a unit from the point of view of ownership, giving in all about 820 net items. I did not take a complete census of net holdings for all the community, but from the census of over 300 fishermen in the central part of the area it appears that about two-fifths of the fishermen own no nets or net components at all. In some cases this is due to their concentration on line fishing (accompanied often with cultivation of rice), but in most cases it is simply a reflection of their poverty. These are the men who form the bulk of the lift-net crews, and whose state is described in the words " he has not a single thing; he only helps other people ". Of the remainder, another two-fifths own a net apiece or a component of one, and one-fifth own more than one net. There is thus a larger proportion of fishermen with some net capital than with boat capital. This is due, on the one hand, to the utility and cheapness of the small drift-net, which costs new only about $12 to buy and $10 to make oneself, and which may be bought fairly readily in used condition for about $5, a sum which many men can raise. It is due also to the convenience of putting capital into the deep gill-net ; one section or component of this costs about $10 new, and the yield is likely to be good.

The attraction of the deep gill-net is seen by considering the distribution of ownership of about 450 sections which make up about 20 of the nets in use. These sections are spread among 85 separate owners, mostly from 2 to 5 sections each. One man has one section only, but 60 people have from 2 to 5 sections, 13 people have from 6 to 10 sections, 7 from 11 to 15 sections, 3 from 16 to 20 sections, and 1 has more than 20 sections. Only one net is owned completely by one man, and the sections of this are mostly old ; he prefers as a rule to go out with another net. This distribution of ownership in these mackerel nets contrasts strongly with that of lift-nets, where nearly every one is the property of a single owner. In drift-nets the same principle operates, but to a much less degree, the 8 nets being made up of components from about 18 owners ; each section here, however, cost about $16.50 complete with cords (in 1940).

As in the case of boats, there are a few nets or components that are owned by persons not fishermen ; these are mainly invalids, old men or widows, though one prominent fish dealer—who, incidentally, is reputed to have an interest in eight boats —is an importer of drift-nets, and always has at least one at sea under the management of someone else.

LEVELS OF INDIVIDUAL BOAT AND NET CAPITAL

The question of the distribution of capital equipment by units —rather than by amounts of capital invested—is important from a practical point of view in understanding the economy of these fishermen, since distribution of the yield is done on a unit basis. Each boat, each net or component of a net receives a share in the takings as such, irrespective of its quality or the amount of cash investment it represents. An old boat that has cost $20 receives just the same share in a net group as one that has cost $200 ; a section of deep gill-net that has cost $5 gets the same share as one that has cost $10. The poor man who cannot afford better equipment thus gets an equal chance with the wealthy man who can buy the best. But the disadvantage of poor equipment is first that it endangers the efficiency of the group, and may be dropped because of this, and secondly that it demands much more labour in repairs. There is thus an incentive to supersede poor equipment by better, and the level of capital of different individuals is largely an index to their success.

One of the striking features in a Malay fishing economy in an area such as we are studying is the great difference in levels of capital at the command of different fishermen. Apart from the men with neither boat nor net, whose fishing capital is comprised only in essential gear worth a mere few dollars, the levels range from $5 or so in a *pukat tĕgĕlang* or $8 or $10 in an old boat, to $500 in a lift-net, some sections of *pukat dalam* and a new large boat, or even higher in a few instances. A detailed census of actual cash property values for the area was not feasible, but some idea of the range of differences is obtained from the following table, which embodies the estimates of boat and net capital for

TABLE 10

RANGE OF CAPITAL INVESTMENT IN FISHING EQUIPMENT
(Excluding hooks, lines, etc., and personal gear)

Levels of Investment. $		Number of Fishermen at each level.
0		96
1 to 50		85
51 ,, 100		15
101 ,, 150		17
151 ,, 200		5
201 ,, 250		8
251 ,, 300		3
301 ,, 350		6
351 ,, 400		1
401 ,, 450		3
451 ,, 500		6
501 ,, 600		3
601 ,, 700		4
701 ,, 800		2
801 ,, 900		0
901 ,, 1,000		1
1,001 ,, 1,100		1
Total fishermen in sample		256

256 fishermen, the calculation being based partly on actual valuations given or purchases noted and partly on estimates of cost less depreciation, as indicated earlier. The number of cases is about five-sixths of that of the total fishermen in the census

area (Table 3), and the figures are fairly representative of the distribution. Though, since most of the more wealthy fishermen were included, by accident of residence near the beach, in the census area itself, the proportion in the higher levels of investment is rather greater than it would have been if a complete census for the whole fishing community had been taken.

In the aggregate, then, only about 30 per cent. of these men have fixed capital of more than $50, less than 12 per cent. have capital of more than $250, and only about 4 per cent. have capital of more than $500. These figures, of course, refer to fishing equipment alone ; some of the men, especially those in the higher levels, have considerable capital in rice lands, coco-nut orchards, etc. But included in the figures are estimates of capital invested in the making of nets for sale as well as for use. Of the 18 lift-net experts included in the sample, only 4 have a capital of less than $250, and 9 are above the $500 level. Of the 4 with small capital, one has a boat but got his net on a small cash payment, and afterwards sold it when his group broke up ; one operates with a borrowed boat and got his net for $38 cash down out of a total purchase price of $230 ; and the two others have boats of their own but no nets, relying on kinsfolk who own the latter. On the whole, the tendency is for the fishing capital of a *juru sĕlam* to be correlated with his relative success. But in a few cases there seems to be some over-capitalization if regard be had to the level of income which the equipment yields. Their comparative lack of success at lift-net fishing, and their heavy investment in nets of other types as well, seem to have resulted in their being chronically short of liquid capital.

MANAGEMENT OF CAPITAL BY INDIVIDUALS

We have now to see how these fishermen as individuals mobilize and dispose of their capital resources, how they calculate the comparative utility of different items of fixed capital to them and substitute one for another. Some account of their preferences for different types of equipment, and the rapidity with which they exchange them, has been given in the previous chapter. Here we are concerned with the actual finance of these affairs.

We may first consider how a man gets a. start as a possessor of capital equipment. Some have the help of their kinsfolk, especially their fathers. One fisherman explained that he and his eldest son were the joint owners of a lift-net, his son having

put $30 into it.   I asked where he got the money.   He replied :
" His money, but my money ; with a child it's not definite "—
in other words, it was a gift from him.   Help is often received
from a father-in-law, especially where this man is the head of
a net-group and the son-in-law is a member of it.   Money is
commonly advanced to finance the purchase of a boat, repayment
being made in instalments out of the boat's share of the yield.
Help to a son-in-law may take the form of a gift to the daughter.
The Malay attitude is well illustrated by some remarks of an old
man with two sons-in-law.   He had accumulated about $1,500
by trading in copra years before.   To one daughter he handed
over about $500 in the form of an orchard and about $250 in
cash to pay for a boat ; to the other about $120, also put into
a boat and land.   In each case the property was put in the
daughter's name so that if she were divorced her husband would
not get it, but otherwise he had the use of it.   The old man had
also lent between $40 and $50 to the first son-in-law, and was
mending a net of his, without payment, as we talked.   He
explained the position—" One just helps ; one is an old person
not up to working . . . one gives to one's child, *gives* to her, to
help with food, before one dies ; she has a hard job . . . it's a
Malay custom if a child finds it hard to make a living, and the
father is alive, the father helps . . . the child has not yet got
property, one has come a little before, one has some . . ."
I asked him whether if he had had a son there would not have
been trouble over these gifts.   " No," he said, " he would calculate like this : that it is difficult for a female child ; that she
wants to accumulate property, but it's hard for her to do so."

On the other hand, various men complained to me that they
could not get a proper start because their fathers were not
wealthy ; or spoke enviously of others who had received help
from their fathers ; or boasted of how they themselves, with no
rich father to back them up, had built up their capital from
nothing.   But in general, emphasis was laid on the importance of
saving, and its difficulties.

Independent enterprise in accumulating capital takes many
forms.   The two commonest are to go out as an ordinary member
of a lift-net crew—which means, in effect, the hiring out of one's
labour—or to go out line fishing, with the minimum expenditure
of less than one dollar—which means a greater reliance on one's
individual skill.   A third method is to work with borrowed
equipment.   The general principle here is for the borrower of

boat or net to pay the lender a half of the share which the equipment receives in the general distribution of the takings. The
proportion which the borrower retains in addition to his own
" body " share varies, for boats, from about ¾ per cent. for an
ordinary lift-net boat, to 2¼ per cent. for a boat used as fish
carrier in lift-net work, and 5 per cent. for a boat used for line
fishing or light drift-net work. The proportion with nets varies
more. But a man may well increase his capital thus by 50 cents
to $2 a week in good seasons. The limitations here, however,
are the comparative scarcity of equipment available for borrowing,
and the likelihood of its being in demand by the owner when the
season is most profitable. But in some cases the borrowing of
equipment is a step which leads ultimately to its purchase.

As a rule the cash gained in these various ways is put into an
old light drift-net, the net of most utility, or into an old boat.
With a net one's individual earning powers are broadened ; with
a boat not only is one's mobility greater, but one gets an additional one-tenth share as well as a personal share in the takings
of any joint fishing. The purchase of an old boat or net means
that one must be prepared to put in time and labour, and a small
amount of extra capital, in heavy repairs. A fairly common
sight during the monsoon is of men taking to pieces old boats
that they have bought, replacing worm-eaten or cracked planks,
putting on new topstrakes or false keels, caulking and puttying
seams. Some boats are almost rebuilt in the process. There is
a constant tendency for equipment on the margin of employment
or of output to filter down from the more wealthy or less energetic
fishermen to the less wealthy and more energetic, allowing
opportunity for the investment of small capital and the capitalization of labour and time. It is significant that even with the
kolek buatan barat, the most prized boats in the Perupok area,
about three-fifths held in 1939-40 had been bought second-
hand.

In this community there is equipment to be bought by all
except those—usually married men with large families—on the
absolute level of bare subsistence. It is true that in a number
of cases men or their wives spoke of not having the cash to buy
boat or net, of being " broke " (the Malay word. *sĕsok* being the
equivalent). But in all except a very few it was almost certainly
lack of initiative as well as lack of cash. The prices of some of
the poorer items of equipment are extremely low. Light drift-
nets, costing new $10 to make or $12 to buy, are commonly

sold second-hand for \$4 or \$3, and I recorded several as having been bought for \$2, and one for which only 50 cents had been paid. Boats are more costly, even very old craft normally costing more than \$10, though I recorded one purchase at \$8.50 and another at \$8. In one case, where I saw a lad rebuilding a battered old small boat, it turned out that he had paid nothing at all for it ; it was in such bad condition that the owner made him a present of it as an alternative to using it for firewood. Sales of such old equipment are usually for cash— the situation of the buyer and the condition of the goods mean that the seller would have poor security in any credit transaction. This comparative inability of the poor to get credit is one of the difficulties in the way of an improvement of their economic position, though perhaps less so than in many other countries.

The poverty factor is also an element in the supply of used equipment, by forcing realization of assets in time of stress, especially during the monsoon. Several men I knew sold casting-nets and small drift-nets then, when money was short, hoping no doubt to be able to get others when income improved as fishing began again. And a certain amount of equipment tends to come on the market from young unmarried or divorced men who bought it when they were flush of cash, and realize it when they get short of funds, for coffee, snacks, cigarettes or women. Older men wag their heads at this propensity of the young who when hard-up will sell anything—usually at a loss—to get some ready money.

The economic problem of choice enters into the calculations of a person with slender capital in several ways. He can continue to save, with a view to investment on a larger scale later on ; he can invest in either a boat or a net, with their respective advantages ; and he can choose size or quality. The problem can be sufficiently illustrated by examples of the last point in the case of boats.

A lift-net crew man in Pantai Damat bought a large *lichung* in Besut for \$20 ; he estimated that it was over 10 years old, and that it would last for about 3 years longer. He said that he bought such an old boat deliberately because he wanted a big one rather than a smaller one in better condition. On the other hand, another man who also went out with lift-nets as crew, but was a skilled line fisherman, sold his boat, which had cost him \$32, for \$25 after about a year, because he wanted a smaller one which would be better for his specialist work. He thereby lost some time, because line fishing was then in swing, and he could not get a place in a boat. But he said that he expected to be able to buy another one before the squid fishing came on, at the beginning of the next moon.

A fisherman with five children bought an old tarred *lichung* in April for $10 and repaired it, in order to go out with the lift-net of a near-by expert. He formerly had a small *kueh*, which he sold for $6, and a light drift-net which he sold for $4, thus making up the price. It was explained by others that he had many children, and that if he didn't have a boat things would go hard with him. But he couldn't afford a good boat, so bought this. It was a large craft, suited to lift-net work, and his purchase meant a conversion of his capital from one form of fishing to another. His venture, however, was unsuccessful, because the net-group he joined had poor takings, and in June he sold the boat again for $8, thus losing $2 of his capital.

An example of more successful investment is that of a man who specialized mainly in line fishing and light drift-netting. He built up his capital thus. He bought first a very old boat, about in its last year of life. The owner had been offered $3 for it and a section of gill-net for catching sprats as bait for dorab, but had refused this. Our man bought boat and net for $5.50. He hadn't much money to spare, but dorab then seemed to be fairly plentiful, so he was willing to pay out the cash. On the first day alone he took $2 worth of fish, and in one week had got his capital back. He said that the gill-net alone was worth $2—he still had it— and he bought another section for $3 to make it more effective. Then he bought a larger boat, a good all-weather craft more suitable for drift-netting and other work, for $11. He had this about three years and got his capital back, in one year he reckoned, by going out with a lift-net group. He had also acquired one light drift-net and had another nearly completed, and also a section of *pukat hanyut*, which, however, was in store and now unfit for use. When I talked to him first on these points he was having a small *kueh* built for him as well. He had given the builder $20 to buy timber, and it was agreed that another $25 should be paid later as payment for the work, making $45 for the cost of the boat. He added that the $25 need not be produced as soon as the boat was ready, as the builder trusted him. I asked him where he got the $20—at Tumpat, where, as I knew, he had spent some time fishing during the monsoon? He said not, that he had been able to save only a little there, but that he had " sought for it at home ", in other words, had dug it out of former savings. His principle, as he made it clear, was to buy old cheap equipment in the early stages, and he was scornful of men who weren't " smart " in this way. When I last saw him he had sold the large boat for $17—incidentally making a straight profit of $6 on it—and was going to concentrate on work with his new small one. Without going into further details it is clear how in a few years he had built up his working capital from about $6 to $60 or more.

In determining investment in a boat there are many technical considerations apart from those mentioned above. Stability, kind of timber, thickness of planks, line and flair, shape of prow and stern are all important to these fishermen, who naturally are conversant with all the points that go to make a good craft. A *pĕraih's* boat, for instance, should be easy to sail and to paddle, but deep and broad to hold plenty of fish, and yet light to pull up on the beach with the catch.

Apart from these technical factors, however, there are others which might not be so apparent to an observer. One is the economic factor of comparative supply and demand, which the fishermen themselves realize clearly enough. As one man put it when his wife was urging him to buy a small boat for line fishing —in the monsoon season when boats were cheap he had no cash, and now (in April) when he had the cash such boats were dear, and hard to get. And in regard to nets another man said that he had a *jaring* which was worth $30—or, he added, $25 if he had to sell and the fish were not plentiful.

Among other factors also is the influence of the wife on a man's decisions. The rôle of women as keepers of the household cash resources has already been mentioned, but they are equally important as advisers and often leaders in matters of investment. Quite a number of cases came to our notice where men refrained from selling their boats owing to their wives' opinions, and, apparently quite sincerely, gave this as the reason to the prospective buyers.

One further factor in choice lies in the sphere of the personal equation between man and material. Here, as elsewhere, a boat will seem to suit one man and not another. But bearing in mind the elements of ritual and magical belief sketched out in the last chapter, one can realize how among these fishermen this suitability of a boat tends to be expressed in more irrational, even anthropomorphic terms. Awang Lung said that my boat, which he had borrowed for line fishing, suited him. He held that if in measuring a boat the concluding arm-stretch meets the end of the boat at the palm, or in the middle of the forehead, then the boat is "in agreement" with the person who is measuring it. He said that men like a craft which is thus " in agreement " with them— they will pay $10 more for it in the case of a large boat. But if in the last arm-stretch the end of the boat is reached at the shoulder, or along the arm, then the boat is not " in agreement ". To " agree " with a man means that the boat will probably do well for him in catching fish. Again, in the case of a *pĕraih's* boat, I asked the expert of the net-group why it had been sold. He replied " it didn't kill—it wouldn't work the *pukat dalam* ; the boats of other people were full ; his boat wasn't full ". In a way, the boat is conceived as being an active partner in getting fish, and if one's own catches are poor while others are not, it may be the boat's fault.

In studying the acquisition of equipment we have been considering, so far, mainly the smaller transactions in which cash is paid for the goods on the spot. As the scale of values involved increases there is more tendency for credit to enter in, partly because of the difficulty of the buyer in raising the ready money, and partly because the items of higher value provide better security should the buyer be unable to fulfil his bargain and they be taken back. The complexities of the credit system are examined in the next chapter. But it may be noted here that even in large transactions the necessary cash is sometimes found at once by the purchaser, either out of his savings, or from current income, or through the sale of other assets, which in their turn have demanded the production of cash by the earlier buyer. It is not possible to estimate on any general scale the proportion of cash to credit in the purchase of boats and nets, but I recorded sales in which $70, $100, $200, $210 and even $300 were paid in full for boats at the time. In the case of the $200 the money came from the sale of a former boat. In that of the $300 the major part of it came from the sale of the boat for which $210 was paid, this money apparently being the result of a very successful season by a *juru sĕlam*. The total amount of new capital invested in such cases is therefore not great, but the tendency is for the higher-priced items of equipment to gravitate towards the more successful fishermen, who have the necessary margin of free capital over and above that realized from the sale of a similar asset of their own.

HISTORY OF ONE MAN'S INVESTMENTS

In all these transactions the concepts of capital (*modal*) and profit (*untong*) are kept well to the fore, and most men know fairly closely at any time how their boat or net stands, how far they have got their initial capital back on it, whether it has yet earned any profit, and if it is sold how far they will have made or lost on that particular investment. Many of them, though quite illiterate, remember the profit-and-loss position of a series of their past investments. As an example, which also illustrates a number of other points of boat and net finance, I give a synopsis of an account which I received from one man, Saleh-Minoh.

As a child he was in the household of a Malay noble. Then for about five years he worked on a European plantation, where he accumulated some capital. Then for ten years, from about 1920 to 1929, he took up trading in dried fish and copra, going to Singapore with the goods. One season he lost about $400 when the cargo sank in a boat, but recovered the amount in a year's trading. He gave up this work on his marriage, having made no great profit. His wife objected to it, saying that he would probably lose everything he made.

He then took up lift-net fishing, and in six years had five boats, all *lichung*, and was partner in five nets. His boat transactions were as follows :

Boat (i) cost $150, and was sold for $120 after 3 years' work.
  (ii) cost $50 and was sold for $40 after a year's work. It was 10 years old when he bought it.
  (iii) cost $80 and he sold it for $100 after 2 years. It was 8 years old when he bought it, and a fine boat.
  (iv) cost $40 and was sold for $35 after 3 years. This was a very old boat, perhaps 20 years in service when he bought it.
  (v) cost $90 and was sold for $100 after one year. It was 2 years old when he bought it.

On these transactions he thus lost only $15 by depreciation of his capital. His net transactions were as follows :

Lift-net (a) cost $400 (nets then being very dear) to the combine of 5 mén. They paid $100 down, and later $150. But after the second year the net came to pieces, the thread not being good, and they sold it for $30, with an additional $20 for the ropes and drying rack. On this net, he said, they lost $100, or $20 per man.

(b) cost $100 ; it had been in use for more than a year already. The same combine took it, giving no cash down at all, but paying instalments on it bit by bit till the purchase price was cleared off. On this net they made $50 profit at sea, and another $50 by the sale of the net, thus getting $20 each per man clear.

(c) was bought for $160, also in second-hand condition. This was used for six months and then sold for $60, the same combine being the owners.

(d) cost $80. This also had been used for a year, and the combine, this time consisting of 6 men, made it last for another 3 years. It was a fine net and gave a large profit, the six men getting about $200 among them. They finally sold it for $40.

(e) was bought for $190, this time in new condition. The combine, by now enlarged to 7 men, paid $50 down, and the remainder little by little out of the takings. The net was used for two years and then sold for $180 ; it yielded as profit at sea about $30 per man.

After this Saleh gave up owning boats and sold the last of them. This meant in effect also his retirement from lift-net work, of which he had had enough. He then entered into partnership in seine fishing. He was concerned in two of these nets :

Seine (a) a net bought for $400 in partnership with one other man. The net was already about 5 years old when they acquired it, and they used it for 3 years. They sold it for $250, and they had taken with it at sea about $100 per man.

(b) a new large net bought for $1,200 by 7 men, one of whom, who acted as *juru sĕlam*, had made the net from his own capital. Saleh and the others participated by paying $50 per man, after which the net went to sea. In six years the complete cost of the net had been paid off. At the end of this time Saleh was tired of the work, and did not want to continue ; the others, however, wished to do so. Saleh told the *juru sĕlam*, who offered him $25 to buy out his share. This Saleh refused, wanting $50. The *juru sĕlam* was unwilling to give this, so Saleh (following the common practice of proposing an exchange of rôles when the two parties to a bargain cannot agree) said that he would pay the *juru sĕlam* $50 to buy *him* out, and would carry on with the net with the other partners. The *juru sĕlam*, ashamed—since he was the expert—did not wish to let the net go, so paid over the $50 to Saleh, who then withdrew.

After this Saleh took up heavy drift-net (*pukat hanyut*) work, at which he is still engaged. He has had an interest in three of these nets.

Drift-net (*a*) He bought a section and ropes for $11 and used it for one year, selling it then for $5, and getting a small profit from it at sea.

(*b*) He then ran a net which had been used for only one month ; there were 6 sections, for which $50 was paid, the money being found by a fish dealer, who did not go to sea himself, but received $\frac{1}{7}$ of the takings—the crew of the net being 6 men. They used the net for one year, and it then broke up (" died ") and they got only the ropes out of it. The initial outlay was recouped by the fish dealer but Saleh himself got no profit out of the net.

(*c*) He then bought one section complete with ropes for $14, from the fish dealer, who is also an importer of these nets. This net he still uses, in its second year. He has got his capital back (at the beginning of 1940) and is now getting a profit from it.

During much of his time with these various nets he also had various others of smaller type—he couldn't remember how many. At the present time he owns, in addition to the heavy drift-net, a light drift-net which cost $5 a year before ; a sprat net which he made himself and which he estimated cost him $40 ; and three sections of mackerel net which cost him $27. He has no boat now, and his total fishing capital, including hooks, lines, etc., is therefore between $75 and $80. Asked how much cash in hand he had at the present time, he said that he didn't exactly know ; his wife kept it, but he thought it was about $150 in paper currency. He added that it would probably be largely dispersed for his funeral expenses, and that his son would get the house, the coco-nut orchard he has and his nets.

This account is not complete enough to enable one to get a picture of his whole financial history, but it shows how partly because of the short life of their nets, but mainly in search of profits, these fishermen are continually changing the form of their investments. It shows also the lines along which a fisherman calculates his capital position, balancing against the cost of net or boat the amount of its earnings plus the amount received from its re-sale. The term " profit at sea " refers to any excess of earnings of the item of capital over the sum set initially as its price. One of the difficulties in weighing up a fisherman's financial transactions is that he often fails to specify, when he is speaking of " profit " from a net whether he means " profit at sea " alone or total profit including any sum derived from the re-sale of the net.

A summary profit-and-loss account drawn up for Saleh's successive investments in lift-net and seine work gives the following results, taking his proportionate shares in each case.

Investment.	Total Cost.	Total Re-Sale.	Depreci-ation.	Total Earnings.	Profit.	Annual Profit.
	$	$	$	$	$	$
Lift-net boats .	410	395	15	114 [1]	99 }	31.50
Lift-nets . . .	172	65	107	197	90 }	
Seines . . .	250	175	75	150	75	9.37
	832	635	197	461	264	18.85 (average)

Apart from these investments, Saleh of course obtained an income from his share in the takings of the nets as an ordinary crew-member, and from his work from other forms of fishing. Reviewing his position as a whole, then, and assuming that the

[1] The earnings of his lift-net boats have been calculated as $\frac{1}{10}$th of those of the nets during the same period.

$170 he put into his original boat and net was practically all that he had, he has added about $50 to his total capital in the fourteen years—though he has reduced the amount of it invested in equipment and holds more in cash. He has taken out as profits nearly $20 per annum, the most of which has apparently been absorbed into his current expenditure.

To attempt to strike any general level of profits for the fishing community as a whole is difficult. The more efficient fishermen, especially with lift-nets and mackerel nets, reap a much higher return than in the case of Saleh. But his annual profit of roughly 30 per cent. may well represent the attainment of fishermen of moderate means and capability.

As an example of building up capital on a higher level may be mentioned the case of a member of the crew of a lift-net. This man owns a large boat, and has been at sea for over twenty years. In addition he is a rice cultivator. His wife does the ploughing, and he leaves the sea at planting-time to do his share of the work. He is also an energetic vegetable grower, selling from $50 to $60 worth of vegetables a year. He told me that in all he had bought about $2,000 worth of land, nearly all of it an investment of money saved from his work as a fisherman.

FINANCING THE PRODUCTION AND MAINTENANCE OF EQUIPMENT

So far we have been discussing mainly cases of purchase of equipment already produced. But an important feature of the fishing industry in this area, especially of lift-net fishing, is the way in which much equipment, nets in particular, is produced by the enterprise of some of the fishermen themselves. This involves new problems of finance.

The position can be most clearly seen by taking as an example the production of lift-nets. The primary costs involved here are for yarn, spinning, netting process (done in sections), joining the sections, edge-ropes and their attachment, and finally, hauling-ropes. I followed the making of several nets fairly closely, and got a number of estimates of cost, in which there was little variation. In the two sets of estimates given below, *B* is that of a *juru sĕlam* who was getting the net made for his use in the coming season ; *A* was that of a fish dealer who made several nets each year for sale, and who was therefore probably in a position to give rather lower rates for his more regular work.

The yarn is bought in packages of about a score of hanks. It is given out to women to spin, and then the thread is given out to other women to make up into sections of different mesh for the various parts of the net (Fig. 16). The entrepreneur goes about once a week to the surrounding villages with his balls of thread and bamboo mesh-gauges and hands the work out to the women, returning on his next round to pay them and collect the work. The rates are on a piece-work basis, and vary according to the size of mesh required, a section of larger mesh being paid at a lower rate than an equivalent one of smaller mesh. One net usually needs about a score of women workers in all, each woman doing a section in about a week or ten days, though some fast workers do one in four days. The same woman often does 3 or 4 sections, getting for this from about $1.50 to $2.50, and averaging probably about 7 to 10 cents a day.

The task of joining up the sections of the net is always done by a skilled man. For this he normally receives $5 or $6 for a week or ten days' work, but when the net-owner or his partners do part of it his fee is proportionately less. The other operations of fitting out the net are done by the people who will use it, and their labour receives no direct payment.

Two estimates of the cost of lift-nets are as follows :

	Net A.	$	Net B.	$
Thread :	15–20 packages @ $4 each	60–80	17 packages @ $4.20	71.40
Spinning :	,, ,, ,, $1.20	18–24	,, ,, ,, $1.50	25.50
Netting—Various sections :				
Pěrut	1 section	7.00	1 section	7.00
Kape pěrut	8 sections @ 70 cents each	5.60	8 sections @ 80 cents	6.40
Měduo	10 ,, ,, 70 ,, ,,	7.00	10 ,, ,, 80 ,,	8.00
Měnigo	13 ,, ,, 60 ,, ,,	7.80	12 ,, ,, 60 ,,	7.20
Měngepa'	15 ,, ,, 40 ,, ,,	6.00	14 ,, ,, 50 ,,	7.00
Mělimo	16 ,, ,, 40 ,, ,,	6.40	18 ,, ,, 30 ,,	5.40
Pělěroh	4 kati ,, 20 ,, ,,	0.80	5 kati ,, 25 ,,	1.25
Joining sections		5.00		6.00
Edge-ropes		10.00		10.00
	Total cost of net	$133.60–159.60		$155.15

Thus about the end of 1939, the time when these figures were obtained, the cost of lift-nets was in the region of $150 apiece, the precise amount varying according to size and piece-rates paid. Since that time, however, their cost has risen, owing to a war-time rise in the cost of the cotton yarn, which is imported. Whereas before the war it was $4 a package, by December 1939 it had risen to $4.20, and by the middle of June 1940 to nearly $4.50 a package.

Having a net made in this way is important from the point of view of finance. It involves finding the capital before the net can be used, instead of getting a large part of it out of the proceeds of the net in use, as is the common method with a ready-made purchase (which is mostly on credit). The edge-ropes may be bought on credit, and the fee for joining the sections of the net may be postponed, but payment for yarn, spinning and net-making must be made on the spot. How is the money found? Almost any source of funds may be utilized, including savings, part of current income, borrowing, realization of other assets, or mobilization of credit in the form of a feast (*krejo*, see Chapter VI). The method most commonly employed by the fishermen is to proceed by instalments, beginning to buy a few packages of yarn and sending them out to be spun and made up in preparation for a new net as soon as a good yield comes in from the current net. They are thus continuously putting their capital back again. And should the return of the current net during the fishing season have been too low to allow of the virtual completion of the new net, the old one may be sold and the proceeds devoted to finishing off the new one. The impossibility of fishing with the nets during the monsoon stimulates this turn-over of capital, and results in the appearance of a number of new nets as fishing begins again. The less efficient fisherman may have to wait two seasons before he can replace his old net, but the richer and more successful one aims always to have a new net ready each year.

Some details from the transactions of two fishermen will illustrate this aspect of their finance.

1. In the case of net *B* above, the owner had bought 13 packages of yarn first, and then a further 4 packages later. He estimated that it had taken him about five months to save up the first $50. In the middle of November 1939 he sold his old net, which he had used for two seasons, for $50 to a man from the Redang islands, off the Trengganu coast. He received $10 down, and it was agreed that he should get another instalment in three months' time. But it was not paid, so early in May 1940 he sailed over to the islands to collect some of the balance, either in cash or in turtles' eggs, which are abundant there. He returned after five days, however, with only a few eggs, which he had paid for in cash.

Net *B*, which had been finished on January 11th, 1940, went to sea on January 14th. By about the middle of April it had received about $50 as its share of receipts, and the owner said that the ropes, etc., were nearly all paid off. About $10 was still owing on these, and it was expected that this would be settled at the next division. By the middle of June, $75 had been received as the net's share of the total yield ; the owner had bought six packages of yarn at $4.45 per package, and had given them out to be spun as the first instalment of work on a new net. Thus about $35 of capital was already being put back into replacement of equipment when this was only six months old.

2. In 1939–40 Awang Lung was engaged in manipulating his capital in three lift-nets.

(a) The net which he had used in the 1938–9 season had been sold for $90 to a man of Senok, who had paid $40 by the end of December 1939. At this time the wife of Awang Lung and one of his partners had each failed to get any money out of the buyer further by successive visits, and he had spoken angrily to them. Awang Lung therefore went himself, determined to take the net back if he did not get satisfaction. It was arranged then that $25 should be paid at the next week-end and the remainder when the new season opened again. The man duly came, but paid only $20 instead of the promised sum. But Awang Lung said that the man had acted correctly ; that since the man had appeared and explained, he would have been satisfied if only $15 had been paid. By the middle of April, however, the balance of $30 had not been settled, and Awang Lung, hearing that the man had been seen in Bachok, and had not visited him, was angry and spoke of sending a messenger soon to ask for the money.

(b) Meanwhile, at the end of December 1939 he was thinking of selling the net he had used during the past season, possibly for $130–140, but more likely for $120. His original capital of $150 in this net had already come back, the net having earned $179 so far. But $21 of the price he had set upon it in the terms of his partnership (see p. 155) had still to be gained. If he sold the net for $120 he would keep $20 as remainder on the agreed value of the net in partnership, and divide the $100 among the two partners and himself. Towards the middle of January a prospective buyer came, but Awang Lung told him he wouldn't sell for another fortnight—he wanted to take advantage of the fine weather to take the net out, since his new net (c) was not ready. He entertained the man to coffee, but no price for the net was fixed. In the third week of January the net was sold, but still no price was fixed. Awang Lung wanted $130, the buyer wanted to give $120. Awang Lung told me that if he could get a few days' fishing and some cash therefrom before the net was taken away he would take $120 ; if he fished without result he would keep to $130. The net was later sold for $120, but by the middle of April no cash had been handed over. A member of the buyer's crew came to Awang Lung to say that all the takings so far had been absorbed by the buying of ropes and other accessories. Awang Lung appeared to be satisfied.

(c) In the meantime a new net was being made. By the middle of December 1939 Awang Lung had spent $15 on yarn, five weeks later the sections had been made and were being joined together, and at the beginning of February the net was made ready for sea.

In this case each new net was financed primarily from the income from the current net, the sums accruing from the sale of used nets coming in too slowly to allow of operation on them at the turn of the season. By the time net (c) was in use Awang Lung had received only $60 from his past two nets, and had rather more than $80 of personal capital outstanding in them.

The position of the net manufacturers who are not fishermen themselves will be discussed later.

Not every lift-net expert can afford to produce his own net in this fashion. The poorer ones, especially those who are just beginning their career, have not the initial capital, and are forced to buy ; it usually takes several seasons before they can begin to manufacture. So much is this so that the distinction between those who " buy " and those who " make themselves " is practically a division between the poorer and the more wealthy expert fishermen.

This situation is the reverse of what one might expect—in a peasant community it is more often the poor who make their own equipment and the rich who buy. We must examine the reasons why this is not so in the Perupok area.

The first reason is the structure of the credit system in which the seller of such large-scale equipment stands out of the bulk of the purchase price, allowing the buyer to pay by instalments, as already described. (See further at the end of this chapter.) The second reason is that the cost of a lift-net to make is roughly from $50 to $70 less than its cost if bought. And the third reason is a combination of the first two which involves an analysis of the procedure of maintaining the net when in use.

Although practically all the lift-nets in the Perupok area are owned individually at the present time, there is a practice whereby each net has its combine of from two to about six men. These men who enter the net combine (*masok konsi*) are usually boat-owners in the group, or other men of substance. They put in no initial capital, and are not responsible for the costs of manufacture of the net. But once the net is in use they come in as contributors of labour, in particular assuming the work of repairing, dyeing and generally caring for the net. Moreover, the financial basis is agreed upon with the net-owner, who puts a " price " upon the net, as a rule about $50 above its actual cost to him. Thus a net which has cost $160 will be assessed at $210 or $220. All proceeds from the net's share of the takings each week go to the net-owner till this figure is reached, but after that he and the members of the combine divide the net's share equally. This system means in the first place that the *juru sělam* (assuming he is the net-owner) has an additional incentive to make rather than buy his net. Not only does he save an outside manufacturer's profit, but he himself gets an entrepreneur's profit and an assured supply of labour for net maintenance in return for foregoing a large percentage of the net's future income. The system is an entirely voluntary one as far as the other members of the combine are concerned—if they do not like the figure the *juru sělam* has set they need not join, and they can always drop out if they wish. But that it is satisfactory to all parties on the whole is seen by the fact that nearly every net has its combine, the more successful nets tending to have the larger number of members.

In the case of net *B* above, for instance, which had only mediocre success, the combine consisted of three men—the *juru sělam*, his sister's husband (who had actually put some capital into it for the purchase of yarn) and one member of the crew. In another net of poor yield the only partners were the owner and his son ; he explained this by saying that it was useless to have others—because of wrangles, " You talk like this, I talk like that ". But other people held that it was because there was some suspicion of his honesty. Awang Lung, on the other hand, had five men in his combine, and it was said that Japar, most successful of all the lift-net experts, had probably at least six.

A further aspect of net finance comes up with the sale of used nets, a common practice with lift-nets in the Perupok area. The ordinary life of a lift-net is about 3 years, though this varies with quality of the thread and amount of use the net gets. Nets are often sold after two and even after one season's work, fetching roughly $50 and $100 respectively. If the *juru sělam* has already received the assessed price of his net out of its share of the takings, then on its sale he normally divides the proceeds with the members of his combine, who thus get a bonus for their work.

The question here arises as to the merits or demerits of selling a net after its first or after its second season. At first sight it might seem more profitable to keep it for two seasons and then sell it, because its takings during a second season should ordinarily be considerably above the $50 difference in price that it would fetch after the first season. This however, does not seem to be the opinion of the Perupok *juru sělam*, who tend to turn their nets over after a season's work if they can. Unfortunately, I did not discuss this point with the fishermen and I did not ever hear it raised by them, so it is possible that there is little to be said either way, when the uncertainty of fishing is considered. But I think that some economic reasons may exist which urge the *juru sělam* to try and sell his net after the first year. In the first place, by so doing he gets an annual profit instead of a biennial one on the price which he fixes to his " partners ". Secondly, a net demands less expenditure on thread in its first than in its second season, and also less work in repairs. Thirdly, in its weakened condition it is liable to tear more easily and lose fish. For all these reasons it is possible that his return may be less in the second season than the first.

The two constant items in the maintenance of most types of nets are mending and dyeing (*pukat těgělang, jaring* and the small nets used on the beach are not dyed). We have already seen in the case of lift-nets how provision for labour in these tasks enters deeply into the sphere of net finance. The cost of the materials used falls on the *juru sělam*. Though he can put forward some of the more permanent equipment (such as the oil drum used as a dyeing vat) as a capital charge upon the takings of the net, to be deducted from the net's share before he begins to share with his partners, the actual thread and dye used are his own responsibility. The cost of these is heavy. Awang Lung once said, " What's the use of going out (with my net) ? Awang (my brother) has not got a single cent for two days. If one goes out there is the expense of thread and of dye—one throws money away." Another fisherman, listening, said : " He is right." Dye alone at $3 to $4 a picul is used up at the rate of about

$1 a week.   The expense of thread varies according to the fortune
of the net, but probably runs to half that sum weekly at least ;
sometimes it is much more.   Every now and then when the
current is strong, especially when a crew is small, a net will get
caught up in the coco-nut-frond fish-lure, or it will strike an
unsuspected coral pinnacle, and be badly torn.   Several such
cases came under my observation, in one the damage being
estimated at about $20, taking several days to repair.   With
pukat dalam, which get ripped on coral, in one bad night in May
a total of 33 sections in 4 nets were damaged, many of them so
badly that they were useless, and the remaining sections had to
be distributed elsewhere.   On this latter occasion the damage
must have been at least $150.   In all such cases the cost of
repairs falls on the owner of the net or section as the case may
be.   Even in ordinary times the work of mending the nets is one
of the most regular occupations on the beach in the evening
(Plates IIB, VIA) or among the coco-nut palms by the sea every
Friday.

In the finance of net maintenance there is often no specific
payment made to labour ; mending and dyeing are part of the
general services rendered by members of the group which use
the net.   With boats, however, the case is different.   In the
monsoon season and the slack mid-year months a great deal of
repair work, puttying and painting is carried on.   The repair
work, if carried out by a skilled craftsman, requires a fee, which
varies according to the magnitude of the task.   For instance, in
fitting a false keel of nibong palm wood to enable the boat to
be hauled up more easily and with minimal damage, a fee of
50 cents is usual in the case of a large boat, the work taking
probably one or two days.   With a small boat, a craftsman may
" simply help " and receive a meal as his wages.   In such repair
work, as also in puttying, painting, and general overhauling,
members of the crew usually take part.   They get no money as
payment, but food, tobacco, betel materials and sometimes coffee
are due to them as a customary reward.   A common sum needed
for this is about 40 cents, in addition to 75 cents to $1 for materials
in puttying.   For painting it may be as much as a dollar or
more, the cost of materials then being $3 or $4.   These sums
may seem small, but to an ordinary Malay fisherman they can
be a serious item.   When one of our neighbours was puttying
his boat he spent 40 cents for resin and 40 cents for oil, borrowing
3 men one day and 2 men the next to help him.   He fed them

with salted fish and rice, a frugal meal, but the best he could afford. Then he went on to finish the job himself. As he was doing so a man passed and asked : " Why are you puttying all alone ? " The worker replied rather tartly, " Puttying alone because I can't put up the expense " (of feeding others). He then remarked to me that if he had had the funds the work would have been done the day before, and stressed the cost of tobacco and betel materials and the difficulty of working single-handed.

It is not necessary to bring forward more examples to show the constant pre-occupation of these fishermen with problems of finance both in the acquisition and in the maintenance of their equipment.

THE ENTREPRENEUR IN NET MANUFACTURE

There is still one question, however, which must be briefly discussed in connection with capital in fishing in this area. That is the rôle taken up by a number of the more wealthy men in producing nets, especially lift-nets, not for their own use but specifically for sale. A number of these nets are sold locally to the poorer experts, to those who are starting new groups, or to those who for one reason or another have no net of their own immediately on hand. The entrepreneurs work in the manner already described as far as the manufacture of nets for sale is concerned, and with some of them the production of nets for sale is really an extension of their own activities as fishermen. They are drawn from several sections of the community. In 1940 there were about a score, of whom I knew 17. Of these, 4 were practising lift-net experts, 2 were retired experts, 2 were *pěraih laut* of successful net-groups, 3 were dealers in fish, copra, etc., and the remaining 6 were men with capital, nearly all fishermen, but having their wealth from their fathers or through their wives. In all they get made and sell from 40 to 50 lift-nets a year. Some like the two *pěraih laut*, the two retired experts, or one of the dealers (a comparatively poor man) manage about one net a year ; others sell two or three, while the most enterprising and wealthy, such as Japar the lift-net expert, Saleh-Esoh the fish dealer and Mat Saleh the rich fisherman and land-owner, dispose of half a dozen apiece.

The market for these nets is a wide one. As a rule only

about four or five are bought locally ; the rest are bought by men of Besut and Semerak to the south, and of Kubang Golok, Senok, Sabak and Tumpat to the north, and even by men as far afield as Menaro in Siam. The last-named market, which is a Malay one, is supplied mainly by one entrepreneur, a fish dealer, who in 1939 told me that he had sold 15 nets there so far, representing a capital turnover of about $3,500. His wife goes periodically to collect the payments on the nets—being away for about 5 nights, her fares on each occasion being about $3. The reason for the demand in these outside markets seems to be the lack of a tradition of net-making there. The dealer said of the Tumpat men : " They have the cash, but they are not skilled." And another entrepreneur said of the Semerak men : " They don't understand making them ; they always buy."

The sale of such a net is a transaction involving credit, ranging from one-half to the full price. The actual price itself varies according to the amount of credit required, and the size of the net. One entrepreneur, estimating the cost of the net to him at about $180, sells at about $230, when half the price is paid down. Another, estimating the cost of the net a little lower, sells for $230 if $50 is paid down, and for $250 if no money is paid down at all. Still another, for $40 or $50 down, sells on about the same margin ; three nets of varying size were sold by him for $200, $235, and $240 respectively. Those entrepreneurs who make a regular business of selling nets to other areas may cover themselves by having a document of sale formally drawn up and " signed " by the thumbprint of the purchaser. In case of dispute this can be brought before a magistrate and an order for payment or for return of the net obtained. Cases of this kind sometimes occur. But it is often difficult to prove what payment, if any, has actually been made, since where there is a document, subsequent instalments may be handed over without written acknowledgment of receipt. In one case that took place while I was in the area the seller claimed return of a net from a man of Besut on the grounds of failure to pay. The sale had been made on verbal agreement only, for $240, and it was alleged that only $35 had been paid, during a long period. But I was told by other people locally that the man of Besut had paid altogether $100 on the net, and that since he then did not complete the payment the seller had taken the net back and sold it again to a man of Semerak. Recently he had told this man to hold his tongue, and was claiming only the original sale. Whatever be the truth of the matter in this case, the lack of documentation complicates efforts to sift the evidence when such a case occurs. Transactions within the locality are normally made without any written agreement, and the sum paid down is apt to be smaller than when the net is sold abroad. Here the personal contacts between the two parties and the general local knowledge of the situation make disputes more rare. In three cases of local purchase of new or comparatively new nets the prices were $200, $230 and $220. In no case was anything paid down, though in two cases the buyers were well able to have made such a payment. In the third case the sale was made to a local group without cash advance after an offer by a man from a southern district had been refused ; he had wanted to buy the net for $230, paying $50 down at once, and the seller had wanted $100 down. The general principle here is that the local buyer is trusted more than is the buyer from another area, and so is allowed to pay for the net out of his takings with it. Should it not give an adequate yield the net is taken back. Moreover, it is easier to keep a check on the local buyer, and to know whether he is actually getting an income from the net or not.

This practice of manufacturing nets for sale depends upon the existence of a stock of liquid capital in the possession of some members of the community. The practice appears to have increased in recent years for two reasons. One is that the amount of available capital has probably increased with the development of the fresh-fish trade from the Perupok area. The other is that the price of lift-nets was formerly much higher than it is at present, and that this tempted the possessors of liquid capital into the business. Formerly, I was told, lift-nets cost $300, $350, and even upwards to $600, one of the reasons being that the rates paid to the net-making women were higher. One of the features of this expansion of net-making for sale has been the entry of the fishermen themselves into the business. A fish dealer complained that the modern practice of *juru sĕlam* engaging in net manufacture had reduced his own outlets for capital. Formerly he himself used to get nets made for sale, and even supply the *juru sĕlam*. But now that they had taken not only to making nets for themselves every year, and selling off the second-hand ones, but also to making new nets for direct sale, his trade in nets and that of other fish dealers had been damaged.

The evidence for the change is slender, but it does appear as if we have here an extension of the functions of the more wealthy primary producer into the rôles of trader and financier as well. We may be seeing the phenomenon of the growth of an embryonic capitalist class in this coastal peasant community.

CHAPTER VI

THE CREDIT SYSTEM IN FINANCING PRODUCTION

The analysis of capital in the fishing industry has brought us to a point where it is now necessary to examine in more detail the whole question of the organization of credit in the community.

From the details given it is clear that these Kelantan fishermen have not only a developed cash economy, but also operate a credit system of some complexity. There are three main spheres in which the credit system is important in the fishing economy : in financing the poorer workers, especially during seasons of unemployment or low yield ; in facilitating the purchase of capital equipment ; and in facilitating the marketing of the product. A problem of economic and sociological interest which arises therein is the taking of interest on loans. The study of this brings out the difficulty of applying Western economic ideas directly to peasant institutions.

SEASONAL ADVANCES

The financing of the poorer fishermen is primarily a matter of providing for their subsistence and that of their families during the monsoon period. This is done either by loans of rice or by loans of cash. Loans of rice are obtained by them either from their more wealthy friends or kinsfolk, especially those who have rice crops of their own, or from the small retail shop-keepers with whom they do business in the village. No interest is charged on these loans, which in cash value rarely amount to more than a couple of dollars. Repayment is normally made when fishing begins again after the monsoon. Loans in cash are most commonly made by *juru sĕlam* or *pĕraih laut* to the members of their crews, and here also no interest is taken. These loans tend to be regarded as a definite obligation, and are thus an integral part of the organization of a regularly established group.

Some examples of such small loans to crews may be given. In 1939, at the beginning of the monsoon, Awang Lung told me that he had lent to members of his crew varying sums of $2 and $3, making $15 to $20 in all. None of it yet had been repaid. He said that he had not lent more, partly because he was himself a poor man (this was an exaggeration) and partly because if he gave his men more they would only raise their standard of consumption. " Malays are stupid ; if food is there they eat largely of it, and if I lend them much I think it may be difficult for them to pay it back." He said that he lent the money because they were men of his own crew, and that he did not press them to repay the money. " If a man went to be a member of a different crew, I would ask him to pay me the money back." A month later his brother Semain, *pĕraih* of the same net, stressed to me the way in which crews ask for support, and run off to other *juru sĕlam* if they don't get it. He said that he had lent about $10 in the monsoon. In some cases he gave rice instead of cash—men of his crew had come to him and said that their children were crying at home for food. Money he gave in small amounts—$1, $1.50 or $2. Some of his crew, he said, did not ask for loans at all. " They know they have to pay it back afterwards, and they don't want to borrow."

Another *juru sĕlam* raised the same matter of loans as an explanation of why it was hard to save capital for buying net yarn. He said that he had lent money to 15 men of his crew, in the following amounts : to 4 men $5 apiece, to 4 men $2 apiece, to 7 men $1 apiece. He defended this total of $35 by saying : " If one does not give money, they won't go to sea ; they go along with some other *juru sĕlam.*" He added that they will say : " Awang Lung—or some other *juru sĕlam*—has money ", and will go to him.

Since the organization of a net group is one of free association between capital and labour, " running " of crew members to another *juru sĕlam* is no breach of contract. The effect of this system of small loans bearing no interest is then that it helps to maintain the poorer fishermen during the time of stress in the monsoon without allowing them to be exploited by outsiders, and is a means whereby the *juru sĕlam* keeps his group together. The sanction which leads him to grant the loans is his need for a full group when work starts again. The sanction for repayment is that an accumulation of unpaid debts will tend to debar a fisherman from being accepted in future seasons as a crew member. And since the *juru sĕlam* controls the division of the proceeds he can bring pressure to bear on a reluctant debtor once the season starts again. There are, of course, cases of men who run up a debt with a *juru sĕlam* during the monsoon and then " run " to another when the fishing season starts. But such deliberate evasion is not easy, since as Awang Lung said, the *juru sĕlam* would begin to worry him for the money when the other net began to pay out. The system means also that the strain is thrown more heavily on the less successful *juru sĕlam*, since their crews have not been able to accumulate as much reserve for the monsoon. Judging that the prospects of the forthcoming season may not be too bright, they require more in the way of loans to keep them contented.

A custom which does not come under the head of loan but is germane to it, is that of charitable gift. The practice of charity is enjoined upon good Muslims, and contributions to funeral expenses, gifts to the poor and to religious teachers are common in Malaya. In addition there is a certain amount of private charity by men with means. The interest of this in the present connection is that it sometimes takes the place of loans directly, and tends to reduce the amount of loans sought. Moreover, it frequently has some consideration behind it. Thus Awang Lung said that he gave away from $15 to $20 per annum as gifts to kinsfolk and others. Among the recipients of his occasional bounty is Nik Rung, a teacher who also at times performs magical rites for success in fishing. Nik Rung performs the offerings over the net of Awang Lung ; for this he takes no fee, since he gets fish constantly from the net. But Awang Lung has given him a dollar occasionally, and in virtue of this calls upon him to perform the ritual without offering him the usual honorarium. Awang Lung also said that when a poor man comes to him for a loan of $1 he often does not *lend* it to him, but *gives* him 50 cents. The reason he gave was that if he lent the money he might never get it back, since the man was so poor. It was better to give outright—besides Tuhan Allah will probably take account of it. He thus minimised the material loss while at the same time getting spiritual credit.

FRIENDLY LOANS

There are other types of small loans to assist the purchase of productive equipment where also the benefit derived by the lender is indirect, and no interest is charged. Such loans are known as *bĕrutang mundung*. They are made normally only by kinsfolk or friends, and to people who can be trusted to repay. Considered as an investment of liquid capital the loan may yield a return through the general social benefits which accrue to the lender, or through his association with the borrower in some enterprise which is facilitated by the latter's use of the capital.

One such loan, not to a fisherman but by a fisherman, was of $15 to a man who opened a coffee-shop in Pantai Damat. The borrower was not a kinsman of the lender, but he had rented the shop from the lender's brother. Here the loan was partly a matter of assisting the brother's tenant—and so was of indirect help to the brother—but was possibly motivated by an expectation by the lender of getting favoured treatment for himself or his children in the shop.

Another example, this time concerned with fishing equipment, was of a man who needed the cash to buy a small net for $3 and an old boat for $17, to go dorab fishing. He borrowed the money here and there from friends, a dollar or so from each, paying no interest. He told me that he was able to pay off the debts in one year.

Such friendly loans are often made by a lift-net expert to an actual or potential member of his crew to enable the man to buy a boat. The loan is free of interest, and the incentive to the lender is that he thereby reinforces the supply of fixed capital for his net-group, since it is implied in the loan that the borrower will use the boat in company with the lender. But no period is stipulated for their association, nor is there of course any written agreement. Consequently it frequently happens that the boat owner later goes off to join some other group. The lender then usually tries to get his money back as soon as he can. The following two examples illustrate the situations which are apt to arise.

In December 1939 Awang Lung told me that he had lent $20 to Yusoh-Seripo to assist him in the purchase of a boat. The loan was without interest, and it was understood that Yusoh would join the net group of Awang Lung. This was done, and they fished together for some months. But in April 1940 Awang Lung was getting restive. Yusoh had become slack about fishing ; for three days, as I myself had seen, he had spent most of his time in flying kites, and had not gone near Awang Lung, nor sounded any of his crew about their readiness to go to sea. Awang Lung had other complaints about him too. He said that he feared that in the morning, when the net was going out again, when a man was sent to wake Yusoh, he would declare that he had a headache, a belly-ache or fever, and would not be willing to go out. Then it would be too late to get another boat, and the day's fishing would be lost—since Awang Lung was netting with 5 boats only at this time. This complaint of Awang Lung to me led him on to a discussion of interest-taking. He accentuated the fact that he had made the loan of the $20 interest-free, and had not asked for a cent extra in return. All he wanted was coöperation.

Another loan to secure coöperation in fishing was in the case of Haroun. Awang Lung in the 1938-9 season had lent him $20 to pay off a debt on his boat to Yusoh Panar. The consideration here was that Haroun should come out with Awang Lung. He did come out for three days, but no catch was got, so he deserted to Awang Kelechen. The latter got nothing the next day, while Awang Lung got two boat-loads! In the present season Haroun went out with Yusoh Panar. By April 1940 I was told that Haroun had paid back all but about $6 of the debt to Awang Lung. He had deposited with the wife of Awang Lung a gold ornament, the property of his own wife, saying : " If I don't pay, you can sell the ornament." Awang Lung at this time did not know the exact sum owing. The matter was in the hands of his wife—a common practice among these fishermen. Here again the loan was interest-free.

The lending of money raises the question of the security for the loan. In the small seasonal and other loans mentioned earlier, the lender relies on the personal security of the borrower, whose character and economic situation are well known to him. The loans given by *juru sělam* to help in the purchase of boats are also made on personal security in the last resort, though if the borrower remains in association with the lender the latter relies upon his own skill and capacity as part security, and through his control of the distribution of the takings can exert some pressure toward repayment. Failure of the borrower to pay may result in bad feeling, but involves the borrower in no direct economic loss.

But in cases where the sum involved is more than a few dollars, and there is no association in production between borrower and lender, it is common for security to be given in the form of an item of jewellery. Jewellery, sometimes gold, but often gold-dipped or pinchbeck among the ordinary fishermen's wives, is commonly worn by women in the form of coat-buttons, brooches, lockets, ear-rings or hair-ornaments. It is acquired either by gift of the husband or as part of the dowry at marriage. As in most Oriental peasant communities, it serves as a means of storing capital, and has the advantage of being easily realized if cash is required. A common custom among the Kelantan peasants is to pledge a piece of jewellery as security for a loan—the pledging being known as *gadar* (*gadai*). *Gadar* may also be described as pawning, but the term must be understood in a rather wider sense than in a European community. Between kinsfolk, or persons trusting each other, the sum advanced may be the full value of the article pledged, or even more. In such case the deposit of the article is rather an earnest of repayment than the provision of a purely business cover. No interest is normally charged on such a loan, between Malays at least. But while the article is thus in pawn it is common for it to be worn by a woman or a child in the family of the lender of the money. The use and display of the article is thus in effect a kind of interest, though it is not viewed as such by the people themselves. The pledging of jewellery usually meets the borrowing of sums of $10 to $20.

There are other cases, however, usually where larger amounts are affected, where the pledging of goods does give the lender of the money a definite economic return on his loan. This brings us to the question of specific interest.

INTEREST-BEARING LOANS

With small seasonal loans and friendly loans, where no interest is paid, local Malay custom is in agreement with the religious rule, that exaction of interest from fellow-Muslims is proscribed.

By the Kelantan peasant cash interest on a loan is described as *aseh pero'* (*hasil perak* in standard Malay), a " tax on money ", or *ano' pero'*, the " child of money ". To take interest is termed briefly *makan aseh*—literally, " to eat the tax ". The orthodox Muslim position is well expressed in a statement by one fisherman who tended to set himself up as a moralist on such matters. He said : " People who take interest are people who don't listen to the rule, but they already know it. The rule of Mohammed is, one must not eat the tax on money. To eat the tax on money, to eat the child of money—it is the same expression. The Prophet Mohammed has spoken in the Kitab ' A tax on money becomes a flame of fire, which roasts us '." [1]

---

[1] Cf. " The taker of usury and the giver of it, and the writer of its papers and the witness to it, are equal in crime." (No. 408 of *The Sayings of Muhammed*, translated by Allama Sir Abdullah Suhrawardy, new ed., London, 1941.)

It is interesting from the economic point of view to observe, however, that as with the medieval Church in Europe, while Islam in this community exercises a restraining influence upon returns from the investment of capital, evasion is practised. Among these Kelantan Malays, this evasion follows two lines. One is that of simple breach or disregard of the rule. The other is more ingenious, and rests upon distinction drawn between taking interest in cash upon a loan, irrespective of its function, and taking a return in cash or kind in a form which combines interest with profits, as a proportion of the general yield.

The direct taking of interest in the form of a regular cash increment, calculated by reference to the amount of the loan, definitely occurs. In a few cases, where the lender of even a small sum of money, perhaps $10 or $20, is unwilling to make the advance without some incentive, the borrower consents to pay something extra when the loan is returned. For a loan of $10 a repayment of $11 may thus be made at the end of a couple of months. In general, however, local opinion draws a distinction between small and large loans. The general formulation often expressed both to my wife and to myself, by women and by men, is that for loans of less than $50 no interest is expected but that on loans of more than $50 interest is paid. The rate is calculated not per annum but per month, and consequently, as is common in a peasant community, it is apt to be high. It is not a standard rate, but depends in part upon the personal relations between borrower and lender. In one case a loan of $50 by a fisherman's wife to a Malay shopkeeper obtained interest at the rate of $1 per month. This rate of 24 per cent. per annum is commonly said to be that which the Indian Chetti takes in Kelantan. Moreover, when articles of jewellery are pawned to local Chinese, according to one fisherman, about half the value of the article is advanced, and interest is paid at the rate of $3 per $25 per six months, which gives the same figure. But higher rates appear to be charged in some cases. Another fisherman, who himself was an occasional lender of money, said that $2.50 per month on $50 was the rate charged. And in the early days of my inquiries when I raised the question of interest on loans, adducing the European practice as an illustration, a group of men, after looking meaningly at one another, admitted that interest was taken, and gave the rate as $3 per $100 per month.

The existence of an interest-charge, despite the religious rule to the contrary, indicates some scarcity of liquid capital in the community. The high rates exacted are due partly to the strength of the demand for capital, but may be also linked with the poor nature of the security offered. For in cases where interest is taken in this way the borrower often has only his personal security to give. In theory, when such a loan is made and there is default in payment, the lender can legally recover through the Courts only the capital sum advanced. But when a written document has been used it is possible for interest also to be claimed by writing down as the loan a sum larger than that which was really handed over. The borrower has to agree if he wants the money.

INTEREST DISGUISED AS PROFIT-SHARING

A commoner method of obtaining what is in effect interest on a loan without using the direct method just described is for the lender of the money to take over a piece of the borrower's property as a pledge, and to get his increment from the usufruct of this.

A summary of the position as regards security for such loans was given me by a fisherman, a *pěraih laut* of a lift-net group, who himself had invested money in several directions. He described four forms of lending.

i. If a man has rice lands, then these can be *pěreto* or *pěgang* by the lender of the money, i.e. " governed " or " grasped ". (In theory, I believe, such lands should be registered as having their title deeds charged upon the Mukim Register. But in practice this seems to be rarely done in the area described.) From the lands thus taken as security the lender gets half the crop ; the owner of the land works it and gets the other half of the crop for his labour.

ii. If the borrower has rice lands, but does not wish to give them as security, then it will be agreed that at every harvest he will pay 10 gantangs of padi (worth about $1) for every $10 borrowed.

iii. If the borrower has no rice lands, then, my informant said at first " We do not lend money ". Questioned further he said that if the borrower has a coco-nut orchard, the lender takes the fruit—usually collected 5 times a year. The lender gets the whole of the produce, less only the small expenses of climbing the trees or using a monkey to do so.

iv. If the borrower has no lands at all, but has a boat, then half the boat's share is taken to the lender of the money each week—though if the share is *nil* or very small, the lender gets nothing.

In all these cases interest on loans is in fact secured, though it is first of all interest in kind, in the form of an actual increase from the capital. It could be termed " natural interest " in the sense that one of its characteristic features is its conformity to productivity.

It should be emphasized that in all cases the original sum borrowed must be repaid in full, irrespective of what has been obtained by the lender from the product of the fixed capital. This latter is not regarded as a return of part of the principal. But the widespread system of pledging property between Malays, with or without the taking of interest, means that these people do not get into the hands of the money-lender through the accumulation of unpaid interest-charges at high rates, as seems to be the case in some other parts of Malaya, and is common in peasant communities as a whole. The security pledged may later have to be sold, and the original capital thus lost, but the usufruct system of taking interest relieves the debtor from some of the worst evils.

The situation with regard to the lending of money on rice lands and boats must be analysed in more detail.

In the case of rice land pledged (*gadar*) for a loan there are other courses open to the lender than that described under i. If the lender is a rice planter he will probably work the land himself and take the whole of the produce. If he is not a rice planter, then he will get someone else to work the land in accordance with the customary system called *pawoh*, under which the worker gets half of the produce, and the owner—or in this case the person to whom the land has been pledged—gets the other half. But according to the ethics of the system, if the man who has pawned the land is a rice planter, it is the right thing to allow him to *pawoh* the land himself, so that he gets half the produce in return for his labour on it. The system differs from the simple and direct taking of interest in that the half-share of the produce goes annually to the lender of the money, irrespective of the proportion which the size of the loan bears to the value of the land. Moreover, considered as return on capital, this half-share of the produce varies according to the particular season.   In this, the system shows the same features as the return obtained from lending money on a boat or a net.

The system of pledging land differs from the system of mortgage as we know it in two main respects. Firstly, the interest obtained is at a variable, not a fixed rate ; secondly, the productivity of the goods on which the loan is secured passes over to the lender, and the borrower can be deprived of it altogether.   The terms of the contract may seem severe, but they have this advantage to the borrower—that the getting of interest is a task of which the onus falls primarily upon the lender.   The borrower is not put in the position of having to find a fixed sum periodically from sources the productivity of which may, from causes beyond his control, have seriously decreased.   The removal of the threat of accumulated interest charges at a time of depression of markets or failure of production—a threat before which so many unhappy mortgagors have quailed—is one great advantage of this Malay system.

The lending of money on the security of a boat, or, more often, to assist in the purchase of a boat, is a very important feature in the economy of the fishermen.   When it is not done as a friendly gesture by, say a kinsman, or by a *juru sĕlam* to secure additional fixed capital for the working of his net, it follows the same general lines as in the case of rice lands.   But there is this exception, that the boat is not explicitly pledged to the lender of the money—though this may sometimes be done in effect by a change of the owner's name officially.   The borrower of the money uses the boat, and gives each week to the lender half the boat's share of any takings, as described above. This is interest only, not part of the repayment of principal. Even if the money lent is less than half of the cost of the boat, the half-share of the takings is still paid.   When I asked why, the answer was " Because the boat owner has not enough money ; he must pay heavily."   Here again this indicates a comparative shortage of finance capital in the community.

There were three major lenders of capital in the Perupok area : Pa' Che Su, fish-trader and copra-exporter ; Saleh-Esoh, fish-trader and net-manufacturer ; and Japar, the most successful *juru sĕlam* and also a wealthy man in ricelands and other goods. But other people also lent money in the same way, as did Che Daud, trader in fish, poultry and copra.   Awang Lung described the procedure as follows.   Pa' Che Su, for instance, lends $100 to help buy a boat worth, say, $240.   The borrower affixes his

thumb-print to a form, agreeing to pay half the *bagian* of the
boat, and to repay the principal later.   Pa' Che Su has his name
registered as part-owner of the boat in the Customs Boat Register
—because he has put $100 into the boat.   So far this is like a
partnership.   But as the boat depreciates in value it is the owner
and not Pa' Che Su who loses ; the latter has the paper stating
that $100 has been lent, and insists that $100 shall be repaid.
If the value of the boat goes down to $150 or even to $100 in
the course of time, Pa' Che Su will still get his $100 complete
should the boat be sold again.   " The boat-owner is the loser ;
Pa' Che Su cannot lose."   Thus if the association of borrower
and lender be viewed as a partnership, it is one in which there
are no preference shares as regards interest, but the capital of
one partner is a first charge upon the assets.   If it be regarded
as a mortgage, it is one in which the interest rate is fluctuating,
and normally very high from our point of view.   As Awang Lung
pointed out, half the boat's share in a year may amount to
practically the whole of the value of the sum originally lent, while
still leaving that sum to be paid off.

It is important to observe that the main lenders of capital in
this area are not Chinese or Indians, as in some other parts of
Malaya, but Malays.   The question then arises. What is the
position in relation to their religious rule ?   The answer is that
taking half the *bagian* of the boat is not regarded as in the same
category as taking interest at a fixed rate—" a tax on money ", or
" the child of money "—because " it is uncertain ".   One week
the provider of capital may get a good share, the next he may
get nothing.   To the Malays, then, it is classed as a share in
profits, not true interest.   But to us the failure of the provider of
capital to share in the depreciation of the assets, the fact that the
money borrowed is not necessarily used for productive purposes,
and the fact that the repayment of the principal is entirely at
the discretion of the user when he has the cash, would seem to
remove it from the profit category—even though the provider of
the capital shares the risks of the undertaking from which he
draws his increment.

But consideration of these mechanisms for the provision and
reward of capital in this economy suggest the difficulty of applying
rigidly the concepts derived from European economic structures
to the phenomena of other economic systems.

The same principle of irregularity of return is invoked in the
classification of other types of return for the use of capital.   The

function of the " catcher " of a net (p. 208) is to collect the cash from buyers of the fish, and if necessary to put up his own money until he can do so. I asked Awang Lung whether the return of $\frac{1}{20}$ of the takings were not interest, i.e. *aseh pero'*. He replied : " I do not know, but I think it's really not interest ; I think it is of the same kind as a fee only, not a tax on money." He added that it was not like getting the " child " of a loan. If no fish were caught, then the " catcher " got no percentage ; it was not like lending money where one got a regular monthly return. He can be said to be correct here, in that the percentage of the " catcher " is a reward for labour at least as much as for use of capital. But the basis of his argument shows the line along which the classification is made.

The summary analysis of this topic given so far has concealed some of the complications which often occur in practice.

Two examples will show how the system actually works.

The first concerns the boat of Awang Muda, which he bought in 1938 from Besar of Bachok, for $170. He paid $110 down, making up this sum partly from the sale of a smaller boat, for which he had paid $40, and which he sold for $17½ ; partly by savings ; and partly by borrowing $50. This last sum he repaid in instalments of $20 or so at a time. Towards the end of December 1939, Muda left the net of Bakar, with whom he had been fishing. He still owed $60 on the boat, and wanted to pay it off. He asked Awang Lung to lend him $15. The latter replied that he had no money then, but said that if he could go to sea for a few days he would have some. He told Muda to ask someone else for the cash, and later, when he (Awang Lung) had got money from fishing he would advance it to Muda, who could then pay off his interim loan. On January 5th, 1940, Muda came to see me, bringing with him the fish dealer Che Daud. They asked me to type out an agreement by which Daud, in consideration of lending Muda $30 for six months, would receive half the takings of the boat for that period, in addition to the return of the principal. Both men stressed the fact that it was only Muda's wish to have the matter in writing that brought them to me for the document ; Daud said that he and Muda were friends, that he trusted Muda, and did not want a document, but that Muda insisted. (He was an honest man who wanted to acknowledge his debt.)

It transpired later from information gained by my wife from Lijo, wife of Muda, that this $30 went to pay, not Besar, but

Seripo, wife of Bakar the *juru sĕlam,* from whom Muda had previously borrowed this sum to pay off an instalment to Besar. And it was because he was leaving Bakar that he had to settle the amount.   On March 29th Lijo explained further that in the interval Besar had been pressing for a further instalment and that she and Muda had paid him $25 of the $30 now still owing to him.   This was made up as follows : $5 she had borrowed from Che Mbong, the wife of our servant, with whom she was friendly, and the other $20 she alleged they had saved.   It was evident from our figures that Muda, who at this time was fishing with Selemen, a kinsman of his, could not have saved this amount. On further questioning by my wife on April 11th Lijo then stated that she had herself saved $9—which was correct as far as we could see—$1 had been given her by her mother's brother Pa' Che Hen, and $10 she said her husband had produced, saying that he had saved it secretly over the monsoon.   My wife asked about this last and she said that she did not know the source, since he had simply produced it with this explanation.   She evidently made inquiries of him for on April 13th she said he answered : " How could I have saved it, since I give all the money to you ? " and admitted that he had borrowed it from Selemen.   By April 22nd Muda had parted from Selemen, whose net had done poorly, and was line fishing with Awang Lung, with whom he was intending to go out when the lift-net fishing began again.   Selemen, hearing of this, was annoyed, and demanded the return of the $10.   On April 30th Lijo told my wife that Besar had come again and asked for the final $5 of the debt to be paid him.   In the interval she had repaid Che Mbong her $5, which as she had said to my wife, she regarded as the first debt to be met—though the latest incurred, it was more a debt of honour than the others.   But now she went back to Che Mbong and re-borrowed $3—not liking to ask for the full sum back—and got $2 from Klesung and Po' Yih, who had been living in her house while theirs was being re-built and were therefore under an obligation to her.   With this she paid off Besar in full.

The debt on the boat was thus ended, but only by incurring a set of other smaller debts, viz., Daud $30 (on which interest of half the *bagian* of the boat was being paid) ; Selemen $10 ; Che Mbong $3 ; Klesung-Po' Yih $2; and Pa' Che Hen $1.   Thus in 4 months Muda had reduced his debt from $60 to $46 by saving, and transferred and split up the debt among five people instead of owing it to one person.   The saving of $14 was made largely from line fishing.   We were unable to follow the later transactions concerned with this boat.   But it is worthy of mention that when Selemen at this time asked again for the return of his money Muda said to him : " If your net had earned only even enough for us to eat, there might have been some chance of my paying my debt.   But as it is there is no chance of my paying back the $10 yet."   Muda announced to us also that he was sick of the matter and was going to sell his boat. He said that he had bought it for $170 because—he alleged— Besar wanted strongly to sell it, and wanted money ; and that it was somewhat against his better judgement.   He said that he had just been offered $165 cash for it by a man in Kubang Golok.   This, however, did not come to anything.   But Lijo was apparently keen to sell it, and this influenced him greatly.

This case, which we followed in detail, seemed to be fairly typical.   The conversion of long-term large debts into short-term small debts is a common way of handling credit finance among these Kelantan fishermen.   In the case of boats working with nets which are more successful, however, the financial operations are apt to be rather less complicated, since the debtor has more chance of larger savings and can handle his obligations in larger blocks.

Another *motif* which emerges clearly from the above example is that of the financial interest taken by the *juru sĕlam* in the boats of his net group, as described earlier.

Another example of the working of this system is seen in the financing of the boat of Po' Su, *pĕraih* of the net of Awang-Yoh. The details were given me by Awang Lung, who said that he had simply heard them in talk, and would not vouch for their truth. But they were probably accurate.   The boat of Po' Su was really not his own, but had been the property of Deromen, a man recently dead, and was now held by his widow in trust for her children.   Po' Su received half the *bagian* for acting as captain. But a complication arose when a Kubang Golok man wanted to buy the craft, the price, according to Po' Su, being $170. (I do not know, but imagine that legally the estate would have had to accept this offer unless a similar arrangement had been come to elsewhere.)   Awang-Yoh said that if the offer were accepted he would not be able to take his net to sea—he had not enough boats without this one.   Po' Su was married to Awang's sister, and they had an interest in keeping together.   So Pa' Che

Su—the capitalist—was approached, and put in $70, and Meriam, the widow, stood for $100, thus making up the value of the boat. In the *bagian*, out of every dollar Pa' Che Su took 50 cents ; Meriam 30 cents ; and Po' Su, for running the boat, 20 cents. Awang-Yoh said that about 3 months later he would pay off Pa' Che Su and Meriam, and would then halve the *bagian* with Po' Su. But Awang Lung thought that he would not have the cash to do this. He said that he had heard that Awang's own boat was financed to half its value by Pa' Che Su, who got half the *bagian*. He added that he had not verified this ; he did not ask Awang-Yoh lest being a fellow *juru sělam* (with a leader's pride) he be ashamed.

This instance, incidentally, was given me by Awang Lung in answer to my question about the *bagian* of boats in case of multiple debts. Obviously, where there are several lenders of capital on the same boat, each cannot get half the *bagian*. Awang Lung said that there is no case where the man who owns or runs a boat foregoes *all* the boat's share, and gave this as a case in point.

This instance and that of the boat of Muda shows what happens. Small creditors, and creditors such as *juru sělam* or others with an interest in the boat apart from its takings, receive no *bagian*. Creditors whose primary interest is in getting a yield on their capital, such as Che Daud with Muda, and Pa' Che Su with Po' Su, take half the *bagian*, or an appreciable share of it. It will be noted that Pa' Che Su's financial interest in the latter is less than that of Meriam, but he gets a larger proportion of the takings because he is a money-lender whereas she is more of a partner.

Where a net or boat is bought on credit from the seller, he does not take half the *bagian* ; it is only when money is borrowed from a third party for the purpose that the half share is taken. The reason for this is that in the first case the buyer is in a position of advantage ; he can refuse to buy the boat at all. Moreover, the price of a boat or net is higher when credit is given. But when he is borrowing money he is at a disadvantage ; if he refuses the half of the *bagian* the other party will refuse to " help " as the Malays euphemistically put it.

MOBILIZATION OF CREDIT THROUGH CAPITAL EXPENDITURE

We have now to consider a phenomenon which, though familiar to the anthropologist in general form in many other communities, is to most Europeans a strange and unexpected way of calling upon one's credit. This is the feast, a customary Malay method of celebrating an important social occasion such as a wedding, a circumcision, or other high point in family life. At first sight it appears that such a celebration, with its heavy expenditure of capital upon goods consumed on the spot, has nothing to do with the use of credit for purposes of production. In fact, the general European interpretation of a Malay feast in economic terms is that it is a wasteful affair, a drain upon a family's resources which should be curbed as far as possible, always having regard to Malay susceptibilities. But closer analysis shows that while it cannot be regarded from the economic point of view as a particularly efficient method of mobilizing capital resources for production, it does, nevertheless, perform this function to a significant degree, quite apart from the wider social satisfactions which it provides.

How is this function of mobilization of capital or credit achieved ?

Among the Kelantan peasants a large feast is colloquially known as *krejo*, (*kĕrja*) " a work ". The essence of every feast is that while the giver of it—the *tuan krejo*—has to accumulate in advance a considerable amount of capital to pay for house extension, food, entertainment and some labour, the guests who attend, usually numbering some hundreds, must each bring individual contributions, in cash or kind or both, which are paid into the exchequer of the host. The host thus reimburses himself for his outlay to a greater or less extent, depending upon the size and success of the feast. So far from being simply a wasteful expenditure of capital by the host, the feast is conceived by these Malays as designed to pay for itself. It is not a success from the host's point of view if he is greatly out of pocket by it, and a really successful feast is one in which he has covered his expenses with a considerable margin to spare.

This, however, is only from the short-term view ; the long-term view introduces complications.

The contributions to the feast are of two major types. One, termed *pĕngelen*, and made either in cash or kind according to convention, is a straight-out gift to the host, a return for the hospitality received. The other, termed *derā*,[1] is also a gift, but carries with it the obligation to make an equivalent return to the donor on a similar occasion later on. Whereas the *pĕngelen*

---

[1] *Bĕrderau* in standard Malay means simply " to co-operate ".

are of small value, not exceeding \$1, the *derā* range from \$2 to \$10 and occasionally more per person, varying according to the wealth of the donor, his kinship or friendliness to the host, and other factors. Each item of *derā* with the name of the donor is announced in loud tones by a master of ceremonies, and commonly written down in a notebook by a clerk as well, so that both host and guests have recorded for them the various contributions. (A few contributions, made privately to the host's wife by her kinsfolk, or handed in the day after by people who have not been able to attend, are not entered in the book, but are remembered.)

I was able to take down the major details of these contributions from the notebooks of three feast-givers, which gave the following range :

Contribution.				Number of contributors to each feast.		
\$				A	B	C
1	.	.	.	89	63	41
2	.	.	.	17	29	21
3	.	.	.	3	10	4
4	.	.	.	1	2	1
5	.	.	.	0	7	8
6	.	.	.	0	1	0
7	.	.	.	0	0	2
8	.	.	.	0	0	1
9	.	.	.	0	0	0
10	.	.	.	4	4	7
15	.	.	.	0	2	1
20	.	.	.	0	0	1

In each case various other sums, mostly in small denominations, were contributed but not noted down, bringing the total cash receipts to about \$250 in *A*, \$296 in *B*, and \$340 in *C*. And in addition somewhere between \$50 and \$75 worth of rice was given in each case, apart from the small sums in 10 cents and 20 cents given by the women guests which are not counted in the total.

The profit-and-loss aspect of the feast is strongly to the fore, not only in the mind of the host, but also with other people in the neighbourhood. The latter speculate beforehand on the probabilities, and are full of gossip the day after as to how much money was taken and whether the *tuan krejo* made or lost on the affair. In each of the three cases cited above the host made a profit, that of *A* being small, that of *B* being just over \$100 (apart from about \$60 worth of rice) and that of *C* being probably between \$100 and \$150. In other cases that occurred during our stay, or of which we received details, the position varied. In one the feast-giver, a woman, got receipts of about $140, and came out about even ; the Malays expressed it thus : " She didn't lose, but she was ' empty ' ; there was no profit," or more simply : " She didn't get anything out of it." One explanation given here was that her husband was not a local man and so few kinsfolk of his came to support the feast with heavy donations. Another, perhaps more revealing, was that since he had two wives perhaps people were afraid to contribute much to one wife's feast lest she be later divorced and their gifts be lost. In another case, a feast brought in about $250 in cash, with about $60 worth of rice as well ; this probably yielded about $150 profit. And in another, a much larger feast, the expenses were about $500 and the receipts were about $800 in cash and about $200 in rice and coco-nuts in addition. It is probably fair to say that an excess of receipts over expenses of $100 or $150 is common.

The economic aspects of this method of feast-giving are worth examination.

In the first place the feast-giver, by this custom of contributions, is enabled to meet an expenditure on a scale which would be impossible to him (or her) from ordinary capital resources. Relying on the contributions which will accrue he is able to spend all his immediate capital and even raise loans beforehand to pay cash for the many items of food, decorations etc. which must be incurred. Occasionally, he even pledges a piece of land to help meet the cost of the feast. In effect, his circle of kinsfolk, friends, neighbours and villagers combine to pay the bill. Without their active assistance the affair is a dead loss.

In the second place, when a profit is made on the affair, this capital sum is not normally put to use in meeting the everyday household expenses. After any loans for the purpose of the feast have been met, the balance is invested in some item of fixed capital. Sometimes this is a consumer's good, such as a better house, but often it is a boat, a net, or a piece of land. In this way one sees the paradox of abnormal capital expenditure becoming an occasion for accumulating funds which may be used to finance production. This may be documented by examples. When I asked the man who obtained an excess of receipts of about $500 over expenditure what he did with the money he replied that he had bought some land, and a boat, and had built a new house. And when I asked the man who had about $100 in excess what he would do with it he said that he was going to use it as capital—he wasn't quite certain yet how, he might buy rice land, or a coco-nut orchard, or buy yarn for net-making.

Thirdly, what has so far been termed " profit " from the feast is really only a gross profit.   The contributions received, apart from the straight-out gifts, are of two kinds.   One (*bayar derā*) consists of sums which are reciprocal payments for contributions which the feast-giver himself has made to the various guests on former occasions when they have been the hosts.   The other (*derā*) consists of sums which he will be obliged to repay to the givers at some future time when their turn comes.   The gross profit must therefore be reduced to a net profit by an amount corresponding to the total of fresh obligations which the host has now incurred.   The amount of these new obligations varies with the size of the feast and the age of the host—an elderly man tends to have a larger proportion of guests owing *derā* to him.   But the man whose gross profit was about \$500 estimated that about \$200 of his receipts was in cancellation of *derā* that he had previously made, and about \$500 constituted fresh obligations which he would have to repay.   The man with about \$150 profit estimated that about \$100 of his receipts represented repayments to him, and about \$200 constituted fresh obligations.

The essence of the system is this.   On the one hand the feast-giver has placed deposits of capital through the community in small sums, at earlier periods, sometimes years in advance, giving credit which he calls up in bulk for his feast.   On the other hand, he mobilizes his own credit at the feast by getting simultaneous small advances from a large number of other people, to be repaid scattered over a long period of years.   Then, if he wishes, he converts this mobilized credit into fixed capital of a productive nature, the income from which may well exceed the annual commitments which he has incurred for the future.

To test how far the people themselves appreciated the economic implications of the whole process my wife and I discussed the financial aspect on many occasions, sometimes deliberately stressing one feature to see what reaction followed.   After a feast which we attended shortly before we left, Awang Lung on the way home said that he heard that the total expenses were \$180, that the receipts up to 2 a.m., when we left, were \$270, and he estimated that the host would get about \$300 in cash and about \$50 more in rice and coco-nuts.   My wife commented, to draw on the conversation, " Then he'll get a profit ".   Awang Lung answered : " Yes, but he's got to pay too ; it's like borrowing." He added that the host might possibly have two or three people in one month to pay off, and that it wouldn't be easy. Then he said again : " It's like raising a loan." This point of the contribution establishing a claim to future repayment was made on another occasion from a different angle. The people with whom we went to a feast asked how much I was going to give as a contribution. I said " Four dollars for the two of us ", regarding this as reasonable in the circumstances. But several of them expostulated, saying that one or two dollars was enough —that local people, if they have no *derā* to make or repay, give only that, and since we were shortly going to leave the district, we should have no chance to get our contribution back. And later, when we gave a shadow-play performance as a kind of farewell entertainment to the neighbourhood, it was suggested semi-seriously that if we would only kill a bullock and invite people to a feast we should be able to recoup ourselves for the various contributions we had made elsewhere.

On the other hand, in discussing feast *C* in the examples cited above, I was told that $400 had been taken by the host (this was an exaggeration), and my informant added laughing : " He has no need to go to sea to-day." I replied : " But he must pay most of it back." The answer was : " He can carry on work first ; he can buy yarn and make a net, and he's only got to pay slowly—about two people a year ; he gets a little profit."

So far from the feast being regarded by these Malays as a simple celebration, or " cutting a dash " by thriftless expenditure, or a means of raking in a profit—which are various angles from which it might be viewed—they are quite cognizant of the more subtle issues : of gross profit to be set off against obligations incurred, and of the productive use of the capital accumulated to yield a profit before these obligations mature.

There are other factors in the situation which may complicate it further. A wealthy man may give a larger feast for display, in which case his expenses mount though his receipts are not proportionately increased. Some of the obligations incurred by a host may fall through owing to death of the creditors. Some of the sums he has previously disbursed may not be repaid to him through poverty, or death, or simple failure of the recipients to fulfil the convention. A larger contribution may be given for the prestige to be got from it even though the prospects of equal repayment may not be too good ; or with the definite object of piling up credit for a specific end in the distant future, as a circumcision ceremony. But such complications are still part of the general picture given here—though they emphasize that the feast has social as well as financial implications, and that from the financial angle the risks are therefore considerable.

CREDIT IN THE MARKETING OF GOODS

The preceding chapter has shown how important in this fishing economy is the acquisition of fixed capital, especially boats and nets, on credit. It is comparatively rare for a large item of such equipment to be bought for cash. The initial payment is usually about 20 per cent. or 25 per cent. of the price, the remainder being handed over in instalments. When the parties to the transaction are kinsfolk, or living in close proximity, the initial payment may be nil. The subsequent instalments paid are a matter of arrangement between the parties, but the principle that they should broadly be proportioned to the actual takings of the boat or net is fairly generally recognized. The transaction thus assumes the character of a coöperative enterprise, the seller supplying the fixed capital and the buyer the labour and management, with the prospect of taking over the fixed capital out of the earnings. Lack of yield from a net is often given as the reason for failure to pay off instalments on the agreed date, and is usually admitted by the creditor as a valid one. In cases of evasion of obligation, or of continued lack of success, the creditor has the resource of taking back the net. In this case the debtor loses all the payments he has made so far.

This system of credit is a vital element in any expansion of the fishing industry since it allows enterprising fishermen with small resources in liquid capital to undertake production.

The credit system is fundamental to the fishing economy in another way, in facilitating the marketing of the product. Direct contact of consumer and producer is, in most cases, not feasible, and the entry of middlemen is essential to secure the wide dispersion of the fish produced. Few of these middlemen have at their command any large stocks of free capital, so it is necessary for them to receive the cash for the goods they sell before payment can be made by them to the fishermen. The onus of supplying the credit thus falls upon the original seller of the goods, as it does in the case of sales of capital equipment. And in all the forms of production where groups of men are engaged, the individual working fishermen participate in granting the credit.

The coöperative system of association does not provide for payment of the workers at once out of the cash stocks of the owner of fixed capital or entrepreneur-in-chief, the *juru sĕlam*. Each member of the group depends directly for his reward on the periodical receipts from the sales of the catch. As will be seen in the following chapter, the physical organization of marketing in the fishing industry means that, in the majority of cases, the middleman receives the cash for his sales of fish to the consumers or to secondary middlemen so late in the day that it is not possible for him to turn over the payment to the producers that evening, and by the following morning they are again out at sea. The obvious course is therefore adopted, of making payment at the end of the Muslim week, Friday being a day on which the lift-net groups and many other fishermen are not out at sea. The middlemen who buy the catches do so on the understanding that they will pay at the end of the week, so on the average, three days' credit is allowed them.

It is recognized that the middlemen are in effect operating on the capital of the fishermen during this period. But they pay no interest on this—just as in European trade no interest is charged on a customer's monthly account. In some cases, however, the catch is sold for cash. A distinction is made implicitly between a credit and a cash purchase by the fact that for " cash at once " the buyer can get a better, that is, lower quotation. The difference between a cash and a credit quotation rests upon the elimination of risk in the former. By the end of the week the buyer may have his capital tied up in other transactions and be unable to meet his debt just then, or more likely, he may have lost on his purchase and wish to " cut " the price, as by custom he may claim to do. Cash down avoids these risks. There may be also an element of short-term interest involved in the higher quotation for a credit sale, but it must be small. It is difficult in such cases to separate out the risk element and the interest element in theory, and impossible in practice. But there is one field where a test is applicable. An ordinary crew member of a lift-net group supports the lower quotation for a cash sale though he himself does not receive his share in the cash until the end of the week in the general division. Since he does not handle the money any sooner, it is clear that as far as he is concerned it is the *certainty* of the cash in hand that is the primary appeal of it for him, and not a foregoing of interest on the use of the cash for the remainder of the week.

A more detailed appreciation of the credit system in the marketing organization is given in the next chapter. But in conclusion here one may stress the flexibility of the system of credit finance carried out by these Malay fishermen, and its comparative efficiency in allowing scope for development and for individual enterprise in a community which is, on the whole, short of free capital. The Malay is sometimes represented as having little economic sense, as unskilled in the intricacies of financial affairs, and since he lacks capital, as being unable to undertake enterprises on any scale. The material given in this and the preceding chapter should show that for the Kelantan fishermen at least such a judgement is too simple.

CHAPTER VII

MARKETING ORGANIZATION

The system of marketing is the most intricate and the most fascinating of all the aspects of the Malay fishing industry on the east coast. The industry is too large to allow a direct market between consumers and producers to operate on a broad front. The intervention of middlemen is necessary. And the individual producing units are too small and too scattered to allow the middleman-trade to be concentrated in a few hands, apart from the fact that the capital at the command of middlemen is, on the whole, quite inadequate for any such concentration. Only in the vicinity of Kuala Trengganu, where the export trade in cured fish is highly developed, is there any tendency for a single individual to assume unified control. Even here many small middlemen operate independently of him, and others, who act as his agents, may employ their own capital in transactions apart from his. In the present condition of the industry the existence of a large number of small middlemen is an essential feature.

THE MIDDLEMEN

The functions of the middlemen are primarily those of providing an immediate market for the fish when it is brought to shore, supplying the mechanism by which the bulk catches are made available in small lots to individual consumers, often at a great distance, and to some extent acting as a cushion for prices in times of glut and scarcity by their preparation of cured fish which can be kept and sold off more slowly. In Kelantan, by their demands for credit, they affect the organization of the capital resources of the community. In Trengganu and Pahang, though only to a small extent in Kelantan, they commonly provide the working capital of the fishermen. In these two latter States, as described in Chapter II, the return to the middleman is of a complicated kind ; in Kelantan it is normally on a simple profit-or-loss basis.

The general term for a middleman on the east coast is *pěraih*.[1]

---

[1] See Wilkinson, *Malay-English Dictionary*, under *Raih*. The form *pěrais* (transcribed as " *price* " in the *Annual Reports of the Fisheries Department of the F.M.S. and S.S.*) appears to be a variant in which the *h* sound is pronounced as *s*, in common with such words as *lěbes* (*lěbeh*), *putis* (*puteh*), etc.

Thus *pĕraih ikan* is a dealer in fish, *pĕraih daging* a dealer in meat, *pĕraih itek* a dealer in ducks, *pĕraih sayur* a dealer in vegetables. On the coast the term *pĕraih* when employed without qualification refers naturally to a fish dealer, and various descriptive additions indicate more precise functions. Thus *pĕraih kandar* and *pĕraih bisikal* are fish dealers using carrying-pole and bicycle respectively ; a *pĕraih hidup* (literally a " live dealer ", is one who trades in fresh fish ; a *pĕraih kĕring* (commonly called *tauke kĕring* from his command of capital), literally a " dry dealer ", is one who trades in fish for curing. The broadest distinction is between *pĕraih laut*, " sea-dealers " and *pĕraih darat*, " shore-dealers ". The former operate from boats, the latter on land. The functions of the *pĕraih laut* in the Perupok area have already been described, and in this chapter we are concerned with the *pĕraih darat*.

The " shore-dealers " are of several types, distinguishable by the scale of their operations, the product in which they primarily deal, the market which they serve, and the nature of their relations with the fishermen. There are the dealers who buy in bulk from the boats, go off by bus and sell the fish fresh on the same day to other dealers in the inland markets. There are those who also buy in bulk from the boats but who concentrate on curing and long-term sales or export of the fish. There are those who do not come on the beach at all but buy from the former and sell retail in the inland markets. There are the " carrying-pole " and " bicycle " dealers who buy on a smaller scale on the beach and sell retail inland. There are the " little dealers " who operate with a small turnover only on the beach itself. The interconnection between their functions, and the tendency of some types to assume the rôle of others as circumstances change, render the whole middleman organization one of considerable complexity.

The analysis may most conveniently begin with an examination of the rôle of the dealers who buy in bulk on the beach.

The first feature that strikes the observer is the number of these middlemen. According to the census of fishing taken by a Revenue Officer in 1937 there were 325 *pĕraih darat* (apart from 105 *pĕraih laut*) handling fish in Kelantan. This number, while probably not complete, gives an idea of the middlemen regularly engaged in fish dealing on the coast alone ; it does not include stall-holders in the markets and other retail dealers inland. In the Perupok area, taking simply the census field examined in 1940, there were 30 men whose primary occupation was fish dealing, representing more than 5 per cent. of the total male population at work. That this number was unwarranted by the volume of sales was an opinion advanced by some of the dealers themselves, and apart from their natural desire to see as little competition as possible, it does seem that the occupation tended to be over-crowded. In January 1940, one dealer said that he had not bought fish for some days ; that he was ashamed to rush into the sea to secure a good place when the boats came in. It is true that he was short of cash at the time, since he was re-building his house, but other men said much the same. Another dealer, a man of more substance, complained both then and later in the year about the intense competition. He said that the dealers used to wait while one bargained for the catch, and that then they would divide it ; now they were bidding against one another. (As will be seen later, this may react unfavourably on all parties to the transaction.) He contrasted the local situation with that at Tumpat, where, he said, there was less pressure. This tendency to over-crowding is facilitated by the ease of entry into the ranks of the dealers. Men with practically no capital at all can participate, and share in the profits, with the result that in times of financial difficulty or unemployment elsewhere fish dealing is the occupation towards which many men gravitate. In their eagerness to secure even a small return some of these men tend to force prices up and thus create trouble for the more regular dealers, whose responsibility is greater.

The second feature which is apparent, then, is the lack of constancy in the numbers of fish dealers. There are first the men who because their net-group has broken up, or because they have sold their equipment, or because they have lost their money by gambling, " go on the beach " for a period to earn a living and try and accumulate some capital. When they have got some money, or been driven out by a succession of losses, they find other occupation again. One such case of a " marginal dealer " was a lad, already divorced, who lived either with his mother or his mother's brother. When I first knew him he was a fish dealer in a small way, but seemed to make little out of it. Then he began to go to sea in his uncle's boat, primarily as a kinship obligation. A month or so later he was on the beach again, equipped with a little travelling cooker, selling gobs of meat on skewers, bought by the fishermen after their day's work. He seemed to do moderately well at this for some days, and said that it was better than going to sea ; he complained that the

competition among the fish dealers was too keen to give any
profit, but later again he was back among their ranks.   Another
addition to the fluctuating numbers of the dealers is made by
men who have a regular occupation, but go in for fish trade as
a side-line.   Among these were several lift-net experts, especially
during the monsoon.   I chaffed one of them about turning
pĕraih, but he replied : " If there is work at sea we go to sea ;
if not, one must seek one's food ; one buys fish to sell."   Some
of them take up dealing seriously, on some scale ; others look
on it more as an occasional diversion which gives them enough
fish for a meal, with perhaps a small profit.   Awang Lung said
on one occasion : " I bought on the beach, but I am not a
fish. dealer ; I was just amusing myself—just my inclination."
This attitude is epitomized by the action of one *juru sĕlam*.
Having come in with an empty net he immediately waded out
up to his neck in the sea to another boat just entering with a
load of fish.   He bargained for the catch for $7, sold it at once
on the beach to the dealers for $8, and went off chuckling " a
dollar profit ".   An old man commented : " He's being a
nuisance," while one of the boat's crew said : " He couldn't get
fish, so he's looking for cash to pay for his coffee."   Ordinarily,
fishermen do not turn dealers on a working day.   But one man
made quite a habit of strolling over to other boats to buy fish
after he had bathed and changed on days when his own boat
had come in early.   This fluidity in the ranks of the fish dealers
rests partly on the good relations which normally obtain between
producers and middlemen, so that occupational jealousy does not
enter, and partly on the facilities afforded by the credit system
of buying.

A third characteristic feature is the way in which regular
fish dealers combine this work with other occupations.   One
man, trading in fresh fish, goes to sea from time to time, manufactures lift-nets for sale and draws an income from lending out
his three boats.   Another, dealing in fresh fish, has also a large
business in dried fish, rents several coco-nut orchards and prepares
copra, and deals also in cattle, meat and poultry.   A third, an
exporter of dried fish and a dealer in fresh fish in a small way,
prepares and sells large quantities of copra, manufactures lift-
nets for sale, imports drift-nets for lease or sale, and advances
money for the purchase of boats and nets at high rates of interest.
A fourth, a retired fishing expert and now a fish dealer, is the
local importer of rattan for the net-ropes needed by the lift-net

groups. A fifth, dealer in fresh fish and curer of fish, earns a subsidiary income by going out as a medicine-man. Most of the other regular fish dealers have some interest in copra manufacture too. Less than half a dozen trade in fish alone. Three of the dealers were formerly fishermen who could not stand the heavy work, but many had not been at sea at all. Occasionally a fish dealer turns fisherman if times are bad on the beach or the prospects at sea are attractive. One, with a steady business in dried fish, went out with a casting-net in the surf during the monsoon when other fish were scarce. Later he bought a section of drift-net and went fishing in a boat of a neighbour. He said to me : " If one doesn't go to sea and has no work on shore there is nothing to eat. If a man has a boat like that "— indicating a small *kueh*—" one can sit comfortably ; one gets a share from the boat, and enough fish to eat." Shortly afterwards, however, fish from the lift-nets became plentiful again, and he hardly used his new equipment.

WHOLESALE BUYING ON THE BEACH

The actual technique of buying fish in bulk on the beach is intricate.

At Perupok the carrier boats from the lift-net fleet are most important. They begin to come in, as a rule, shortly after midday, about which time groups of dealers start to assemble under the coco-nut palms above the beach. Each dot on the horizon is scrutinized, and speedily resolved into a line fisherman, a passing junk or a lift-net carrier. When I was first studying the fishing I was puzzled by the confident and accurate way in which people on the beach could tell not only which were the lift-net boats, but also which had a good catch and which had not when they were still miles away from shore. (When they are close inshore they can be told by their draught.) The mystery was solved when I discovered that partly from convention and partly for speed in reaching the market, a carrier boat with a good catch hoists two sails—which make a distinctive pattern on the horizon —instead of the single sail which is customary on other occasions.[1] This provides a very useful signal, allowing the dealers on the beach to save time by going only to the boats that are worth while, and getting the first attention for the people with the good

[1] But the position is complicated in the late afternoon. Returning boats, full or empty, often hoist two sails in order to get home before dark.

catches. With possibly scores of boats coming into a half-mile stretch of beach at the same time some such convention is necessary. Another factor which assists rapid assembly of the dealers at the spot where the fish will be is that each group always lands at the same place, where its boat-skids are. So when by the shape, colour or patching of sails boats have been identified in turn, the dealers waste no time moving from one to another.

As each carrier-boat approaches the shore its sails are lowered, masts taken down and stowed on the arm at the prow. The crew paddle in, and as the boat grounds the dealers wade out to it and help to drag it up the beach. They gather round the sides as the gear is removed and the floor-boards are taken up to reveal the catch (Plates XIA, XIIA).

Then the bargaining begins. After a few moments' inspection the captain is asked : " How much ? " He replies, naming a figure that is almost certain to be higher than that he will finally accept. One of the dealers then names a much lower figure, often about half only of that named by the seller. The latter refuses, and further bids are advanced. In this bargaining process the figures named by the seller are referred to as what he *kata*, " says " ; those by the buyers as *tawar* " bids ". After some chaffering, in which the seller is induced to lower his initial figure, he is asked to name a bed-rock price. In this he is said to *mati*, " to die ".[1] This bed-rock price may or may not be accepted by the buyers ; it depends upon the keenness of their demand. If it is still too high they may ask again : " *Mati molek* "—" Die properly "—that is, name a reasonable limit. To this he may retort, " *Tawar molek la !* " " Give a reasonable bid, then ! " For large catches the bargaining goes by $5 units as a rule ; for smaller catches by $1 units. Normally the buyers do not bid against one another, but either leave the bargaining to one of their number or act in concert. The captain of the boat is the only seller, though members of his crew occasionally give a word of advice. During the bargaining the fish are not removed from the boat, though some of the buyers turn them over to gauge their size and thrust a hand down to estimate the depth of the catch. While the haggling proceeds, a crowd usually gathers, and members of it, mostly dealers, take out of the mass of fish any odd ones of other than the bulk species ; by custom these are the perquisite of those who get them. It is common also for dealers, whether they are buyers or not, to extract one

[1] From the more complete expression *harga mati*, " dead price ".

or two fish from the bulk and take them away for food, without payment.

When a price has been settled the next procedure is for the buyers to divide the fish, for it is rare for a single dealer to take the whole of the catch. The division is usually made according to compartments, which are created in theory by the ribs of the hull. The ribs, and the resulting compartments also, are termed *kong*. Five or six *kong* is the usual number sold ; the others are kept for food or other special purposes.

When it has been settled who are to take the various compartments, and at what prices, the fish are scooped out, with hands or a flat basket, and put into large cylindrical baskets for transport to inland markets, further beach sale, or gutting for curing. A sale is sometimes made for cash, but usually it is on a credit basis. No written record is made, either between seller and buyers, or among the buyers themselves ; the public nature of the transactions obviates dispute at this stage.

SAMPLES OF BARGAINING TECHNIQUE

As part of my records I took down in as much detail as possible the bargaining in several hundred cases of fish-buying. I give here three samples of these records to indicate some of the more typical features and to show the empirical basis for my analysis. In the confusion of talk not every individual bid could be recorded, but the progression of the sale at different stages was accurately obtained.

Case 1. Four compartments of *sĕlar kuning* were for sale, a medium catch. Fish were plentiful that day, and this boat was late in. The seller asked $40. Initial bids, lightly given, were $12, and then $15. The seller said ironically : "Aren't fish saleable ? Have they no price ? "—" How much bedrock ? " he was asked. " *Mati* thirty dollars."—" Twenty ? " —an intentional mishearing that raised a laugh. " Thirty ! " replied the seller with emphasis. One dealer offered $17. " Won't give ! " answered the seller. " $18 ? "—" At $20 I wouldn't give them ! "—" Twenty " was then bid. " Won't give, and $25 won't get them, either ! " He was then asked again : " *Mati* how much ? "—" Bid ! I have *mati* already ; less than $30 won't get them," the seller replied.

Baffled, the dealers then began to ask about individual compartments. One dealer offered $4 for one of the rear compartments. " Won't you dare $2.50 ? " the seller asked sarcastically. The dealer then bid $22 for the whole catch. " Won't give," the seller replied. The dealer then commented : " A small boat, and no bus " (meaning that the catch was not so large as it looked and with no means of transport the fish might have to be dried, at a lower profit). The seller said to this : " If you want it, $30 ; at $25 I won't give it."—" Take them then ; I don't want them," answered the dealer in disgust. Offers were then renewed for various compartments. Three small dealers who bid went off, disgusted with the stiff price the seller demanded. One larger dealer was held up for some time on the first two compartments ; he was offering $11, and the seller insisted on $13. Finally these were sold for $12 ; the third compartment was sold for $7 and the fourth for $3. Thus the total reached was $22, a figure rejected previously by the seller. This is an example of keen bargaining by the seller in a dull market ; he was noted for his stubbornness.

Case 2. Five compartments of *sĕlar kuning*, moderately full, with some mackerel in addition in the fore-hold. This was the first boat of the day, in before noon.

For the bulk catch $25 was bid, cash down. " You can bid $26, $27, I won't give . . ." replied the seller, adding " at $30 I won't sell."—" At $28 you'll give or not ? " the dealers asked. " No ! "—" *Mati,* how much ? " they inquired. " *Mati* at $42, cash down," he answered. " $40 we don't dare," they commented on this ; and one of them added : " Speak decently, a little less." Then they asked again : " *Mati ?* "—" Forty-two," he replied firmly. " Less ? " they inquired. " Not a bit ! " was the reply—but a moment afterwards " forty, cash down ".

The dealers ignored this, and began to discuss the possibility of taking the catch at $35. There was talk of telling the seller to take the fish himself. At the time no bus was in sight. Then there was a lull in the bidding. At last the seller roused himself " You don't want them at $40 ? " The dealers said no, so he began to remove the fish. " Two less than forty I don't wish to take," he replied to a question. He and the crew unloaded the fish, which made five full baskets. A bus then arrived, and the fish were carried up to it by the crew. One dealer had gone, and no fresh ones had arrived to stimulate the bidding, so finally the seller agreed on $35 before the bus left. When that evening the net-owner heard from me the other prices for the day—fish were scarce—he said, " I think the dealers were frightened, thinking that the fish behind were plentiful " (that is, that the early arrival of a good catch presaged a glut later). " If he (the seller) had come later at Perupok he would have got $50." On the day's prices this seems probable (see *Case* 3, on the same day).

For the mackerel the seller wanted $3 a hundred. An offer of $2 a hundred was made, but the seller said : " Won't give . . . at Perupok they're $2.50." A lad of the village bought them at this figure—but later actually paid only $2.

Case 3. Six compartments of *tamban běluru* and *anak sělayang*, with a few mackerel and *sělar kuning*. This, a very good catch, came to shore in mid-afternoon, after many other boats had arrived empty. There was great shouting from the dealers as the covers were taken off, and keen interest was shown.

At once sixty dollars was bid, and then sixty-five, both offers being refused by the seller. Then someone said : " *Mati* now, plenty of people want to buy at once. *Mati* properly." The seller named $90 as the bed-rock figure. A dealer bought at this immediately, for cash. " Cash must be paid " said another dealer in warning tones to some of his colleagues who wished to come in. Others stood back, not participating in the purchase, some because they thought the price too high, others because they had not the free cash. The total catch was at once taken off by bus. It made about 11 baskets, and the gross takings in the inland markets were $109. The transport costs were about $10, and the dealers then divided about $9 among them. This example illustrates the speed with which a sale is concluded when supply is short and demand keen.

FEATURES OF BARGAINING

The samples bring out most of the basic features in these transactions.

In the first place, the sales are a public affair, with a crowd round, of crew, buyers and potential buyers, and bystanders.

All are free to interject, to comment on the fish and the bidding, and to joke with the principals. This publicity has two important effects. It helps to provide a flow of information from one boat arrival to another, putting each transaction into focus with others that have taken place and tending to assist both parties to a finer appreciation of the market situation of the day. It also helps to make the final price agreed upon a matter of common knowledge ; with sales on credit and verbal contract only, this provides for a sanction against later disputes.

Secondly, a decisive rôle is played by the captain of the carrier-boat, the *pĕraih laut*. He has the responsibility of naming the initial price and the (theoretical) bed-rock price which crystallizes negotiations, and of concluding the final price. The income of the whole net-group of 25 men or so rests largely upon his business acumen and knowledge. It is important to note here that though I occasionally heard grumbles that the captain had not sold so well as he might, there was never a single case in which the crew of his boat refused to accept his decision and called for further bids. Moreover, though occasionally there was a dispute between seller and buyers later about the exact price agreed upon, there was never any case in which the *juru sĕlam*—who, it will be remembered, is the man with the largest capital interest—revoked the price agreed upon by his representative.[1] The remedy of the *juru sĕlam*, if he is dissatisfied with a series of bargains, is to " throw away " his *pĕraih laut* and get another.

Again, the examples show the essential coöperation that exists between the buyers. Bids are advanced by different *pĕraih*, not as competitive but as representative offers up to what is regarded as the value of the goods. If opinion differs as to this value the more cautious can always drop out. This convention of coöperation has several effects. From the point of view of any individual it gives a check upon his judgement in buying. But more important, it helps to spread his risks by giving him a small share in several boat-loads. And an individual with little capital can earn a living with practically no outlay of cash, though he must take some risk. From the point of view of the seller, it means a lower level of prices ; but it also means increased security, a factor of great importance. Moreover, it means a more even distribution of prices.· Since a number of dealers can

---

[1] I recorded one case, however, where the *juru sĕlam* entered to let the catch go for less than the *pĕraih* was asking at the time.

participate in the purchase of each boat-load there is every incentive for many to assemble at each sale, and the later arrivals of fish suffer less than they would in a competitive system of individual buying. In brief, the coöperative system of bidding and purchase seems fairly well adapted to a situation where the total capital at the command of the buyers is small.

An outstanding feature of this bargaining system is the common Oriental procedure of dual quotation—both seller and buyer quote prices, and the margin between them is narrowed down. The width of the initial difference is partly a simple convention—if you name a price, I offer half automatically. It is partly also in the hope of being able to take advantage of the ignorance or weakness of the other party. But in these fish sales it has a further function—it allows both parties to test the market. The seller of the fish has been at sea since before dawn, and does not know how prices are going to-day ; he may not know even whether the buyers of yesterday's fish gained or lost in the inland markets—an important factor in determining to-day's prices. The potential buyers, particularly at the arrival of the first boat-load or so, do not know how the fishing has been going at sea—whether this boat-load is likely to be followed by many more, or whether they can bid high in confidence that fish to-day will be scarce everywhere. The initial seller's demands and buyer's bids, wild though they seem, give both parties more inkling of the market situation.

A further feature of the bargaining is the convention of the *mati* price set by the seller. In theory this is the bed-rock figure below which he will not go—it is the " dead-point " in the negotiations. Sometimes it is actually this, and buyers are forced to rise to it. But often it is not the real bed-rock, but the point of crystallization. It represents the seller's view of what the price ought really to be, as distinct from his optimistic quotations of what he would like it to be. The serious bargaining normally lies just below this point. The obvious function of the *mati* price is in giving a focus to the bargaining, in concentrating the differences of view between seller and buyers upon the narrowest range after preliminary variations have been explored.

On the buyers' side, an equivalent, though less precise, is the statement of what they " dare " or do not " dare ". Like the *mati*, this figure of what they " dare " is meant to be a serious quotation. As often as not, it is invited by the seller when negotiations are dull, and it is put forward by the buyers in consideration of the future market in which they will have to sell the fish, either fresh or dried. Like the *mati* price, it is subject to revision, and it is more elastic in that one buyer will sometimes " dare " what another will not.

One element in the dealers' calculations is transport charges, since most of the traffic from the fishing centres to the larger inland centres is by motor-bus. Carrying-pole dealers and bicycle dealers avoid these charges, but their market is correspondingly limited in range. Bus charges are roughly 5 cents or so per basket of fish per mile, but vary according to quantity carried, and at times according to the amount of transport available. For instance, from Perupok to Melor or Ketereh, when the ordinary charge for a single basket with a man travelling with it may be a dollar, the rate may be reduced to 80 cents or so if, say, 10 baskets are taken. Again, the normal charge for a basket of fish alone to Kota Bharu is 70 cents ; on a holiday, however, when hardly any buses were running, the driver of one wanted a dollar, and agreed to a little less only after much wrangling. (Since fish weigh so much less when dried, a dealer who buys for curing can reckon on proportionately smaller transport charges per picul of wet fish at the boat than can his colleague who is buying for fresh sale inland.)

Another characteristic of this type of bargaining is the mechanism whereby a failure in the adjustment of seller's and buyers' prices is met by the seller assuming the rôle of " buyer-in " on behalf of the net-group, or of buyer on his own account. This tends to prevent the dealers from taking too great an advantage of their combination and keeping prices at a very low level. It tends also to check the exercise of monopoly by a seller on a day when supplies are short. The expression used for " buying-in" the catch is *kau' sĕndiri*—unloading themselves. Resort to this is not regarded by the dealers as breach of contract or unfair tactics ; it is the legitimate exercise of rights. But it does involve additional trouble to the seller and his crew. Instead of being able to go off home they must unload the fish and carry the baskets up the beach, and some of them must go off in the bus to the inland markets, not to get home till long after dark. Moreover, they run the risk of losing money. If, therefore, the seller insists on a price higher than that which the buyers can honestly dare, the threat of having to take the catch himself often brings him to terms. Sometimes, however, when the resistance of both parties is stiff, but not quite unyielding, the

matter takes on amusing complications. The seller, refusing all
offers made at the boat-side, proceeds to unload the fish and
carry the baskets to the bus. There the argument begins again,
and this time a sale is made, the precise figure depending on who
feels in the stronger position. Or again, the seller mounts the
bus, and proceeds along the road with his baskets, the potential
buyers accompanying him. *En route*, they chaffer, and finally
come to an agreement; he gets off and catches the next bus home,
while they carry on. In these proceedings some element of bluff
may enter. On one occasion the seller mounted the bus, and
on it got also some of the erstwhile bidders, saying that they had
business along the road. But once aboard, the matter of selling
was broached again, and an agreement was reached. The
seller had a shrewd inkling of what their " business " was, but
was content to wait until, by opening it again, they gave him a
chance to get better terms ! A variant of the procedure occurs
when, though the seller continues to take the catch to market,
some of the erstwhile bidders come in with him as in normal
buying, and they share the risks of the undertaking. This
re-alignment of some of the buyers with the seller is not
uncommon, and is an index to the good relations that
obtain between them outside the formal opposition at the boat-
side.

Another feature which emerges in the bargaining is the effect
on demand of the time of day at which the fish appear. This is
seen concretely in three statements made by dealers as arguments
for taking the fish at a lower price. They say : " It is dark " ;
" There is no bus " ; or " I am going to dry ". The point of
all this lies in the distinction between the fresh-fish market and
the dried-fish market. For the former, naturally, the later the
catch reaches the inland market, the greater the risk of loss. And
moreover, the later the time on the beach, the less chance of
getting a bus, the only means of bulk transport. Hence the
anxiety of the bidders as the afternoon shadows creep down the
beach, and their complaints of " dark " as the sun begins to get
low. The presence or absence of a bus at the time of selling
may make a difference of five or ten dollars in the price. By
about 5 p.m. a waiting bus is rare, and the catch will almost
certainly have to be dried. This means that the buyer's capital
will be tied up for some time, his labour will be greater, and his
profit, though probably more certain, quite possibly may be less.
A lower quotation is thus demanded. Moreover, only some of

the dealers have the equipment for fish-drying, so the demand
tends to be less keen.

The most striking example of this that I observed was one
when all the boats came in very late because there had been no
wind in the early afternoon. The seller for one net landed at
6.30 p.m., when it was already dark. He had 4 compartments
of sĕlar kuning half full. Only three dealers were there to buy.
Previously one of them had said to me : " When it is dark we
buy fish, but cheaply, since few pĕraih are there." Negotiations
were conducted in almost total blackness. The seller asked $25 ;
and $10 was bid. Then there was a long interval, in which the
dealers argued with him for a while, but then turned away and
began to discuss general subjects, joking and apparently quite
oblivious of the fish. They were simply waiting for him to come
down. At last he said : " If you don't want to buy, all right,"
and began to cover up the fish. They then said : " Don't be
angry ; we have to make our living ; we have bid ; the mackerel
this morning took our cash ; fish are very cheap." They then
asked him to mati. He said angrily : " Bid properly first."
They said : " We have bid—$10." He was very annoyed, but
at last mati at $15. One dealer asked : " Will you take $11 ? "
—" No ! " Then another offered $12, which was also refused.
Finally the sale was made at $13.

In this case the buyers had it practically all their own way.
But there are occasions when fish is scarce, and the advantage
lies with the seller. Here at times I observed an interesting
variation of the usual procedure—an attempt by the seller to set
a fixed price, and allow no haggling. I saw this procedure
followed only by Awang, pĕraih of Japar, and it did not meet
with the approval of the buyers. On the first occasion Awang
came in early with a good catch, flush with the deck. He was
asked to give a mati price. He replied : " Mati—bed-rock and
no haggling at fifty." There was some murmuring at this, and
then Japar, the juru sĕlam, who was present, said to the buyers :
" Won't you dare $45 ? "—" Awang hasn't yet given them for
$45," answered one of the dealers, with a grin. But a bystander
called out : " The juru sĕlam has mati at $45 ; at $45 he lets
them go," and they were at once taken at this figure, cash down.
Awang seemed rather put out at being over-ridden, but did
nothing. On another occasion he was more successful. The
catch was thirty enormous horse-mackerel of the types known
as gĕrong and bĕrka'. They were taken from the boat and laid

on the sand while the crowd gathered round. Awang, asked to name a price, did so—$4 for 3 *gěrong*, or $3 for 2 ; and $1 apiece for the *běrka'*. He added : " I don't want haggling ; it can't be less. Whoever doesn't want them, just keep quiet "—this in gruff tones. The dealers stood silent after this blunt statement. Then one of them asked rather plaintively : " I'm not bidding, but are we allowed to ' invite ' less ? " This meek subtlety was refused, and the affair languished. It was a good catch, fish were scarce, and this was the first boat in, but to be deprived of their " natural rights " of chaffering left the dealers at a loss. Then the former spokesman, in mock despair, laid his head-cloth on top of the fish, put his head on them and pretended to go to sleep. He said a little later to another man : " Can we bid ? One can't know ; the owner's angry." Then Awang proceeded to put the fish into baskets and take them off himself, but at the bus they were bought by a dealer at $11 for 10, the original price.

The fixed price is a device which though attractive to the seller, is not likely to have much success. The reason is that in his lack of knowledge of conditions elsewhere the seller is likely to set a price above the market value, and so be left to carry the fish himself. Since owing to competition from Tumpat and other areas the seller can rarely have an absolute monopoly, he tends to lose, and so is ready to accept bids made on a future occasion.

Another feature in the bargaining is the attempt by the dealers to break up the bulk sale into one of separate compartments when the resistance of the seller is strong. The motives for this are twofold. One is by getting the seller to name prices for individual compartments to break his resistance by challenging his price on each. It is bargaining for the whole catch from another angle, the total bid being often the same as before. Another motive is that an individual dealer hopes to get a lower quotation for that section of the catch he has chosen than if it were considered as a proportionate part of the whole. When all the bargaining for the separate compartments is over it sometimes turns out that the seller has agreed to a set of prices which total a figure identical with or even less than one he has previously refused to accept (see *Case* 1). This chaffering for independent compartments, which takes up more time, is usually only with catches that come in fairly late in the day, or when fish are very plentiful, so that the dealers feel no great urgency to complete the sale. In any case, it merely pushes forward by one stage the division that they have later to arrive at among themselves.

A practice occasionally followed late in the day, either when seller and buyers cannot agree on a bulk price for the catch, or when a buyer does not wish to take the whole of it, is to *běli bilang* instead of *běli borong*, that is, to " buy on a count " instead of " buying in bulk ". A price per 100 in the case of *sělar gilek* or *kěmbong*, or per 1,000 in the case of the smaller *sělar kuning* or *lechen*, is agreed upon, and the fish are then counted out. This gets rid of one variable in the bargaining situation, and so makes agreement simpler. In one case, where there was no hurry to buy, it being admitted by all that fish weren't selling well, a buyer came up and after a few inquiries said : " Take them out, take them out, I only want to buy a few ; ten, twenty cents' worth." They were then sold in small lots, by count. In another case, near sunset, the seller wanted $30 for the catch, but agreed to $3.40 per 1,000. There were two buyers, who were going to cure the fish. When they were counted there were 6,900 altogether. The result was greeted by shrieks of laughter from one of the partners. When, mystified, I asked the reason, it appeared that the pair of them had a bet of 20 cents on the result, one arguing that there were 7,000 fish in the catch, the other that it wouldn't reach that figure. I had been asked to calculate the exact sum owing—it saved them trouble—and when I announced the result, $23.46, the loser said with a smile : " Better lose 20 cents than buy the fish for $30." Incidentally, this example demonstrates two points about fish dealing. One is the fairly high degree of accuracy achieved in calculating by eye the volume of a catch ; from other similar counts I should say that this error of less than 2 per cent. is quite typical. (I found that my own error was 25 per cent. or more at first and even after some practice was usually about 10 per cent.) The second point is that loss due to this error of estimation does not normally fall on the wholesale dealers, since they also dispose of the catch by bulk in baskets ; it is the retail dealers who make or lose by their estimates since they sell the fish by count.

When the final margin between what the dealers will give and what the seller will take cannot be adjusted, bargaining is sometimes cut short by the dealers taking the fish and leaving the ultimate price to be settled later. They may tell the seller on the spot they will only give him so much, disregarding his protests, or may leave their offer till after they have sold the fish. This practice occurs normally only with small catches, those that come in very late, or those that are to be cured, and depends upon the existence of good relations between seller and dealers. It really rests upon the fact that for such catches demand is not keen, and the decision of the dealers can be easily enforced. The seller does not want to throw away any advantage by actually agreeing to what they propose, but at the same time wants them to buy, as an alternative to having to take the fish himself.

I have not tried to describe all the subtleties of bargaining here, but simply to outline the major factors which tend to determine the final price.

There is, however, one factor less precise in its operation but of considerable importance, which stands at the back of the negotiations in a general way. This is the concept of a just price. It emerges most clearly in the frequent injunctions to bid (or quote) " properly ". The bargaining is not conceived simply as a matter of taking advantage of the other party wherever possible—though a little deception is quite permissible if one can manage it. A factor of " business ethics " also enters in, and men who always drive the hardest bargain and will make no concessions are unpopular. If sellers, some dealers do not go to their boats ; if buyers, some sellers do not welcome them. The reasons for this are based partly on a rather vague feeling of companionship between buyers and sellers—they are all members of the same community and some of them are friends and kinsfolk and they all have to get a living somehow ; and partly on a more real economic interdependence. Should the buyers combine to force prices down to the minimum not only would the fishermen take to acting as dealers themselves, but the marginal groups might drop out of business. The dealers would also probably lose all the small concessions they now get in free fish for home consumption. Should the sellers force prices up to a point at which the dealers failed to get a reasonable living, then their own receipts would inevitably fall off, and they would be forced to market the fish themselves. The practice of " cutting " the price, which is already in existence, is the dealers' present remedy against unprofitable wholesale buying. This question, however, is bound up with the broad problem of credit, which is discussed a little later.

GRAPHS OF PRICE-DETERMINATION

Having analysed the bargaining, and the factors which go
to determine wholesale prices of lift-net fish, we may now examine

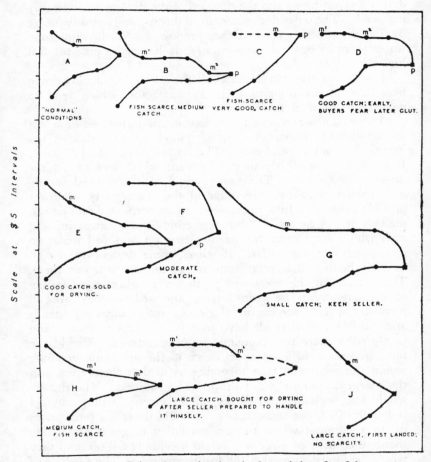

FIG. 17.—Price determination in bargaining for fish.
All graphs have been adjusted to a seller's opening quote of forty dollars.

the actual course of the bargaining more closely, to see what
relation the final market price arrived at bears to the earlier bids
and quotations. This can be most easily done by diagram.
In Fig. 17 each graph represents the course of bargaining in a

sample case from my records, chosen to illustrate a situation of supply and demand. For comparison I have adjusted the prices in each case in proportion to a seller's opening quote of $40. In the graphs the upper line of the curve represents the seller's quotations, and the lower line the buyers' bids ; the final price agreed upon is the meeting place of the two lines. Each point on the curve represents a successive price named by one party, and it is the shape of the curve, not the final level, that is most significant here. The bed-rock (*mati*) price quoted by the seller is indicated by the letter *m*, and cases where the final price was for cash down (*pitis sělalu*) are shown by the letter *p*.

It will be seen that in all cases there is a wide difference between the opening quotation of the seller and the opening bid of the buyers, but that there is considerable variation in the speed with which the gap between them is narrowed. Moreover, in only two cases, A and H, does the curve approximate closely to the shape of the normal curve, with each party making concessions by regular increments. In all the other cases one or other party shows resistance, which may be sustained, or may collapse suddenly to meet the price of the other. In each case the specific shape of the curve can be explained partly in terms of the supply and demand situation of the moment, including calculations of futures, and partly in terms of the stubbornness or complaisance in bargaining technique displayed by the individuals concerned. But primarily it is the knowledge of the situation possessed by both parties that determines the outcome. This explains a frequent form of the curve in which the seller, maintaining considerable resistance for some time, finally yields, having known all along that he would probably have to do so despite his assertions (see cases F and G). In case C, on the other hand, scarcity of general supply gives the seller a strong position, of which he takes full advantage, while in case B a medium catch in a situation of scarcity allows the seller to assume a modified firmness. In case I, a catch sold for drying, the position is somewhat similar ; here the seller carried his resistance to the point of beginning to handle the fish himself. Since the catch was a good one, the buyers had to meet his price to some extent before they could obtain the fish.

The interest of these graphs is that they show how in this situation, given the technical conditions and relevant factors in supply and demand, market expectations, etc., the processes of price formation conform to ordinary economic principles. They show also, however, that the bargaining situation itself, with its conventions, is a factor in the final determination ; some people take more advantage of the conventions, others less. Moreover, while the direction of price movements can be inferred from general analysis, their magnitude cannot, and the empirical data such as are given here are essential for this. Examination of my full material shows that while the opening quotation of sellers is apt to vary widely from the price finally attained, the opening bids of buyers bear a fairly constant relation to it—from 25 per cent. to 30 per cent. below it, as a rule.

Though only ten cases have been given here, for lack of space, they are drawn from more than sixty originals, which bear out the generalizations here made.

CASH, CREDIT, AND INSURANCE AGAINST PRICE-REDUCTION

In most cases of these wholesale transactions the fish are sold on credit granted by the seller until the end of the week. But in some cases cash is paid on the spot. We have now to consider why this is so, its frequency, and the effects upon other transactions.

The most important reason is as already stated, the certainty of the payment. Not that there is usually mistrust of the honesty of the dealers on the score of final settlement for what they have bought, but there is a definite fear that the amount given in settlement may be less than the price agreed upon at the time of sale owing to a " cut " imposed. This point will emerge clearly a little later. As a complement to this attitude on the part of the seller is that of the dealers, who may offer cash down as an inducement to him to let the fish go at a lower price. A second reason which operates at times is the desire of the crew to have the catch sold for cash so that they may draw upon the proceeds at once. A comment I overheard from a crew-member at the sale of one of the first catches after the monsoon illustrates this. He had been asked if the sale had been for cash. " Cash at once," he replied, " when men haven't been to sea for two months of course they want cash down." But as a rule the cash is not disbursed to the crew till the end of the week.

I did not record all the cases when cash was paid on the spot, but from 45 cases noted in the two best months of fishing it was clear that cash payments were usually made for the first three or four good catches of the day. These were taken by bus to sell fresh. Cash can be paid in such cases because there is a daily turn-over, whereas when fish are bought for curing the dealer must stand out of his capital for some considerable time, and may sell the cured fish from stock perhaps only once or twice a week. The later catches of the day, then, are usually bought on credit, since whatever the wishes of the seller there is not the free capital available to pay for them on the spot. The highest amount of cash I noted as paid out for these wholesale purchases on one day was $260 for a total of five catches. The next day, however, after $135 had been paid out for three catches the seller of the next catch demanded $23 cash down. " Oh! the cash is finished," replied one dealer. But then another asked " Shall we buy for $21 cash ? " and this was agreed to. The largest amount I saw handed over for any single catch was $100, made up by about five dealers. It is difficult to estimate what the total amount of free capital at the command of these dealers is at any one time, especially since the more wealthy of them do not bring all they possess on to the beach. But on one occasion when I expressed surprise at a heavy cash payment a dealer said to me : " How much cash do you think these people here have among them ? I think about $300." Considering the scale of the transactions this was probably a fair estimate. Most dealers work on a capital of about $10 to $15, though I saw several with about $35 in cash, and the most wealthy probably have a free capital of $50 or more. One dealer said that he bought on some days $50 worth of fish, on other days only $5 worth—" It's uncertain ". When cash was demanded then it was " difficult " ; otherwise one needed no capital—two dollars was enough. That day he had bought fish for fresh sale from three boats, paying cash in each case, the total amounting to $30. In addition he had bought on credit a further $11 worth for drying. Of these, however, he had sold $4 worth to ordinary retail buyers.

The desire of sellers to have cash sometimes causes difficulty. On one occasion in a fairly slack season a seller, Awang, *pĕraih* of Japar, had first stipulated " Sixty without haggling " as his *mati* price, to which the dealers' reply was : " If you'll sell for fifty people will pay cash at once." Then he was prepared to let the catch go for $55, but added : " I don't want you to buy unless you pay cash." One of the crowd commented cynically : " They'll have to borrow, then." In this case the cash was found. But the following day the same man sold his fish for $50 " cash ". When the time came to pay some of the buyers said they hadn't the money. Awang was angry. Saying " I don't want you to buy " in a trembling voice, and with flushed face, he threw the floor-boards across the boat with a petulant gesture. The buyers with lowered voices proceeded to divide the catch, while Awang stood sullenly by the stern. The upshot was that half the buyers paid cash and the other half did not. Psychologically, it was an interesting situation, linking up with the efforts of the same seller to set a fixed price on other occasions. It illustrated the friction that sometimes occurs between seller and buyers, with the possibility in the background of the price being " cut " on a credit sale. At times one dealer has to be helped out by another if the seller proves intransigent. A small catch, the first of the day when fish were scarce, was sold for $13. The buyer then explained : " I want to buy at $13, but I want credit, I haven't the cash." The seller insisted on the money, and argument proved fruitless. Then another dealer came up. The buyer explained : " I want the fish, but I'm hard-up."—" I'll buy," said the other, and after some attempt to beat down the price he handed over a bundle of notes. But when the seller counted there was only $12. " One more," he said. " Count again " said the buyer. But the result was the same, and the bundle was handed back. Then the buyer counted each note out separately, with a laugh, dipped in his purse and flung down the complete payment. It was an obvious attempt to get round the agreed price by a trick sometimes followed. If the seller is incautious, or easy-going, he accepts what he is given ; it borders on cheating, but in some cases at least is rather a persistence in the bargaining than an actual deception.

These examples show that the demand for the payment of cash can be both a source of friction between buyers and seller, and an index of underlying friction. It rests upon the seller's knowledge that the result of a credit sale is often a reduction in the price actually received from that agreed upon.

We have now to examine this practice of price-cutting in more detail.

When the buyers have lost on a deal it is a well-recognized practice for them to " cut " the price when payment comes to be made. The " cutting " is not automatic, but depends upon the acquiescence, however grudging, of the *pĕraih* or *juru sĕlam* from whom the fish were bought. The extent of the cut varies, but about 10 per cent. is not unusual ; an average in 10 cases

XIVA FISH-DRYING ON A SMALL SCALE
Dealers inquiring the price of the fish, which are drying on bamboo trays in the sun. It is the off-season for fishing with lift-nets, and the large boats are therefore drawn up among the palms.

XIVB A SMALL RETAILER OF FISH
A small trader with a bicycle loading up before going off to the market.

XVa A STAGE FOR LARGE-SCALE FISH-DRYING
This staging of bamboo, at Beserah, Pahang, is owned by Chinese dealers. Attached to the platforms are sheds to hold fish, salt, jars and other containers.

XVb TAKING DRIED FISH INTO STORE
This staging, with drying fish scattered over it, is at Batu Lipo, Trengganu.

of actual or estimated cuts was a little less than 8 per cent. But while it is regarded as customary, " cutting " is naturally viewed from different angles by buyers and sellers. The buyers' point of view is that it is reasonable to ask the seller to share in their losses, since insistence on the full agreed price would make things very hard for them. Moreover, should *juru sĕlam* and *pĕraih* consistently refuse to cut, then the risk of buying is too great. (I noted this attitude only with a few of the more conservative dealers.) They hold that being friends and kinsfolk, as many of them are, the sellers usually agree without difficulty to the proposition of the dealers. But they admit that confidence must exist between the parties ; that if a seller does not trust a dealer when he says he has lost on the transaction—and there is only the word of himself and his fellow dealers for this—then he will not agree to reduce the price.

The point of view of the sellers is apt to be that cutting is due to the bad business habits of the dealers. As one explained it : " The dealers fight ; one bids twenty dollars, another twenty-one, another twenty-two ; afterwards they lose, and cut." And the reluctance of dealers to pay cash even when they have it is attributed to their desire to safeguard themselves in case of loss. The attitude of sellers tends to crystallize into a cynical view that losing and cutting is a kind of natural characteristic of dealers. On a number of occasions when fishermen asked me what price a certain catch had brought their comment was of this type, " Thirty-five dollars ? Oh ! they'll pay only thirty. They lose and cut. They'll lose ten and cut five." Or of a catch sold for $12, " What do you expect of dealers ? They cure the fish and afterwards they cut ; I think it'll be only ten dollars." Or, " Thirty-five dollars ? Write thirty, Tuan " (as a record in my notebook) ; " afterwards they'll cut."

A cut, while usually accepted in the end, is disliked for a further reason, namely that it may lead to trouble in the seller's group. One *pĕraih laut* told me how he had sold $4.50 worth of fish to a dealer who had promised to pay the next day, but gave him only $4.30, cutting 20 cents because he said he had lost on the re-sale. The *pĕraih* said that he was afraid the *juru sĕlam* would be angry with him, since he would have heard what the fish sold for, and would accuse him, the *pĕraih*, of trying to cheat. " Selling at one figure, and paying at another " is what an unscrupulous agent can do, putting the blame on a fictional cut of the dealer. A case mentioned to me by Awang Lung brings out the same point. He was selling a catch, and gave as the final price $15. The dealer who bought came along at the end of the week with $12 only. " I didn't want to take it ; I was afraid evil would be thought of me," meaning that the crew might think he himself was cheating. So he put off the payment, saying he must ask his crew. They refused the $12, and the onus was thrown on the dealer to pay up.

The whole matter of cutting is then a rather ill-defined one, in which the principle is defended by the dealers but questioned and disliked by the sellers, who in practice usually agree, sometimes disallow it, and try to circumvent it by insisting on cash sales where they can. Viewed from the economic angle, the practice has the effect of evening out the profit-and-loss fluctuations for individuals. The losses of the middlemen tend to be distributed over the general body of producers as well. But on the other hand the knowledge of the middlemen that they can cut if they lose allows them to bid higher than they otherwise would, so that if they get a profit the producer in effect has also benefited.

In the attempt to escape from the system of cutting, sellers sometimes adopt the procedure of sale for a fixed credit sum. On one occasion after the bargaining had reached $45 the seller said that this was a final price ; if afterwards the buyers proposed to pay less he wouldn't take it ; they must accept $45 as the absolute price. They took the catch on these conditions. On another occasion also a catch was sold, not for cash but " *bayar tidak kurang* ", for payment of not less than the agreed sum. This is not a common procedure, probably for the reason that the buyers cannot be prevented afterwards from " asking " to have the price reduced, in spite of their original verbal acceptance of it.

Another mechanism to avoid the incidence of price-cutting is the revival of an institution which was formerly tried when the Singapore dried-fish trade from Perupok was in full force. This is the *tangkap*, the " catcher ", who fills the function of an insurance agent. When a net has a " catcher " it is his job to be responsible for the payment to the *juru sělam* at the end of the week of the full sums agreed upon for the various sales of fish. If he has not collected all the amount from the various dealers by that time, he must make up the balance from his own cash reserves. He is thus debt-collector and supplier of short-term cash to the *juru sělam* for whom he acts. For these services he receives a share (*chabu'*) of $\frac{1}{20}$ of the total receipts for the week. The position was put thus : " The ' catcher ' is the man who fills in the cash ; because he eats a profit. If he can't eat cash it's of no use ; what's the use of working ? . . . Because he supplies cash from his own house, therefore he knocks off $5 a hundred— because it's his money he supplies ; he uses his own cash." It was estimated by a fisherman that since the " catcher " must pay on the nail when the end of the week comes, he must have a capital of at least $200 to $300 for each net which he " catches ". In former times the percentage he received for the job was 10 per cent., because when the fish were sent to Singapore the payment for them could not be received in time for the weekly division, and the " catcher " therefore had to lay out a great deal of capital. But the system did not appear to have worked well, and was discontinued. Nowadays, with the daily cash returns from the sale of the fresh fish in the inland markets, the outlay demanded from the " catcher " is much less, and his percentage is consequently also less. In 1940, when the system seemed not to have been going long in its revised form, there were seven nets with " catchers " in the Perupok area. All these men were fish dealers themselves—partly because they had capital, and partly because as dealers they could keep watch more effectively on their brethren. One of them, who was " catcher " for two nets in the Perupok area and one in Kubang Golok, had only recently taken over the first two. He said that his percentage from them in the first month had amounted to about $25, and he had not yet lost any money in his collections from the buyers of the fish. He commented on another prominent dealer that this man was not a " catcher " because people didn't like his going off and not watching his boat ; the " catcher " must be at the boat when the fish are sold, watch who are the buyers, and if he thinks they may be untrustworthy, get the cash from them at once before letting them take the fish.

The institution of the " catcher " is important from its reactions on the practice of cutting. Since the " catcher " not only gets a percentage of the gross receipts, but is also personally responsible for any deficit from the amounts agreed upon when the fish was sold, he has every incentive to refuse to allow any cuts, and to insist on the buyers paying in full. The economic effects of this are brought out by giving the views of the *juru sĕlam*, Awang Lung. When I asked him if he did not have a " catcher " for his net he said no, he didn't like the practice. Then, surprisingly, if one bears in mind what has been said earlier, he added that if there is no " catcher " the dealers know that if they lose they can ask to cut. But if there is a " catcher " they know they have to pay in full, and so tend to buy more cheaply, and also, they do not flock to buy to the same extent. " If I had a catcher perhaps there would be less of a crowd to buy my fish." Nowadays, he said, with the selling of fresh fish, he had no difficulty in getting the money at the end of the week. Then he went on to say that a Pantai Damat friend of his had said a few weeks before that he wanted to " catch " his (Awang Lung's) fish. He had said " All right, but I don't want one, since if the dealers lose badly and can't cut, perhaps fewer people will want to buy." These statements are extremely interesting. Not only do they show a *juru sělam* defending " cutting " from his own point of view, but they indicate how clear a grasp of the essential economic principles can be shown by an intelligent Malay fisherman.

The whole system in effect is a demonstration of the tendency of returns to an equilibrium. By sharing the risk the convention of cutting helps in the long run to maintain the market. And by attempting to block the lessened returns due to cutting, the " catcher " tends to depress the market by making buyers more cautious, so that the reduction in effective demand merely tends to bring down prices at an earlier stage.

ARRANGEMENTS AMONG THE WHOLESALE DEALERS

So far the dealers have been considered as a group, operating opposite the seller. Now the subsequent arrangements between them must be examined.

The function of the dealers is not only that of transporting the fish to the inland market, to sell it there in smaller parcels to retailers. Some retain it for curing ; others sell on the beach to other middlemen, or even retail ; others spread out among the villages on the bus routes, at Melor, Ketereh, Peringat, Pasir Tumboh, or go on to the town market at Kota Bharu. Their choice is guided by their individual estimate of the profits in each case. Therefore if there is more than one buyer they must split up the catch and come to terms.

There are four methods by which they may do this. The first is to divide the catch by compartments while it is still in the boat, and settle the price which each compartment will bear in the total amount. The second is to take out the fish into baskets and settle the price per basket on the beach or, if they are all going by bus, while waiting for it to arrive or *en route*. The third is when dealers who have stood out of the original purchase on the grounds that it was too dear take compartments or baskets from those who have bought, by making independent offers. And the fourth method is for dealers who have not entered the original purchase to combine with the purchasers, not on an equal basis of putting up a share of the capital, but as " helpers ", getting perhaps a dollar out of the profits for their assistance.

The most common method is arrangement by compartments. Each man stands by the compartment he has chosen, prepared to point out its demerits in order to get it as cheaply as possible. In accordance with a useful convention, prices are normally settled first for the catch in two major sections, before descending to the argument about individual compartments. There is always a wrangle, sometimes acrimonious, but if the fish are to be sold inland time presses and they try to come to a decision as soon as possible. To begin with, suggested prices are usually made on the basis of dollar margins, but except in the case of the higher value catches, margins of 50 cents, 25 cents and even 10 cents are often used before agreement is finally reached. I give here a sample of the various figures suggested and finally agreed to for a catch of five compartments of fish bought by the dealers for $30.

The catch was first considered as two units—the three fore compartments, and the two rear compartments. The successive prices put forward were :

Three fore. $	Two rear. $	Result.
17	13	Rejected
18	12	Rejected
18½	11½	Rejected
19	11	Agreed

The compartments were then considered individually, thus :

No. 1. $	No. 2. $	No. 3. $	Result	No. 4. $	No. 5. $	Result
6.30	6.30	6.40	Rejected	5.70	5.30	Agreed
6.20	6.30	6.50	Rejected			
6.10	6.40	6.50	Rejected			
6.20	6.40	6.40	Agreed			

The cash was then handed over to the seller by each buyer, who then took out his fish, and went off.

The poorer dealers work on such small margins that a difference of even ten cents is important to them, and even the wealthy dealers do not disdain to haggle at this level. When agreement on a proposed division cannot be reached then a common solution is for one of the parties to suggest a change of places. The objector is thus obliged to take that which he declares is the better bargain or to swallow his complaint. In one case where two dealers were dividing a catch they had bought for $9, after rejecting four sets of figures, they came to an impasse between $5.10 and $3.90 or $5.20 and $3.80. Finally one of them said : " If this one is $5.10 I'll take it ; if it's $5.20 I'll take the other." The second dealer first said $5.10, then changed his mind at the last moment to $5.20, and the other man moved over. Both were laughing at their delay, but serious about the division.

The prices for individual baskets, when the division is made on this basis, are determined in the same way (Plate XIIIA).

Despite the keen wrangling, relations between the dealers are usually good ; but occasionally one is left disgruntled. The following example illustrates this, and also the way in which miscalculation by dealers leads to their cutting the price to the original seller.

A small catch of *sĕlar kuning* and mackerel was brought in, and Ali, a dealer, bid $13 for the lot. Then a separate bid was made for the mackerel by two other dealers, and the seller let them have the fish for $5. Ali said : " If they're going for $5 I'm buying them." But the others said : " Keep to your own compartment ; you're here, there and everywhere," and took away the fish. Ali then asked the price of the *sĕlar kuning*. The seller quoted $10. Ali bid $6. The seller said : " People have already bid $8." Yusoh Panar, normally a fisherman but that day acting as a dealer, inquired who ? Ali said : " I did— but only together with the mackerel ; if separately, then only $6." But Yusoh then bid $7, and after the seller had *mati* at $9 they agreed on $8. Then came the division between all three dealers taking part. The compartments were priced by the others at $2.30, $3.30 and $2.40, from fore to aft. Ali, who was at the centre compartment, made a row. But the others would not give way, so he said to the dealer at the rear compartment : " Leave it, leave it," meaning that he should change over. But the other man replied sensibly, " I don't want to ; that one is dear." Ali, by this time furious, said : " Then let yours come up in price ! " But finally Yusoh and Ali took the centre compartment together at $3.25—much against the grain, as far as Ali was concerned. Yusoh said : " Sell the fish at 30 for 10 cents—not less." As a result, selling retail, they took only $2.66 for the centre compartment. I asked Yusoh why he had bid so high. " Fish are scarce ", he said. Then he began to calculate for the catch as a whole. He said to the other two : " If we reckon to pay $7, how much do we cut ? A dollar divided by three, how much per man ? " But when they worked it out they had lost as a whole just over two dollars, and Yusoh said : " One dollar cut is not enough." At this stage Ali, a fairly poor man, could stand no more and with the others alternately laughing at him and trying to pacify him, rushed off shouting : " I won't pay ; I'll be summonsed instead."

A *dénouement* such as that is rare, but I have quoted these cases to show the detailed calculation that goes on over a single catch. The marketing process is still further complicated by the entry of intermediate middlemen and ordinary retail sellers.

INTERMEDIATE AND RETAIL SELLING

During and after the division of the catch just described, a number of secondary transactions are liable to take place, with bulk dealers coming up late from other boats, with other dealers hoping to sell again to retail sellers or act as retailers themselves on the beach, and with the carrying-pole and bicycle dealers who want fish to sell in the inland homesteads or further down the coast. The activity of these last-named dealers, and of the retailers on the beach, is important.

The *kandar pĕraih*, carrying-pole dealers, are picturesque figures. They come on the beach stripped to the waist, clad only in a short kilt. Their job of serving the inland shops and homesteads off the bus routes means that they must travel along the foot-paths, and since they usually can get fish only late in the day, after the wholesale dealers have bought, they must go at a sharp jog-trot to sell their burden before dark. They often carry considerable loads balanced on their pole—two that I once saw carried about 80 pounds apiece—and the bulging muscles of their calves are an indication of their calling. The area that they serve comprises broadly the segment bounded by the arc of the Bachok—Kota Bharu road, with a radius up to 10 miles.

Partly by convention, but primarily because their free capital is very small, the carrying-pole and bicycle men hold back from participation in the bulk buying. They bargain directly for a catch only when it is a very small one, and rely on subsequent purchases from the wholesale dealers and purchases of *makan lau'* fish, supplemented by purchases from line fishermen etc. to fill their baskets. They nearly always must pay cash for what they buy, pulling out their money from an old tin, a purse or a little plaited basket wrapped in a scrap of cloth. They seldom have more than a dollar and some silver as working capital with them, and they operate on small margins, frequently buying a few cents' worth of fish here and there. When they buy from lift-net boats they are to be seen clustered round the rear compartments, which being narrowest are the cheapest. Since the amount that each man can buy is limited by what he can carry, as many as half a dozen buyers may divide a single compartment or parcel of fish.

These purchases of the small dealers from the wholesalers often give the latter a quick turnover and an easy profit. On one occasion I saw a compartment just allotted in the division of the catch to two wholesalers at $4.20 sold on the spot to a group of carrying-pole men for $4.75. When I asked the wholesalers why didn't the small dealers buy the compartment themselves in the original bulk purchase and then divide it, the answer was : " They can't. The people here buy and reckon up ; the carrying-pole men simply stand there, and later take off the fish. If they try to enter earlier the people here are angry." The wholesalers, who are local men, resent infringement of their privileges by outsiders. This group spirit appears also when local dealers who have been arguing fiercely over prices with the fishermen support the latter when an inland buyer comes along. For instance, Pa' Che Su was bidding $16 for 4 compartments of fish in a boat, while the seller, asked to name his price, *mati* at $20. Pa' Che Su said : " How do you get $20 ? $4 a compartment," and refused to increase his bid. But then a small dealer from inland came along and asked the price for one compartment. " Six dollars," said Pa' Che Su. The man protested, but Pa' Che Su said : " $20 is the price for the whole catch." The man said : " But there are many small fish there," which Pa' Che Su denied—though previously he had been arguing the same himself ! Then the small dealer went away, and Pa' Che Su returned to the attack, saying to the seller : " I am going to take out these fish,"—" If you do, it is at $20," the seller replied. Finally the sale was made at $18.

The position of the carrying-pole men has deteriorated in the last twenty years or so with the development of the fresh-fish market by motor transport. Whereas formerly they and the fish-curers divided the catches fairly equally, now they have been thrust aside as bulk buyers by the dealers who operate on the buses. With the circumscribing of their markets, and the increased competition, their numbers have diminished, so that nowadays there are only between twenty and thirty of them usually to be seen on the Perupok beaches, instead of the hundred or more said to have operated there formerly.

In times of shortage of fish they are the dealers who are apt to suffer most, since the local people naturally tend to absorb what there is, and the prices quoted to them are high. On a number of occasions I met carrying-pole dealers who complained of the difficulty. Some days they could buy *sĕlar kuning* only at 30 cents a hundred, whereas the price at which they could sell them in the inland villages was 3 for a cent, which left no profit, and they therefore had to return empty. Their trouble here was due to differential markets. Competition from Tumpat and other areas was sufficient to keep the retail price down inland, while on the Perupok beach fish were scarce enough to make it worth while for the local retail buyers to pay a high figure. Hence as a fisherman observed of the carrying-pole dealers : " They want to buy cheaply here, but people won't give them the fish." Even when fish are more plentiful these small dealers have to be cautious, since competition inland may force them to lower retail prices there. And since they arrive at their destination late in the day their time-margin for seeking untapped markets that evening is small, and they may have to unload at a loss.

The retail market on the Perupok beach itself is the most amorphous of all. It has no unity of time or place, but is comprised of scores of separate sellers or groups of sellers, drawn from a wide variety of interests. There are wholesale dealers, clearing off a balance from their earlier purchases or taking in what cash they can before beginning their main work of gutting and curing. There are small dealers getting a turnover on a basket bought from a wholesaler. There are dealers or their wives acting as sellers for others with jobs elsewhere. There are crew men selling a part of their share of domestic fish for coffee-money. There are line fishermen selling their catch out of their boats. There are women offering a few fish bought here and there in small parcels. There are even children selling a handful that they have begged or stolen. And through and round this mass of sellers scattered up and down the beach move other dealers, carrying-pole men and bicycle men, and the buyers—housewives from homesteads this side of the river and beyond it, rice planters, fishermen who have not been at sea or who have had no luck that day, shop-keepers, old men—all haggling keenly. Prices are not fixed, and though they tend to an equilibrium there are variations due to differences in the quality of the fish, in the bargaining powers of individual retailers, and in the prices originally paid by the retailers. There is, in effect, not one retail market but a series of markets, among which advantage can be sought.

For all retail sales cash is paid, and it is this assurance of ready money, combined with the chance of getting a surplus of a few fish for the evening meal, that brings so many classes of people into the market as sellers. The smaller fish, which comprise the bulk of the sales, are quoted at so much per 100 or so many for 10 cents ; " thirty ", for instance, may mean either cents or fish. A till is made by scraping a hole in the sand, and in the confusion of traffic money is often lost. (A casual monsoon occupation of girls and children is to filter the sand to recover this treasure-trove.) A feature of this retail selling is the ability of children. They show themselves apt traders, and adult buyers, even petty dealers, purchase their fish and bargain with them gravely in the normal way. So they soon learn to acquire the keen bargaining sense characteristic of all people in these fishing communities.

A few examples will show the bargaining methods and type of calculation employed in these retail transactions.

A small dealer, in partnership with another, had bought a compartment of fish for $11. The bulk, filling two baskets, had been taken off by one partner by bus, and the remainder, about 500 fish, was being sold retail on the beach by the other, with a third man as helper. The product from the beach sale was to be counted in with that from the baskets, each bearing its proportion of the total cost. The beach fish were offered at 50 cents a hundred, at which there were no buyers. A lad then inquired the price—" 45 cents a hundred." He offered 30 cents, then after a moment the reply came " 35 cents ". He walked

off. As he got five yards away he was recalled—" Take them "
—and he bought 10 cents' worth. A new buyer then came up
and was quoted 40 cents a hundred, but went off. A woman
hailed to come and buy fish was given the price as 30 cents. She
also refused, and the seller then called to a small dealer " Hey,
hey ! Basket O ! (a short-hand term for itinerant dealer) Come
here ! " And then in general terms : " Come here, come, come,
come, come, selling fish cheap O ! " Then, as a boat was being
hauled up past him and his fish threatened to be buried in sand
or trampled on, " Fish bought with cash ! Fish bought with
cash ! " was uttered as a warning cry. The total profit from the
whole deal, taking fish sold by the basket and on the beach
together, was 18 cents to each partner, after giving 5 cents to
the helper on the beach.

This has shown the adjustment of supply price to demand.
The process is usually quick, but different prices often obtain
for a time at different parts of the beach. At five o'clock one
afternoon a dealer began to sell mackerel and the very similar
sělar gilek out of a boat-load at 6 for 10 cents. He refused to give
7 for 10 cents, though he threw in an extra fish to a woman who
bought twenty cents' worth. A little farther up the beach
mackerel were then selling at 5 for 10 cents, and *sělar gilek* from
another boat at 7 for 10 cents. All were fish of no significant
difference in size. At another boat a dealer and his wife, acting
for the boat captain, were offering mackerel at 6 for 10 cents.
They did not sell well, so near dusk the price was dropped to
13 for 20 cents to a carrying-pole man, and then to 7 for 10 cents.
At this price a crowd of buyers assembled. One woman said :
" Eh ! I bid nine " (for 10 cents). The seller replied : " What
do you mean, nine ? " and took out of her dish the 7 fish he was
about to sell her. After a moment she said : " Give them then—
seven " and the sale continued. She took 30 cents' worth, and
then added : " I ask for eight."—" Can't do it," replied the
seller—but then gave her eight for ten cents, though he picked
them out himself and did not let her do so, as she wished. Someone else then asked " How many for ten cents ? " to which the
last buyer replied " Seven " and then added in an undertone to
the seller : " Another tail still." To this he answered " enough "
and turned to someone else.

Greater variations of bargaining power are to be found in
these retail sales than in the wholesale transactions. The argument of comparative cheapness at a neighbouring site is often

used by buyers—sometimes as a trick—but is not necessarily accepted. To a dealer who was complaining that he was being asked 10 cents for a fish smaller than one he got for 8 cents from another boat a seller replied : " What will you ? There are cheap places and dear places ; cheap fish and dear fish ! " But in contrast to this stout defence some line fishermen selling their catch show bewilderment when confronted by the acute dealers. A number of them complained to me that the dealers were sharp. One man said : " We don't understand when the countryfolk say that $1.10 a hundred is 8 for 10 cents "—he was being cheated of 15 cents by this false arithmetic. Another, who sold 140 squid at $1.80 said to me : " He paid $2.20 ; is that correct ? " In this case the common mode of calculation by addition was turned to the buyer's advantage by sharp practice, thus : " 100 at $1.80, and 40 ' tails ' more are 40 cents, making $2.20 in all."

Another factor differentiating these small sales from the bargaining for the bulk catches is the concession often made to kinsfolk or friends, who are allowed to carry off the fish at a lower rate, or have an extra one or two thrown in for good measure by the seller. Inquiry about transactions cheaper than the current market rates often elicits simply : " She's a cousin " or " He's a friend "' as explanation. Normally such concessions are given only when the fish are being bought for household consumption, but occasionally a line fisherman gives a somewhat cheaper rate to a dealer who regularly comes to take his squid or Spanish mackerel.

THE MARKET FOR CURED FISH

The Perupok area supplies cooked and cured fish for both the internal market and the export market, but the former is of main importance. Shrimp paste (*bĕlachan*), pickled anchovies (*budu*), salted fish paste (*pĕlaroh*), spiced pickled fish (*ikan singan*), grilled fish (*ikan pĕrangan*) strip-cured fish (*ikan talung*, Fig. 18) and ordinary salted dried fish (*ikan kĕring*) are all prepared for sale, the last named being the major item. Some dealers handle only the dried fish, others engage in the preparation of the other types as well.

The only charge to be met by the fresh-fish dealer is the cost of transport (even the baskets used belong mostly to the bus owners). In contrast with this, the fish-curer has a range of charges. There is first the cost of gilling and gutting the fish, an operation termed *chekeit*. This job is usually done by women and girls for about 3 cents a basketful, and it is an important source of income for elderly widows or divorcees. (A basket of 1,000 to 1,200 fish is often done in an hour.) Then there is the cost of salt, of which the dealer uses several hundred gantangs a year; this formerly cost $3 or $4 per hundred gantangs, but by 1940 the price had risen to between $8 and $10. The

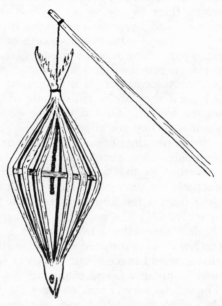

FIG. 18.—*Ikan talung :* Spanish mackerel cured by the strip-method. After the fish is cut and cleaned, a rattan hoop is placed inside to hold the strips open, and it is then hung up exposed to sun and air for several days.

salt is imported from Siam in small coasting vessels. Then there are the overheads, represented by a shed, costing about $15 in bamboos, thatch and labour ; concrete or wooden tubs or earthenware jars costing from $2 to $3 apiece, in which the fish are soaked in brine ; bamboo trays, costing $12 to $14 per hundred from the countryfolk, on which the fish are dried in the sun ; and other accessories such as huge salt baskets, small jars for holding pickled anchovies, kerosene tins for holding pickled fish, baskets for holding the wet fish and plaited bags for holding the dried fish.   The complete outfit of an ordinary fish-curer in the Perupok area normally costs between $30 and $50, apart from the running costs of gutting, salt and spices, and transport.   The labour in curing the fish is usually supplied by himself and his wife, and so does not enter as a cash item.   In fact, the largest element in the " profit " of the fish-curer is normally a return for the labour expended.

A few details of costs and returns will indicate the kind of margins on which these people work.

Fish paste, produced by pounding up fry in a mortar with salt, is sold in jars inland.   One dollar's worth of fry mixed with the same value of salt makes 8 jars of paste, each containing 10 quarts.   The jars themselves cost 25 cents apiece, and the contents are sold at 50 cents per jar if the jars are not returned, or 40 cents per jar if they are.   Thus a sale of 8 jars brings a profit of $1.20 less freight.   Grilled fish are sold in the inland markets by women.   They are clipped between sticks and sold for 2 cents or 3 cents per clip of four *sělar kuning* or two mackerel. A woman commonly takes about 200 clips in a basket, thus getting a gross sum of $4 to $5, on which a profit of about $1 is made.   Sometimes the work of grilling is given to an assistant, payment varying from a few cents per 100 fish for *sělar kuning* to 25 cents for mackerel.   The preparation of anchovies is done on a larger scale.   One curer, who bought a catch for $9, calculated as follows.   He estimated that the volume was about 3 picul, and that it would make about 20 jars of *budu*.   The cost of the salt would be about $4, and of the jars $5, making about $18 in all.   The selling price of *budu* at the time (October 1939) was $20 per *kodi* of 20 jars—it had been as high as $40 per *kodi* the previous year, when anchovies were scarce.   Thus he said, if there were 20 jars in his lot he would make about $2 profit ; if, on the other hand, he had miscalculated and there should be only 15 jars he would lose about $3.   Later he told me that the lot had made 19 jars, so he had a profit of $1.   At that time he had about 150 jars in stock, shortly to be sold.

Mackerel are preserved by various methods.   For salting them with spices in tins a dealer gave the following estimate. He had 125 fish per tin, the fish having been bought for 90 cents per 100.   The cost of a tin was 20 cents, of 2 quarts of salt 10 cents, and of the tamarind and other spices about 5 cents ; in addition there was 5 cents per tin for carriage by bus.   Thus the total cost per tin was about $1.52. He sold the fish in an inland maket, Kedai Laboh, about thirty miles away, either separately at 6 or 7 for 10 cents, or per tin at $2 to $2.50, depending on the prices ruling at the time. He thus took a net profit of 50 cents to a dollar per tin. He said that the fish would last a week or more, if the tins were not rusty at the beginning. A more common method is to put the mackerel down in jars or tins with a much larger amount of salt, when they will last for several months. Here the cost of preparing 1,000 fish was estimated at $2 for salt, $1.25 for jars, and carriage at 50 cents. The profit varies according to the initial price of the fish and the demand at the time of sale, but may be as high as $5 or so per 1,000 fish.

The ordinary process of curing by pickling in brine and then sun-drying is used for horse-mackerel, sprats, and a variety of fish taken by the seine. Here allowance must be made for a loss of weight, estimated variously by different dealers, but roughly about 50 per cent. Thus 2 picul of *sĕlar kuning*, bought for say $6, and with added costs of 50 cents for salt, 10 cents for gutting and $1 for carriage, would yield 1 picul of dried fish, worth possibly $8 or $10, giving a profit of from 40 cents to $2.40.

All the figures given here, however, are merely illustrative of general situations, since there are a number of variable factors which rule out anything in the nature of a steady level of costs and profits. These factors include variations in the initial price at which the fish are bought, in the price of salt, in the quality of the cured product, and the daily and seasonal fluctuations in the price of the product. Moreover, the amount of capital at the command of the individual dealers is important, since the ability to hold on to a stock of cured fish in anticipation of a rise in price at a season of greater scarcity may mean a very considerable increase in profits. Mackerel bought in October at say $17 per 1,000 will fetch up to $30 per 1,000 salted during the monsoon in December or January. But if the dealer is forced by shortage of capital to sell them in November the price will probably be only $20.

The dealer in cured fish finds his best market during periods of bad weather, when fresh fish are scarce. Thus whereas at the beginning of November 1939 the price of dried *sĕlar kuning* was $6 or $7 per picul, at the end of January 1940 the price was $15 per picul, and a fortnight later, after a fall, when a period of bad weather came again it rose to $17 per picul. One afternoon, for instance, a dealer said that the price of these dried fish, sold per 1,000 at his house, had been $3.50 the day before, was $4 that day, and he expected if the weather continued to prevent fishing it would rise to $5 the following day ; that was 15 fish for 10 cents retail, and he would not let them go for less. He pointed out firstly the daily rise in price owing to the bad weather, and secondly that with large capital a man could make considerable profits. But he added that most people had only small capital, and could not accumulate stocks and wait. As the clouds gather and the rains begin to come, however, the fish-curer is exposed to greater risks of loss since he depends to a large extent on sun-drying his product. Here again early buying, which means again more capital, offers advantages. In the broken weather, which indicates the approach of the monsoon, the curing of fish may be so interrupted that in the words of one dealer " they are dry enough to sell, but not to keep ". They must be disposed of quickly, at a smaller margin of profit.

It is difficult to form an estimate of the amounts of capital at command of these dealers. But some idea of its minimum extent can be seen from the quantities of fish they have drying on their trays at any given time. Representative samples for different dealers were : 16 trays, the fish having cost $20 ; 21 trays, costing $32 ; 21 trays, costing $25.60 ; 8 trays, costing $10 ; and for one of the largest dealers 68 trays, costing $60. (Each tray holds about 500 large *sělar kuning*, 8 trays making about 1 picul of dried fish.) Considering dried fish in stock, reserves of salt, and commitments in other types of cured fish, these dealers probably work on a capital of between $50 and $100, at least. It will be remembered from the previous description that a dealer rarely buys the catch of a whole boat, but shares it with others ; on the other hand, he normally buys from several boats on the same afternoon. When he buys for drying he rarely, if ever, pays cash ; he settles on Friday evening or Saturday. It is not uncommon, therefore, for a dealer who specializes in fish-curing to hold back from the cash purchases of the fresh-fish dealers, reserving his liquid resources for the week-end payments for which he has contracted. Should he lose heavily on the fresh fish and not be able to recoup his losses by " cutting ", his credit would suffer. But from time to time, especially on Fridays, he takes a trip inland with his dried fish, coming back with sums of $30 or more from his sales. His turnover is slower than that of the fresh-fish dealers, but on the

XVI MENDING A PURSE SEINE, 1963
The net is different (note the large plastic floats) but the repair technique the same as in 1940.

XVII   HAND-LINE FISHERMEN BARGAINING, 1963
With a catch of two Spanish mackerel, they discuss prices with the fish dealer.

whole, less speculative, because he can hold his durable product a little longer to suit the market. Put simply, as one dealer expressed it, he did not often sell " fresh " because those dealers often lost ; but in selling dried fish one rarely lost.

The position of the fish-curing industry may be summed up by saying that it is the stand-by of the market, waiting to absorb any balance of supply from the fishermen by virtue of being able to hold its stocks, whereas the fresh-fish dealers must turn theirs over the same day. And because the stocks must be held, and the overheads and running charges are much greater than in dealing in fresh fish, the fish-curers buy at a lower price, once, the differential demand for fresh fish has spent itself. Moreover, because their turnover is less rapid, but on the whole their return is more stable, there is little tendency for the fishermen to force them into paying cash down for what they buy.

This study is not concerned in detail with the technical aspects of fish-curing. But from the economic angle it may be pointed out that the methods employed on the east coast are capable of considerable improvement. Examination has shown that the fish are often imperfectly salted and dried, that the conditions of work often lead to the incorporation of sand in the product, and that exposure to flies in sun-drying is apt to result in some infestation by maggots. The consequent poor quality of much of the fish means low prices. One of the problems in improving the fishing industry is to devise better methods of curing or other preservation and—more difficult—to get them generally adopted.

INLAND MARKETS FOR FISH

To complete the picture of fish dealing, a brief reference must be made to the inland markets. The towns and large villages served by the fish trade from Perupok include Jelawat, Ketereh, Melor, Kedai Lalat, Salor, Pasir Mas, Kota Bharu and, less frequently, Pasir Puteh and Kuala Kerai (Fig. 11). These are also supplied by other fishing centres as well. (Kota Bharu, as the largest town in the region, receives fish from very far afield, including dried fish occasionally by boat from places such as Setiu, more than 50 miles down the coast in Besut.) In turn, these inland centres serve to some extent as dispersal points for fish to smaller communities near by.

Each of these larger inland centres has its market place, usually with a large building specially erected and maintained by the government for trading, the capital expenditure and cost of upkeep being recouped by market dues on traders, who display their wares either on stalls or on the ground. The stalls consist of benches of wood or, more often, of concrete, for the sale of dry goods ; wet fish are usually laid out on low concrete slabs. The fish traders are found as a rule in one section of the market, but the market as a whole caters for traders in all kinds of goods, including fruit and vegetables, sweetmeats, rice, cloth and medicines.

As on the beach, there is a broad division between traders who deal in fresh fish—usually men—and those who deal in cooked and cured fish—for the most part, women. Their numbers vary according to the season and the time of day, but in a large market there may be upwards of 50 in all. The traders in fresh fish, whose commodity fluctuates more rapidly in supply, are more mobile ; those in cured fish are usually to be found in regular occupation of their stalls. Both types of trader often handle other items as well as fish, such as eggs. There is of course considerable variation in the amount of capital employed by individual traders, but in general that used by the fresh fish traders, whose turnover is much more rapid, is apt to be smaller than that of the dealers in cured fish. The typical stock-in-trade of a dealer in cured fish operating in a moderately large way in the Kota Bharu market consists of, say, the following : a dozen heaps of dried fish set out on the bench for immediate trade ; half-a-dozen large baskets of the same in reserve beneath the bench ; a basin of shellfish ; a basin of turtles' eggs ; a basin of hens' eggs ; and a basin of dried tamarind. The total value of the stock may be upwards of $100, and the turnover of the basic items about once a week.

Buyers come to the market not only from the town itself but also from the hamlets round about, in many cases from as much as 5 miles, often on foot. Most of them are women, seeking food for their families, but some are petty traders, taking away fish to sell in the small village shops or stalls. Trading conditions, including the system of bargaining, are the same as those described earlier for retail selling on the beach. The prices for the fish are influenced very much by the supplies of fresh fish that come to hand each day ; the demand for dried fish in particular is apt to be sluggish or brisk accordingly.

FLUCTUATIONS IN FISH PRICES

In most of our analysis so far we have been concerned with the processes of arriving at market prices rather than with the fluctuations of such prices from time to time. We must now consider this question, first from the point of view of short-period movements, and then of movements over a longer period, though unfortunately my material on the latter is small.

In short-period price movements there are first of all the changes that take place in a single day. The influence of changing supply and demand here has been already shown particularly in the fall of wholesale prices that tends to occur as the afternoon draws on and the demand for fresh fish for the inland markets dies away. To some extent this fall is reflected in a drop in retail prices on the beach, though any very serious change is apt to be prevented by two factors. One is that in the early afternoon there are few fish available for retail sale, since most are borne off by bus ; the other is that in anticipation of freer supplies later buyers are inclined to hang back rather than compete for what there is at high prices. And a third factor which tends to check a severe fall in retail prices at the end of the day is that though fish are more plentiful then, the possibility of drying the balance takes up the slack in supply, and no frantic competition takes place among sellers to rid themselves of their stocks. As an instance from a day when fish were moderately plentiful—from the first boat in, a few mackerel were sold at 2 for 5 cents, and a few *sělar kuning* were offered, with only an occasional buyer, at 3 for 2 cents. From the fourth boat, a little later, mackerel were sold at 5 for 10 cents, and *sělar kuning* after being offered at 2 for 1 cent, were sold at 5 for 2 cents, or 3 for 1 cent if they were small fish. But from the eighth boat, later still, *sělar kuning* were sold at 2 for 1 cent to housewives, while carrying-pole men were boggling at giving the price, and the dealers who were selling were keeping back the bulk of the catch for drying. For the mackerel from this boat 3 for 10 cents was quoted ; lower prices were refused to most buyers, though to a few privileged people, other fishermen, 4 were given.

At times a shortage of fish later in the day may drive the retail price up high, especially if this shortage has not been anticipated, and early prices have been low. Two days later, the first boat came in at 11.30 a.m., and *sělar kuning* were sold from it at 3 for 1 cent, and *bĕluru* for 1 cent apiece. These were low prices, but sellers and buyers were expecting that the very early arrival of the first boat presaged a glut of fish. Contrary to expectation, later boats had poor catches, and later buyers could not get *sĕlar kuning* for 3 for 1 cent, nor for 5 for 2 cents,

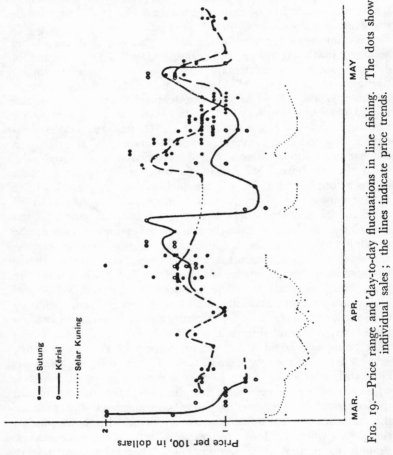

FIG. 19.—Price range and day-to-day fluctuations in line fishing. The dots show individual sales; the lines indicate price trends.

but had to buy 3 for 2 cents, and even 2 for 3 cents. *Bĕluru* were sold at 1 for 2 cents, while some small *sĕlayang* which the dealers began to sell at 20 cents per 100 had the price raised by them to 25 cents a 100, and the 2,000 or so in the heap were all gone in five minutes.

Variations in prices from day to day are apt to be more marked. These are due partly to fluctuations in supply, but also to fluctuations in demand, particularly on the part of the fresh-fish dealers. These men are influenced in their calculations by their estimates of what conditions will be like in the inland markets where they are going to sell, as determined primarily by competition from other areas. Here they use as an index partly their general knowledge of the probable effect of the day's weather on the other parts of the coast, but more particularly the prices that ruled in the inland markets the day before. A series of losses the previous day in Kota Bharu, Melor, Ketereh, etc., tends to depress the Perupok wholesale market. Though as it allows the fish-curers to get fish on more favourable terms, prices are kept more stable than might otherwise be the case.

An interesting example of day-to-day price fluctuations in a combined wholesale and retail market is given by Fig. 19, which charts the prices paid on the beach, per 100, for squid (*sutung*) and sea-bream (*kěrisi*) obtained by line fishing. The squid are bought in part by dealers for sale in the inland markets, and in part by local people for domestic consumption ; the sea-bream are bought almost wholly by the latter. For comparison the prices paid for the *sělar kuning* which are sold retail from the lift-nets are also given. The graph, which covers a period of six weeks when the squid season was in full swing, shows the marked fluctuation in the prices from day to day. On the whole, the three sets of prices tend to move together, the general ruling factor being that when *sělar kuning* are plentiful and cheap, then the prices of squid and especially sea-bream tend to fall. The fall in the price of squid is more slow than that of sea-bream since the former are more esteemed, and even if they are in considerable supply, they can be more easily absorbed by being taken to inland markets. But an example of competition in the inland markets depressing beach prices is given by comparing April 27th and 28th. On the first day squid were plentiful, and were bought at a ruling price of $1.40 per 100 by dealers. The following day the market was very dull. The men who had bought at $1.40 had been forced to sell inland at $1.20, because the supplies from other fishing areas were heavy ; one dealer, who had more capital than the others, put 600 squid in ice and returned with them rather than lose so heavily. So the Perupok dealers were offering $1.20 or less on the 28th, and the fishermen for some time refused suspiciously to accept these prices. Several of them, rather than sell at what they conceived to be an unfair figure, took their squid home to make into a delicacy known as *kětupa' sutung*. One man, offered $1.20 a hundred in the early afternoon, refused and had then to take only $1.10 later on. But in about an hour the fishermen had sold their squid at the dealers' prices. For the next few days *sělar kuning* were fairly plentiful, and the price of squid continued to drop, the lowest price being 90 cents a hundred. But then suddenly on May 4th the lift-net fish failed. So brisk was the demand for squid that the dealers were rushing out into the sea up to their waists to board the line fishermen and get the catch first, and the price ranged round $1.40 again. But squid were also scarce, since the adverse wind which hampered the lift-net fleet had stopped the squid fishers from going over towards the Perhentian islands where their grounds were. Sea-bream were then in great demand, selling at 6 for 10 cents, a very high price.

The lack of any accurate measure of the volume of fish bought prevents any close consideration of day-to-day changes in wholesale prices of most kinds of fish. But since mackerel are normally sold per thousand, an example from such sales will illustrate the fluctuations. The heaviest supply of mackerel came on the market early in May. On May 3rd one boat came in with nearly 2,000 fish, having heard the day before from line fishermen that a shoal had been seen. They were sold at $25 per 1,000 since they were the first catch. The next night many boats went out, and got large catches. The first boat in sold 2,575 fish at $22 per 1,000, a rate obtained also by others, but later arrivals had to be content with $20 per 1,000. The next morning, May 5th, catches were also very large, and were sold at rates ranging between $16 and $10 per 1,000, though most boats sold at $12 per 1,000. This day the market was flooded, and while a large quantity of fish was taken to the inland markets, many of the later lots were salted down. Three boats from Kubang Golok came in to Perupok to sell, the former village having run out of salt.

In Kota Bharu this day the price of mackerel, at the rate at which the retailers took them from the wholesalers, was $1 per 100 or even 80 cents per 100 in some cases. The retail price there started off at 6 for 10 cents, then fell to 7, 8, 10 and even 12 for 10 cents. But the early sales were few, since the buyers, seeing fish come in in quantity by six o'clock in the morning, knew that they must be plentiful, and hung back. Hence the dealers who bought on the Perupok beach and went to Kota Bharu and other markets to sell lost, in some cases heavily.

One dealer estimated that the price on the Perupok beach the following day would be only $8 or $9 per 1,000. He said that dealers who salted the fish could afford to pay $10 per 1,000, or even up to $15 per 1,000 ordinarily—" they can't lose, since in the monsoon salted mackerel fetch 10 cents for 3 fish ". But this demanded capital for salt, etc., and a long wait, and only about half a dozen dealers could manage it. He stated also that the low price of mackerel at the present time was due to the long period still to go before the monsoon—about seven months. The fish do not go bad in the meantime, but they turn black, and are less liked, so fetch a lower price. In October the price of fresh mackerel cannot fall so much as now since there is a stronger demand for them for salting. I asked why Kubang Golok had run out of salt. The answer was because it was the opening of the season. When the first boat from Siam arrived a few days before, the dealers there did not buy much salt—they were hoping that more boats would soon come, and that the price would fall.

On the following morning catches were also very large, and though the price did not fall to my informant's estimate, the rate was $10 or $11 per 1,000. The next morning catches were somewhat smaller, and the rate varied from $10 to $14 per 1,000. One dealer who took a lot at $1 per 100 sold them in Ketereh for $1.50 and $1.60 per 100. The next morning prices were about the same, though the same dealer, buying at $1.40 per 100, was able to sell in Ketereh at only the same figure, thus losing his transport charges. The next morning still the beach price had risen slightly, fish having become much scarcer, though they were no higher than $1.50 per 100 in the Ketereh market. And on the 10th May, the last day on which there were any catches, they bought from $13 to $15 on the beach at Perupok, and $1.50 at Ketereh, while at Salor, further inland, a dealer who bought on the beach at $15 managed to sell at $20 per 1,000.

This example, besides showing the rapid fluctuations in the day-to-day prices, indicates also the close relation of the fish-curing market to the fresh-fish market, and the effect of long-period calculations on wholesale prices.

The impossibility of obtaining records of the precise volumes of fish sold forbids the analysis of long-period fluctuations in

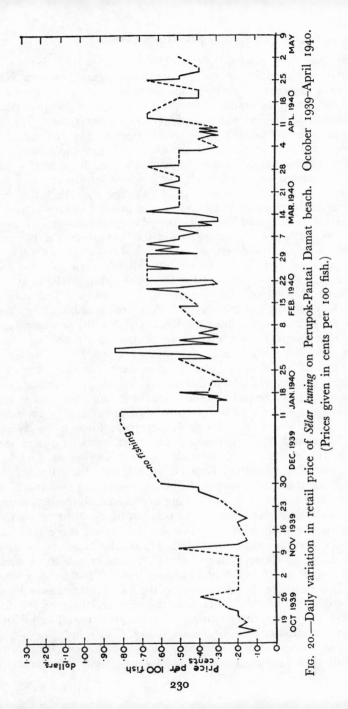

Fig. 20.—Daily variation in retail price of *Sĕlar kuning* on Perupok-Pantai Damat beach. October 1939–April 1940. (Prices given in cents per 100 fish.)

230

wholesale prices. For retail prices, however, some material may be presented. In particular, I recorded the retail prices at which *sĕlar kuning*, the commonest fish, were sold on the Perupok beach during a period of rather more than seven months, 1939–40. The results are given in Fig. 20. Apart from showing the sensitivity of the market from day to day, the graph shows the very considerable rise that took place on the approach of the monsoon, the gap in sales during the monsoon, the fall immediately the new season began again in January, and a rising trend, with sharp fluctuations, in the succeeding months. Owing to 1940 being a bad fishing season the prices in March and April were probably higher than normal, but a rise in price towards the middle of the year must be a common phenomenon because of the seasonal drop in supply as the fish retire for breeding ; the new shoals contain mostly very small fish.

In order to be able to compare retail prices on the beach with those in other centres inland I was supplied, through the courtesy of the British Adviser to Kelantan, with figures taken once a week by local officials in the markets of Kota Bharu and Kuala Kerai. Though these records were irregular and not always complete, and the figures from Kuala Kerai displayed undue steadiness, comparison of the results with my Perupok figures showed similar market trends over the six-monthly period. Comparison of prices in different markets on the same days, however, shows a divergence much wider than can be explained by differences in transport and any regular system of middlemen's charges. It is clear that owing to the incalculable vagaries of production in the various fishing areas, from Semerak to Tumpat, which serve the Kota Bharu market, retail prices there bear no constant relation to the wholesale or retail prices paid on any individual beach. This bears out what has already been seen from other angles, that the profession of wholesale fish dealer is apt to be a speculative one, and that owing to the keen competition of the dealers, there is no systematic exploitation of the consumer either on the beaches or in the inland markets of Kelantan.

The following figures are illuminating from this point of view. On April 8th *sĕlar kuning* were only moderately plentiful at Perupok. They were being bought wholesale at rates from $3 per 1,000 up, some for selling inland, some for drying, and some for disposal retail on the spot. Those which were sold retail made a slight profit, and those taken for drying probably also fetched a profit, if they were kept for about a month, when the price of dried fish had risen. Those taken inland went to Melor, Ketereh, Peringat and Kota Bharu. At Melor, *sělar kuning* began to be sold at 15 for 10 cents, and later the price dropped to 20 for 10 cents. But at the other three markets the price was 40 for 10 cents, because fish from Tumpat had been taken there. The one dealer who went to Melor therefore made a profit; the others lost heavily.

PROFITS AND LOSSES OF MIDDLEMEN

It is a common accusation that middlemen reap the benefits of an industry, exploiting either the primary producers or the consumers or both. This is quite likely true of the Chinese fish dealers in the west and south of Malaya, but does not seem to be so with the Malay dealers in Kelantan. It is difficult, however, to make any general estimate of their returns.

That the Kelantan middleman is not in a peculiarly favoured position for reaping an advantage from the fishing industry may be gathered from the fact that of 35 transactions in fresh fish which I recorded quite at random during my stay, in 17 a profit was made and in 18 the dealers lost money. This does not mean that they were necessarily out of pocket in every one of the latter cases, since the practice of cutting probably allowed them to shed their losses or reduce them on some of the transactions. In some cases, however, where cash had been paid down, this was not possible. Profits varied from 10 cents on retail transactions to several dollars on wholesale ones, with a general average of about one dollar. Losses varied from 10 cents to $5, with a general average of about $1.75. For a total of 10 cases the profit was $10.60 on a capital outlay of $94.70, an average rate of about 10 per cent.

It is obvious that the majority of the middlemen must make a living, so the proportion of losses in the above cases cannot be taken as symptomatic of the situation as a whole. The fact that on a casual record I observed so many losses, however, shows the set-backs to which the middleman is subject in dealing in fresh fish. In dealing in cured fish, for the reasons already stated, the profits are more regular, and probably on a higher scale. One of the most lucrative transactions I noted was the purchase of 300 large catfish for $9; the cost of salt in curing them was $1 per 100, and they were sold for 10 cents apiece, thus yielding a profit of $18 on a capital outlay of $12. But this was very exceptional.

Some idea of the profits of a well-established dealer trading in both fresh and dried fish may be obtained from the partial record of one man's activities during a month. For 16 days of fish-buying during this period his total capital outlay was $136, and on 6 days he did not buy, either because it was a holiday or because he went inland selling his fish. On a group of transactions fully recorded he made a profit of $12 on an outlay of $42.30, and a loss of $1 on a $7 outlay, thus getting a net profit of $11 on an outlay of $49.30. This represented about 11 days' work. On this basis, his income from fish dealing may well be in the region of $7 or so a week. That of smaller dealers would seem to be in the region of $2 to $3 a week, except during the monsoon, when if they deal only in fresh fish, their income drops to perhaps only a $1 a week.

SUMMARY

The system of marketing fish is so complex that it is advisable to emphasize some of its basic features.

The market is a diverse one, with middlemen of several types, each performing a specific function, though not exclusively. Moreover, the system is so fluid that, as there is prospect of profit, people who are normally producers or consumers act for a time as middlemen ; they are not regarded as infringing the privileges of any corporation. Marketing of the fish does not take place on any system of fixed or conventional prices, such as might be expected in a community which attaches great importance to traditional forms of social intercourse, and in which the members are, on the whole, closely interrelated by kinship and other bonds. There are definite conventions, but they apply to the process of bargaining, which is highly developed, and not to the prices as such ; these are characterized by great sensitivity of response to changes in supply and demand, including buyers' and sellers' forecasts. (This haggling in the marketing of goods is in contrast to the procedure in the marketing of services, in which custom regulates the remuneration, and where changes in prices are apt to be very slow.)

Producers sell independently, but wholesale buyers form loose combinations, the composition of which varies from one transaction to another. These combinations inhibit direct competition between buyers, through the system of representative bidding and sharing of the goods. But indirect competition emerges through differential estimation of the profits to be got, with consequent maintenance of bidding or withdrawal from it. The semi-combination of buyers cannot force down prices to the producer too far by reason of the possible entry of the producer himself into the market as a middleman. Moreover, competition between dealers when they sell tends to avoid exploitation of consumers.

While retail sales are for cash, the credit system is well developed for wholesale purchases. And through the process of " cutting " the price, or the possibility of doing so, the risks and the gains of the middlemen tend to be spread over the body of producers as well. The capital of the middlemen is small, and they tend to operate on narrow margins.

The result is a closely-locked system. But its economic organization depends largely on production in other fishing areas, and growing facilities of transport and communication tend also to increase this interdependence of the different local markets.

CHAPTER VIII

THE SYSTEM OF DISTRIBUTING EARNINGS

The system of distributing the earnings from fishing is probably the least interesting to Europeans of all the economic aspects of the industry on the east coast of Malaya. The principles are simple, but they are enmeshed in so much detail that the observer is apt to get bored with the attempt to understand them. Yet it is essential to grasp them in order to be able to understand what are the income levels of the fishermen and what are the relations between those who put in capital and those who provide only labour.

At the outset, one sees many possible ways in which, in theory, the earnings might be divided. It is clear at once that no factor in production gets a constant return—that, for instance, there are no steady interest or wage payments ; each factor has to rely on getting a share of a highly variable yield. But how that share is determined, and whether any factor has preferential treatment is at first no easy matter to find out. For large coöperative undertakings such as lift-net fishing one can imagine possible distributions such as the following : that when fish are few the boats and net, as fixed capital, have the first call on the earnings and the crew share only when the catch is large ; that boats and net receive a share every day, while the crew gets shares once a week ; that when the crew is a regular one the division of earnings takes place weekly and when they are casual labour it takes place daily. These were in fact some of the possibilities suggested to me by the imperfect records of the early stages of my work. Each had some basis, but none were correct. Then I was trying to disentangle the underlying principles from state-ments which each contained some relevant point but which could not be comprehensive, and which often seemed to be in conflict with one another. Their meaning had to be realized slowly by watching the system at work, becoming familiar with the technical terms used and seeing the particular conditions to which each statement applied. These difficulties, which are common to most studies of such peasant economic systems, are noted here because unless they are understood the temptation is to take a brief statement as a near enough approximation to the truth, and to ignore significant variations or special provisions as being just a mass of irrelevant detail.

In general, the system of distribution of earnings from all types of coöperative fishing on the east coast is one of fractional division, the owner of the major capital, net or boat, being the person responsible for the actual allotment of the money. The shares vary according to the type of fishing but in most the net, as the most important capital item, receives the largest share and is the pivot of the distributive system. This is embodied in the laconic statements of principle which the fishermen ordinarily give when asked. To a question " How does the division of earnings in lift-net fishing go ? " the answer commonly given is simply " It divides by three " (*bagi tiga*). This means that in the major sharing-out the net takes one-third, and the crew, etc., get the remainder. Similarly, with the deep gill-net—" *bagi dua* ", the net takes half.[1]

But such an answer, while a good guide to what happens once one knows the system well and wants merely local variations, is no clue to the complexities for the uninitiated. It leaves unstated all the special allowances for the wide range of functions and services which are taken into account. Moreover, different districts are apt to have important local variations of practice.

To elucidate the system the scheme of division for lift-net fishing in the Perupok area may be first taken as an example, and details for other types of net-fishing in the area given later. A brief account of local variations elsewhere on the Kelantan and Trengganu coast is given in Appendix IV.

SHARING OUT THE EARNINGS FROM THE LIFT-NET

The week's earnings of a net are commonly divided on Friday night, the evening of the Muslim sabbath, on which day the lift-nets do not go to sea. The choice of this day is guided partly by the fact that it gives an opportunity to the sellers of the fish to collect the cash from the dealers who have bought it. If they do not pay at least a considerable part of their debts no division can take place, and even at dusk it is sometimes uncertain if the dealers will redeem their promises. At times it happens that the division has to be postponed for a day or more.

The division usually takes place in the house of the *juru sĕlam*.

---

[1] In Kelantan, pronounced as *bagi tigo* and *bagi duo* ; in Trengganu, as *bagi tigær* and *bagi duær*.

A lamp stands in the centre of the floor, and all the crew assemble, often with wives and children. Betel and smoking materials are provided by the *juru sělam* as a rule, as with any ordinary reception of visitors. The *juru sělam* sits facing the crowd, with his *pěraih laut* near him, and probably the wife of one of them with a clay bowl containing a mass of small change, from which she counts out piles of coins. Beside him the *juru sělam* probably has a couple of tins, into which, as perhaps into a pocket of his jacket, he puts certain shares which are to be sub-divided later.

Each day's takings are shared separately, since the number of men engaged is apt to vary.

First, one-tenth of the day's takings is set aside for the *unjang*, the constructions of coco-nut fronds essential in this type of fishing. The remaining nine-tenths are then divided into two equal portions. One goes to the *pěraih laut*, the carrier of the fish to shore. He sets aside one-tenth of this as the share for his boat, and divides the remainder among the crews of all the boats, including himself, but excluding the *juru sělam* and any other men who are partners with the *juru sělam* in the net by the arrangement described earlier. The division is made equally, without respect of age or skill, but only those men who participated in the fishing on that particular day may share. The second portion is divided by the *juru sělam*, and is known as the *bagian pukat* as distinct from the *bagian pěraih*, the carrier agent's sharing out just described. From this sum the *juru sělam* first deducts one-third as the net's share. This is known as the *bagian dalam*, the " share within ". The remaining two-thirds he divides among the crew at large in the same way as the *pěraih* did, but this time the *pěraih* is excluded and the *juru sělam* and his partners in the net are included. Moreover, certain extra shares come to be reckoned. Each of the five boats which engaged in the work of the net gets a share here, frequently the same as that of a man, but sometimes a multiple of a man's share, according to rules which will be described later. And the *juru sělam* assigns to himself and normally to the captain of the boat which actually carries the net an extra share apiece for the special work they have had to do. Thus in this section of the division the *juru sělam* gets at least three shares—one or more shares for his boat, one share for his " body " (i.e. as a crew-member) and one for his " diving " (i.e. for his expert functions in locating the shoal of fish and directing the work).

In the procedure of apportionment these Malays work by a

method of successive fractionating rather than by one of immediate division of the total. Thus, if it is a question of taking out a tenth share of $100 and sub-dividing into two-third and one-third portions for " parent *unjang* " and " child *unjang* " respectively, the method is as follows. The $10 is set aside, and then $6 are put out for the former and $3 for the latter ; the remainder of $1 is then similarly split up, probably first into 60 cents and 30 cents, with the 10 cents final remainder split likewise and the odd cent thrown into one or the other pile. When the initial sum is not a figure easily divisible at sight, a rough trial and error method is followed. Thus at one division I saw, a large sum had to be divided among many men. A number of piles each with eight ten-cent pieces were first set out, but on counting it was found that there was not one pile for each man. " Take off ten cents," said the *juru sĕlam*, so the piles were reduced to 70 cents each and the cash thus available made up into fresh piles, which were then enough to go round.

The *juru sĕlam* and his *pĕraih laut* do the calculation, perhaps helped by their wives, and the crew do not seem to follow every step, or indeed to attempt to follow it. They accept what is pushed over to them, and though the *juru sĕlam* may say " Count it at once " and later ask " Is it correct ? " they are concerned with the accuracy of the amount they receive in regard to what has been called out rather than with the mathematics of the division as such. A man often does not bother to check the sum himself, but leaves his wife to count it for him and take charge of it. But the public nature of the distribution and the mechanics of the calculation, with the actual coins spread out, do allow of a check on any cheating on the part of the *juru sĕlam*.

The large number of shares and the system of fractionating division used mean that a great amount of small change is necessary. To some extent this is met by " making change " from men who, having accumulated several days' earnings, are in a position to exchange their coins for notes. The system takes much time. At one distribution I attended, the division of a total of about $250 took about three hours, but much longer is often necessary. Awang-Yoh, who said of himself : " I am not clever," stated that if he got $200 or so to divide it took him from 7 p.m. till after midnight, and that there were other *juru sĕlam* like him.

In the process of division there are usually small residues of a few cents left over. These are either thrown into someone's share or given to the children, partly as a recompense for their work in laying down skids for the boats, and partly as a pure gift.

This system of distribution is complicated by several sub-divisions of the shares mentioned, by occasional special increments for particular services, and by the fact that there is no complete uniformity among the *juru sělam* on the details of division. Though the main principle is always the same the specific fractions of the total allotted to items such as net and boats vary considerably. The reason for this variation is that the final voice in the distribution is that of the *juru sělam* (normally the net-owner) ; he is the " head " of the productive unit, and to a large extent exercises his own discretion on details. The members of his crew accept his methods ; if they are fundamentally dis-satisfied their remedy is to leave him. One *juru sělam* said to me " There is no custom about it, there are no rules about it ; it is simply at the inclination of the *juru sělam*." This is an exaggera-tion, since it ignores the general principle that runs through the system, and also the regulating factor of the crew's opinion, but it does emphasize the flexibility which is a notable characteristic.

The variations of the system and extensions of it may be now examined in detail by considering the shares in turn.

Share of the unjang. This is known as the *chabu' kěpalo unjang*, the share put aside for the *unjang* head. Normally one-tenth of the total of each day's yield, it is sub-divided into thirds. One-third is allotted to the " child *unjang* " taken out daily in the boat of the *juru sělam* ; since this is made by him he gets the takings therefrom. Two-thirds are allotted to the " parent *unjang* ", the semi-permanent lures on the fishing banks. The " parent *unjang* " from which catches of fish are taken on any day may have been constructed by the *juru sělam*, by his *pěraih laut*, or by a member of his crew, or may be the property of someone in another net-group. Whoever the owner is he should receive the allotted share, though, as explained in Chapter IV, should he live at a distance it may not reach him. Should the day's catch of fish have been secured from more than one *unjang* the allotted share is divided between the owners on a basis roughly proportionate to the amount of fish from each. Since " parent *unjang* " may be made by any members of the crew who are industrious, by implication this is one method whereby an ordinary crew-man may secure an extra increment to his regular income. But since a small initial capital is necessary to make one, this acts as a bar, and the majority of the " parent *unjang* " are the property of *juru sĕlam* and *pĕraih laut*.

Some *juru sĕlam* depart from the normal practice of the tenth share for the *unjang*, though all appear to adhere to the sub-division of it into thirds.   One man, a *pĕraih laut*, described how should the fish sell for $30, the *juru sĕlam* first halved the cash with him, and then each took one dollar out of his part, making a total of one-fifteenth for the *unjang*.   Another man, an old *juru sĕlam*, stated that a tenth share was taken out of the total for the " parent *unjang* " and then an additional five cents in every dollar of the total for the " child *unjang* ", giving three-twentieths of the whole as the combined fraction.   This, however, may not be a modern practice ; I did not hear of it elsewhere.

Share of the net.   While all *juru sĕlam* in the Perupok area, after setting aside the share for the *unjang*, split the distribution into two sections and take the net's share from that which they themselves apportion, the precise fraction allotted to the net varies.   The commonest practice is that of *bagi tigo*, the net getting one-third of the *juru sĕlam's* section and the crew two-thirds.   But some *juru sĕlam bagi duo*, giving half to the net and half to the crew.   Occasionally the practice of *bagi limo* is followed, the net getting two-fifths and the crew three-fifths. Apparently the same *juru sĕlam* may vary his scheme according to circumstances.   Awang Lung said that if he has only a little cash to divide, he takes one-third for the net and gives the remainder to the crew, but if it is a large catch he may *bagi limo* or *bagi duo*.

In recent years, however, some of the Perupok *juru sĕlam* have taken to allotting a half share to the net in their section of the division as a regular practice, on the plea that with the entry of the *pĕraih laut* into the scheme of distribution the third was too low.   The Pantai Damat *juru sĕlam*, however, still keep to the older convention, which naturally is more satisfactory to the crew.   In Paya Mengkuang one net has adopted the new scheme, and since here the only partners are a man and his son, and the takings are small, the crew do not like it.   In the case of the Perupok nets, where the takings are high, the crew do not mind so much.   In this case the higher share for the net is, in effect, a bonus for the successful *juru sĕlam* and his partners.

A further complication arises especially near the beginning of the career of a new net, when the expense of ropes is felt by the net-owner.   Some of these men attempt to recoup this outlay by taking a rather larger allowance for the net than normal, not by subterfuge, but on this plea. Since the actual division lies at the discretion of the *juru sělam* no objection is normally raised, though, if too dissatisfied, the crew tends to melt away later. When I discussed the division of a week's takings of net L with one of the partners, he said that the net's share was $15. I asked why it was so much, considering that the total takings were only $86, and not $100 as they would have had to be normally. He replied : " The *juru sělam* allotted a little more because the expenses were solid "—adding that there were ropes and other accessories not yet paid for. Then he said : " It comes like that because the men of the net couldn't otherwise bear up." He pointed out that there were only the two of them as partners, and that the crew had no special expenses ; there were only two of them since the net was not a very profitable one—the Perupok nets had half a dozen men always willing to enter the partnership. Later I raised this point with Awang Lung. His view was that on no account should the cost of ropes and other accessories come out of the general division ; they should be borne directly by the net's share, even though the takings were small. His own practice, which I checked by observation and which I think was general, was to take the cost of the accessories out of the net's share only, but to give them precedence in reckoning them for accounting purposes for the partnership, apart from the return of capital on the net. When in March 1940 Awang Lung had accumulated $107 as his net's gross share, he had incurred for the cost of anchor rope, hauling ropes, drying rack, dye and oil-drum (for dyeing) a sum of $30—an amount lower than normal since he had some rope left over from his previous net. So far, then, only $77 was " return of capital " on the net, and this was the sum noted by the partners as so much advance towards the time when they would begin to share in the profits.

Most *juru sělam* inform their partners at each division as to the state of the net's finances. Others less scrupulous or more careless may omit to do so. In the case of net L already mentioned, one partner told me early in the season that he did not know what was the net's share to date ; he said that till then it had been used to pay off the cost of net ropes, etc. He added that he still did not know exactly what he had put into the net. This surprising state of affairs was due to the fact that his wife, who was the dominating power in their household, was the sister of the other partner, and had made all arrangements but had not yet informed her husband of the full details.

Shares of boats. A principle commonly enunciated for the division of the proceeds from lift-net fishing is that a boat gets the same share as a man. But in practice the tendency is for boats to receive a larger share. This is admittedly so in the case of the boat of the carrier agent, who takes specifically for his boat one-tenth of the sum he divides. This seems to be a rule without exception.

In the division by the *juru sělam* the boat of the agent does not enter. Of the other boats which actually handle the net, four or five as the case may be, all may get a single share except the boat which carries the net to and from the fishing ground ; this gets two shares. One explanation for this preference was that formerly the net boat got only a single share too, but now it was double because that particular boat was bought for a high price. It was added that the double share would continue for about three weeks and then would revert to a single share. But this explanation cannot be a general one, and goes against the normal principle that every boat of the same function gets an equal share irrespective of its cost and value. A more feasible explanation was provided by another man, who said that the captain of the net boat " worked solidly on the net ". In another scheme of distribution the boat shares are much higher : the two *pěrahu atas haruh* and the *pěrahu bawah haruh* get two shares apiece, the *pěrahu pukat*, the net boat, gets three shares, and the *pěrahu sampan*, the boat of the *juru sělam*, gets four shares. This, however, is unusual.

The most common scheme, as I was told by several experts and other fishermen, is to allow two shares for each boat except the net boat, which receives three shares. Thus Awang-Yoh, who had four boats to consider, apart from that of the *pěraih*, which is allowed for separately, said : " If my crew are thirty, there are 40 shares." Nine of these go to the boats as described, and one is for himself as *juru sělam*. (He himself is, of course, included in the crew and so gets a share for his " body " in the ordinary way.)

In the normal course the boat's share goes to its owner, who is its working captain. But a sub-division is necessary when a boat is taken to sea by a man other than its owner. The return for the loan of this form of capital has been discussed earlier, but two examples will show how it affects the boat user.

In one case a *pĕraih* using the boat of another man works on a half-share basis ; if in the division the boat received $3 the user retains $1.50 and hands over the other $1.50 to the owner. In addition, of course, the user gets his share as a " body " and keeps it. In the other case, also that of a *pĕraih*, the boat's share is divided on the basis of one-third to the user and two-thirds to the owner. The effect is that such a man having little or no capital of his own gets an ordinary income for his labour, plus an increment for his leadership and responsibility as captain of the borrowed boat. The precise allotment between captain and boat-owner is a matter for private arrangement ; there is no definite rule.

Shares of the crew. Unlike the other elements in the distribution the shares of the crew are allotted in two separate acts of division, by *pĕraih* and by *juru sĕlam*. In the share-out by the *pĕraih* each man receives one share throughout ; in that of the *juru sĕlam* all the ordinary crew get one share apiece, but the *juru sĕlam* and usually the captain of the net boat get an extra share each as already mentioned, as reward for their special functions. In each section of the division the amount per man is calculated, and then the captain of each boat is asked how many men he had in the boat on that particular day. As a colloquialism among these fishermen the classifying term *ekor* (tail), properly used for animals and fish, is often employed here for men. Thus one hears : " The boat of So-and-so ; how many tails ? " or " How much per tail ? " The sums calculated are handed over to the boat captain, who is responsible for giving them to his crew. In the share-out by the *pĕraih* those men who have entered the net-combine of the *juru sĕlam* as partners in the profits are omitted. So such expressions are heard as " Five becomes four ", meaning that though the boat has a crew of five, one of them as a net-partner is dropped at this point.

Special increment. The process of division so far described relates to the apportionment of the yield from a single boatload of fish each day. But it has already been mentioned that when more than one boat in a net takes part in the carriage of fish, a boat which is not the regular carrier receives a special share for its work. There are three types of such shares.

(i) *Ikan buritang* or *ikan charu'*, known commonly as *ikan luan* and *ikan bĕlakang* according to position in the boat.[1]

---

[1] Cf. the standard Malay terms : *buritan* and *bĕlakang*, the stern of a boat ; *luan*, the prow or forepeak of a boat ; *charum*, a contribution or instalment ; *gandoh*, to

This arises as follows : The net is cast, a catch is made, and the boat of the carrier agent goes ashore to sell it. If afterwards another cast is made and a catch is obtained it must be carried by one of the other boats. If small, half is sold, the proceeds to be divided among the whole net group. The other half, at the rear of the boat, is at the disposal of the crew of that boat, to be used as food or sold as they wish for their own benefit. Being at the rear of the boat this portion of the fish is known as the *ikan bĕlakang*. If the catch is a large one, filling the whole of the boat, then all in the compartments fore of the mast, known as *luan*, is the property of the crew, and the bulk of the fish are sold in the ordinary way. If, then, the next day or soon after another second cast obtains fish likewise, another boat will take the fish—they take turns, thus spreading the profit. It is the boat which is deputed to take in the bag of the net each time that gets the catch. This taking of *ikan bĕlakang* or *ikan luan* is a matter of custom—Awang Lung described it as " adat ", " hukum " or " undang-undang ". He said that the *juru sĕlam* does not give the crew this bonus of fish ; they take it as by right. Sometimes more than one of the ordinary boats will have fish to carry if the catch is an extra large one, and each boat will take its bonus. Only very rarely, however, is the catch large enough to cause the net boat or the *pĕrahu sampan* to handle fish. The one is burdened with the net, and the other has the coco-nut fronds of the *unjang* on deck, so it is not convenient for them to take the catch. If no provision were made, they would then be at a disadvantage as compared with the other three boats. Hence they are brought into the scheme by special adjustments.

(ii) *Ikan gandoh.* When fish are plentiful, the *juru sĕlam* may hand over one or two compartments full to the net-boat— " because they have worked solidly ". This they sell and divide among themselves, so getting a bonus which is roughly equivalent to the foregoing. They get this about one time in three that there is a catch from which *ikan buritang* has been got. The net-boat's bonus is the *ikan gandoh*.

(iii) *Kayoh unjang* or *duit kayoh sampan.* The crew of the *pĕrahu sampan*, the boat of the *juru sĕlam* himself, rarely get a lot of fish to sell for themselves—perhaps only once a year, said Awang Lung. Their bonus is given them in cash, and is cal-

supplement, to make up the difference ; *tambah* (Minangkubau, *tamboh*), to supplement (Wilkinson, *Dictionary*). The Kelantan dialect has assigned these terms specific meanings and in most cases changed the pronunciation.

culated by reference to the *ikan gandoh*. When this latter is given it means that each of the other boats has a bonus, matched now by that of the net-boat. So at the weekly division the *juru sělam*, knowing that he gave *ikan gandoh* to the net-boat on a certain day, asks what it sold for, and takes out of the total cash an equal sum. Thus in the division of Awang Lung on February 9th, 1940, he ascertained that on the previous Sunday the net-boat had sold their *ikan gandoh* for $5. He then took this amount out of the general takings before any other apportionment began, and divided it later among his crew—taking one share himself as a worker. Even here there may be a further extension. On the Thursday *ikan gandoh* had also been given. On asking what it had sold for, Awang Lung received the answer $3.60. He then handed over to the captain of the net-boat for himself and crew an extra dollar. This was a *tamboh*, or additional gift. It is made with the idea of bringing up the bonus of the net-boat roughly to the level of that received by other boats from the sale of their *ikan buritang*. As such it also automatically increases the bonus of the *kayoh sampan*, which is based on the *ikan gandoh*. So on this occasion Awang Lung took $6 out of the takings for Thursday, to be divided among the 6 crew-men (including himself) of the *pěrahu sampan* on that day. Of this amount $4.60 was the equivalent of the *ikan gandoh* plus *tamboh*, and the rest he took " because they are poor men—I have six men ". He meant that his crew of six would otherwise get less per man than the smaller crew of the net-boat.

It will be clear that these extensions and modifications of the general system do make provision for special functions and extra work, and tend to level up individual labour and individual income. The *duit kayoh sampan* is so called because in working the net the crew of the *pěrahu sampan* have much paddling to do in and out of the net while the crews of the other boats " sit comfortably ". The former " have it a bit more solidly ". But it is also obvious that these special allotments give the *juru sělam* a chance of bumping up his own income in subtle ways.

It may be mentioned that the justification for the *tamboh* to the *ikan gandoh* was held by Awang Lung to be the fact that this fish bonus is given by the *juru sělam* to the net-boat, whereas the fish of the other boats, the *ikan buritang*, are simply taken by them. " I do not know," said he, meaning that he has no control (directly) over what they get.

It should be pointed out also that when there are *ikan luan* in the boat of the *pĕraih laut* they are the property of the *juru sĕlam* for ordinary division. They are not sold for the benefit of the boat's crew since it is their job to carry fish ; they do not work the net as the other boats do, but merely lie off and wait for the catch.

VALUE OF THE SHARES

So far we have analysed the principles of distribution of the yield from lift-net fishing and the processes of sharing at the various stages. Now we have to consider what these amount to in receipts for the different factors of production.

The position may first be summed up by measuring the proportionate returns to the various factors and agents concerned. In a normal distribution, after allowing for any deduction that may have been made for the insurance function of the " catcher " (see earlier), and assuming a full assembly of labour power each day, the aggregate percentage of the total secured by the various elements is approximately as follows :

		per cent.
Unjang, " parent " (2/3 of 1/10)		
" child " (1/3 of 1/10)		10
Net (1/3 of 1/2 of 9/10)		15
Boats, *pĕraih* (1/10 of 1/2 of 9/10)		
other (9/40 of 2/3 of 1/2 of 9/10)		11·25
Juru sĕlam (2/40 of 2/3 of 1/2 of 9/10)		1·5
Crew (9/10 of 1/2 of 9/10) *plus*		
(29/40 of 2/3 of 1/2 of 9/10)		62·25

This assumes that the net takes only one-third in the share-out by the *juru sĕlam* ; should it receive one-half, then its takings will be increased by $7\frac{1}{2}$ per cent. and those of the crew will bear the brunt, being reduced by about $5\frac{1}{2}$ per cent. *in toto*.

In general terms, if we include the *unjang* as items of capital, the scheme of distribution gives to the total capital employed a share roughly aggregating between one-third and three-sevenths of the whole yield, and to the total labour and skill (including " wages of management ") from four-sevenths to two-thirds of the whole.

Considering the scheme from the point of view of the different categories of personnel engaged, on an individual basis, the shares are as follows. The *juru sĕlam*, if the net-owner, receives between 20 per cent. and 35 per cent. of the takings, depending upon whether he owns the " parent " *unjang* used, and whether he divides by one-third or by one-half for his net. The *pĕraih laut*, the carrier agent, gets about 6 per cent. of the takings ; a boat owner who is also a crew man and not a partner in the net gets about 3 per cent., and an ordinary crew man about 2 per cent. These figures are, of course, approximate, since they depend upon several variable factors.

As between individual crew-members, for instance, even with equality of total returns at the end of a week between different nets, there may be considerable variation. Different nets have crews of different size ; because of illness or other reasons the number of men in the crew may vary from day to day ; *juru sĕlam* vary in the proportions they allot to nets and boats ; the number of net-partners abstaining from a full return for their labour differs with different nets ; special increments for extra loads of fish may give higher returns to the members of some boats' crews than to others.

This variation in returns to crew-men may be exemplified by considering a few cases of the actual sums received in specific distributions.

On February 9th, 1940, Awang Lung divided the week's takings, amounting to $257. The shares allotted to the crew emerged per man as follows :

Day.			From share-out by *juru sĕlam.*	From share-out by *pĕraih laut.*	Total.
Saturday	.	.	. 20 cents per man	50 cents	70 cents
Sunday	.	.	. 48 „ „ „	$1.20 „	$1.68 „
Monday	.	.	. 27 „ „ „	60 „	87 „
Tuesday	.	.	. — (no bulk sale)	— „	nil
Wednesday	.	.	. 20 cents per man	40 „	60 „
Thursday	.	.	. 30 „ „ „	70 „	$1.00 „
Total	.	.	$1.45	$3.40	$4.85

It will be noted that the proportion between the amounts allotted in the share-out of the *juru sĕlam* and of the *pĕraih* varied from one day to another.

By no means all the crew received the full total of $4.85 ; one man received as little as $2.40. The crew of the boat of the *juru sĕlam* himself, however, each received an additional $2 from the *duit kayoh sampan*, the special increment recorded earlier. Each boat owner who was a net-partner received $4.50, approximately equivalent to three shares in the division by the *juru sĕlam*. But in addition the *juru sĕlam* handed over to these men at the end about $2.50 to be divided among them as a kind of encouragement or bonus. This act of his, not governed by custom, was received with a laugh and exclamations of pleasure ; it illustrates how the *juru sĕlam* exercises his initiative in the distribution. The share allotted by him to the net on this occasion was $50, representing nearly 20 per cent. of the total takings ; this sum, as often happens, was chosen as a convenient round figure.

It is interesting to compare the rate of return gained by an ordinary crew man in different cases. The following few examples are fairly representative :

Net	Week's Total. $	Crew-man's Share. $	Percentage of Total. %
1. Awang Lung . .	101	2.35	2·3
2. Awang Lung . .	257	4.85	1·9
3. Awang Lung . .	95	1.56	1·6
4. Japar . . .	174	2.60	1·5
5. Jakob . . .	194	2.60	1·3
6. Awang-Yoh . .	6	0.8	1·3

The comparatively low percentages in cases 4 and 5 were attributable to the large number of crew in each. In Perupok village, where the catches are usually larger and more men tend to be attracted as crew, the individual incomes are lower in proportion to the total takings, and to the net's share, than further to the south where the yield is lower and the crews on the whole smaller. When I asked Awang Lung why the yield per man was relatively so much lower in case 5 than for his own catches he replied : " There were many men ", adding that whereas he himself had only 30 men Jakob had 35 or 37, with 7 or 8 men in some boats. On the same theme another expert said : " Men ! because the men are many indeed, it's little they get." And Awang Lung added that the Perupok *juru sĕlam* took a half share in their division for the net. Speaking about case 4, one of the crew said that there were, in all, 45 men in the six boats to be paid. He commented, " The *juru sĕlam* eats much "— meaning that he absorbed a large proportion of the takings—but went on to say that nevertheless he liked going out with Japar since the catches were good.

The distribution of the takings, which in any case rests largely at the discretion of the *juru sĕlam*, may become definitely abnormal through special circumstances, such as a very small week's yield or misappropriation on the part of the *juru sĕlam*. Case 6 above is an instance of the former. The *juru sĕlam* got only two small catches, one of which he sold for $5, the buyer actually paying $4.50, and the other for $1.50. He paid each member of the crew 8 cents, took 52 cents himself and allotted $2 to the net. When I asked him why these latter shares were disproportionate he said that when a total week's takings is less than $10 the *pĕraih* does not receive his usual share of the division. He said that he did not know if other *juru sĕlam* acted as he did, but thought they did the same. (I unfortunately neglected to verify this.)

It is obvious that where the system of accounting and division of earnings is done by memory and verbally alone, there is much opportunity for genuine mistakes, and also for some misappropriation of funds. But the latter is rare, largely, one may think, because of the sanction of the fear of losing the crew, who would simply " run " if they thought they were being cheated. A case of suspected false dealing by a *juru sĕlam* came to my notice when a net did not go to sea one day, though conditions were good. I commented on this to Awang Muda, captain of the net boat and a kinsman of the *juru sĕlam*. He said that the crew didn't want to go to sea. When the last division of the takings was held, though the fish had been sold for $60, the sum that was actually divided was only $43. It was known to the crew that the buyers of the fish had said that they had lost money on the purchase, and had therefore " cut " $10 ; they said they had handed over $50 cash to the carrier agent who was the seller on behalf of the net group. This man denied having received that sum ; he said he had been given only $43. The result was that each member of the crew received $1.30, and Awang Muda himself received $2.60. Since the number of the crew was very small, these sums were regarded by them as inadequate. With buyers and carrier agent in conflict as to how much cash had been handed over, they were suspicious. Muda declared roundly that he believed that a son of the *juru sĕlam* had kept the missing $7. The net had been doing badly, and this combination of reasons later led to a secession of some of the crew from the group.

The position in other types of fishing in the area may be summarized very briefly.

DISTRIBUTION OF EARNINGS FROM OTHER TYPES OF FISHING

Here the same general lines of procedure are followed, though the precise fractions allotted to the agents of production differ, and in the less important types of fishing the distribution itself is a much less formal and lengthy affair.

Deep gill-net (*pukat dalam*) : The general principle here is that half the takings go to the net and half to the boat and the crew.

It will be remembered that the deep gill-net, unlike the lift-net, comprises a score or so of sections, with several owners of these sections in combination. The total share of the net is therefore divided among the various owners in proportion to the number of sections held by each.

The total share of boat and crew is divided according to the following principles :

the boat gets two shares ;

an ordinary crew-man gets one share ;

each of the two men who pay out and handle the net at sea gets one additional share ;

the man who bales gets one additional share ;

the captain of the boat gets one additional share ;

when the net is working by day, a *juru sělam* is needed, and he gets one additional share for his special work.

Thus with a crew of ten men an ordinary man with no special functions would get, after night fishing, one share in sixteen out of the division for the crew, or one thirty-second of the total takings. The incentive to this type of fishing is that not only is there opportunity for a man to get an extra share for special labour, but he may also get a return as the owner of one or more sections of net.

As in the case of the lift-net the principal man of the group, normally the major net owner, uses his discretion to vary the distribution. When Awang Lung used the boat of his brother Semain in order to get Semain and his stepson Hussein as crew he gave Semain a share in the net division representing that of one of his own sections, which he himself gave up. This was in addition to the boat's share and the ordinary share as " body ". Awang Lung also gave the share of another section of the net to Awang Muda, who had helped him in the mending of the net. Shares representing other sections of the net were also given by Awang Lung to other men. The result was that he finished up by receiving only 8 or 9 shares himself instead of the 15 shares to which he was entitled by his ownership of as many sections of the net. By such means the " entrepreneur " gets help in repair of equipment and secures the allegiance of a crew. This is not an arrangement asked by the benefactors, agreed upon beforehand, or part of the rule of division ; it is a voluntary concession made to assist in the efficiency of the enterprise. The sanction for it was expressed by Awang Lung himself to me in the words " perhaps they would like it less " if they did not receive such bonuses.

The division of five days' fishing, using two full nets, of 26 and 22 sections respectively, in November 1939, was as follows. The total takings were about $65, and the time taken to make the distribution was about two hours.

The *bagian dalam*, the net's share, was at the rate of 60 cents per section. Each major net owner, with 14 sections apiece, got $8.40, and the *juru sělam* was given the equivalent of one share in each net, making $1.20. The *bagian awok*, the crew's share, was at the rate of 83 cents per man for five days' work ; those who were out for shorter periods got correspondingly less. Awang Muda, who went out for one day only, would have got 13 cents, but this was made up to 20 cents by Awang Lung who made the distribution as one of the two principal owners of capital. Each of the two boats received $1.66 as its share. The *juru sělam's* " legal " share was also $1.66, but " because he was an old man and a poor man " he was given an extra $3.79 as a bonus.

In terms of capital and labour the proportionate returns were about 46 per cent. to the nets, 5 per cent. to the boats and 49 per cent. to the crew as a whole, including the *juru sělam*. In terms of returns to personnel the two boat owners, who were also the major net-section owners, received 15½ per cent. of the takings apiece. One of them, whose father was ill, did not go out at all, and the other, Awang Lung, who did go out for two days, did not take his share as a " body " but remained content with his other receipts. An ordinary crew-man working full time received 1·3 per cent. of the total, though eight of the crew of about two dozen men received double this, for their special work. And the *juru sělam*, who in this case had no financial interest in boats or nets, received about 10 per cent. of the total takings, partly as a bonus.

Seine (*pukat tarek*) : The general scheme for seine-netting is the same as for *pukat dalam*, half the takings going to the net and half to the crew, with special allotments to those of the crew who perform special functions. But, as a rule, the boat's share is reckoned together with that of the net, and not taken out of the crew's share, since net and boat are in the same hands. There

were, however, only two of these nets in the Perupok area in
1940.

Purse-net (*pukat payang*) : There are none of these nets in
the Perupok area, and the principles of distribution in other areas
are discussed in Appendix IV.

Drift-net (*pukat hanyut*) : There are only a few of these nets
in the area, and the principles of distribution vary according to
whether the net as a whole is owned by one man, or the sections
of it are owned by several men in combination. For three nets
in regular operation the respective schemes were as follows :

(*a*) With the net of Pa' Che Su, a wealthy man who did not
himself go to sea, the principle was that the net received half the
takings and the boat and crew the other half in equal shares.
If the crew consisted of five men, each obtained one-sixth of the
half share.

(*b*) With the net of Pa' Che Mat, the boat (that of his son-in-
law) was given one-tenth of the total takings, and the remainder
was divided so that the net, which was the sole property of
Pa' Che Mat, received approximately one-quarter, and the crew
three-quarters in equal shares. The net owner was himself one
of the crew.

(*c*) With the net organized by Ma'e there were six men, each
owning one section. Here the boat, which was the property of
Ma'e, first received one-tenth of the total takings, and the
remainder was divided into six equal shares (half being for labour
and half for the net).

The relative receipts to capital and labour in these three cases
may be compared in terms of a total yield of $20, with 6 men.

					a	*b*	*c*
					$	$	$
Net .	.	.	.	.	10	4	9
Boat	.	.	.	.	1.42	2	2
Total Crew	.	.	.	.	8.58	14	9

The proportions obtained by labour in these cases are respectively 42·9 per cent., 70 per cent., and 45 per cent., while those
obtained by the principal organizer are respectively 50 per cent.,
25 per cent. and 32·5 per cent. The differences here are significant. In the first case the net owner, doing no work himself,
had an assured labour supply and took a high share for the use
of his capital. In the second case the net owner worked himself
but had considerable difficulty in getting and keeping a crew.
When I asked him why his net had not gone out one night he

replied, " A net of one man is difficult ", meaning that when the crew are not part-owners they have not the same incentive. I asked " Why ? Because they don't get the net-share ? " He replied vigorously : " They get it ! I divide into four ; one share for the net and three shares for the crew." Because of the shortage of labour he was willing to take a lower rate of return on his capital. In the third case every man had a direct interest in the net, and the return to net-capital and to labour was aggregated for practical purposes with no need for concession.

Sprat-net (*jaring*) : Here the value of the net is considerably less than in that of lift-net and other fishing discussed above. One principle of division is to split the takings into as many equal shares as there are men of the crew, plus one share for boat and net together. Here with a crew of from four to seven the return to the net and boat capital varies between $12\frac{1}{2}$ and 20 per cent. and that to individual crew-members likewise. But at times one-third of the takings are given to the net and boat while the crew divide the remaining two-thirds among them, thus obtaining from 10 per cent. to 16 per cent. per man.

Fish-traps (*bubu*) : Here it is a case of a set of fish-traps operated normally by a crew of four men. The principle of distribution is to allot first a share of one-tenth of the takings to the boat used. Then, if the expenses of constructing the traps have been borne by the principal organizer alone, the remainder is divided into five shares, one of which goes to him as trap-owner and each of the other shares to a member of the crew—he himself taking one share in the usual way. If, however, all four men have contributed to bear the expenses of the traps, then a sum of $10 or so is set aside and repayment made, after which all the men share equally in the remainder when the boat's percentage has been met.

Light drift-net (*pukat tĕgĕlang*) : In this organization, where each man of the crew provides his own section of net, and they use the boat of one of their number, the principle followed is to allot one-tenth of the takings to the boat, after which the members of the crew receive equal shares.

Line Fishing : With fishing for Spanish mackerel, squid and most other types, the normal principle is the same as that in the case of the light drift-net ; the boat gets one-tenth, and the crew then share equally. But variations occur.

If the boat is old, or the crew are close kinsfolk, there often is no allowance made for the use of the craft, and the crew simply take equal shares of the total takings. If the boat is new, or belongs to a non-fisherman, or has recently had the expense of new sails and gear, then the percentage is taken. On occasions a reduced rate is applied. Thus one man, who goes fishing with his brother-in-law, takes a share for his boat, but makes it one-fifteenth only. If his partner were not a kinsman he would take one-tenth. And, he told me, when he would use a new boat he had building he would take the one-tenth share, even if his brother-in-law should be his partner.

Apart from the boat's share, there is variation also in the method of distribution among the crew. If all share equally, the crew are described as a " combine " (*konsi*). This means, however, in line fishing, that the less expert reap part of the advantage of the skill of the more expert. When large boats go out hand-lining as a break in their lift-net work the crew is not a homogeneous group of line fishermen. It has been got together mainly for handling the net, and usually has a mixture of lads and experts. Hence the principle known as " individual work " (*masingmasing kreja*) is often applied, and each man sells his own catch and keeps the proceeds. As the boat sets sail the fishermen decide among themselves what rule of distribution they shall follow. " What shall be our work to-day ? " one asks. The expert line fishermen often call for individual returns, and about half the crew usually say : " As you like ; we don't mind." But the boat-captain often says : " To-morrow each man for himself, if you like, but to-day let it be a combine." His word is usually accepted. His interest lies mainly in balancing the different interests of his mixed crew and keeping them together—though if he is the boat owner he may also have an eye on the boat's share of the takings, which will probably be foregone if the principle of independent returns is adopted.

FISH FOR HOME CONSUMPTION AND PETTY CASH

So far we have been considering the distribution of the money from the sale of catches in bulk. But an integral element in the distributive system is the fish withheld from general sale and devoted to the domestic needs of the fishermen ; it is called the *makan lau'*. *Lau'* (*lauk*) means the flesh component of a meal, which is regarded as incomplete without it. The conventions of consumption thus influence the scheme of distribution.

The way of apportioning the *makan lau'* can be seen from what happens in lift-net fishing. The fish for the crew are those normally in the rear compartment of the boat which carries the catch to shore. They are left in place till all the boats of that net have arrived, and then the *juru sĕlam* or his wife puts them out into as many heaps as there are boats. This is done roughly, the contents of a baler being put in turn on each heap, until there are no more fish left ; if the last round would be incomplete thus, the fish are then doled out by handfuls instead. The captain of each boat, or his wife, or someone else appointed by him then divides the boat-heap into shares for the crew, and each man (or a woman from his family) comes and takes his share. If fish are scarce these individual shares are counted out carefully, but otherwise the division is done by handfuls only. There is rarely any comparison of shares by the crew-members or grumbling if unequal numbers of fish have been allotted.

The amount of the shares and the method of disposing of them varies according to circumstances. In the ordinary way each man gets about 20 or 30 fish, but when fish are plentiful the number may rise to about 50 apiece. When fish are scarce but the catch has been big enough to make a bulk sale, then each man may get only 8 or 10 fish. But when the catch is so small that no bulk sale is made, then each man's *makan lau'* may be very large, perhaps 100 fish. The total value of the *makan lau'* for the crew of a net, if calculated at market prices, is normally between about $2 and $5, representing somewhere between 5 per cent. and 10 per cent. of the total value of an ordinary catch.

The primary function of the *makan lau'* is to furnish fish for the domestic meals, and as such it is regarded as a fundamental part of the crew's earnings. If it is small the crew and their wives grumble. Provision of it is, in fact, the first charge on the day's catch. An important secondary function, however, is as a source of petty cash, for giving " coffee-money " to the crew. In all the larger nets they get their major earnings weekly, and if they are married their wives take charge of most of the cash. They rely then on the sale of part of their *makan lau'* to supply them with the few cents they need for a cup of coffee, a cigarette and possibly a cake or sweetmeat at the end of the day. When the *makan lau'* is only a few fish, they take it all home for food, even when fish are fetching a high price on the beach. But when it is moderately plentiful they sell up to ten cents' worth or so apiece, commonly selling half and keeping half back for home consumption.

The *makan lauʻ* has thus an important rôle in the distributive scheme, and any programme of re-organization of the fishing industry should include provision for it, as a customary part of the fisherman's income.

THE BASIC PRINCIPLES OF DISTRIBUTION

This analysis of the principles of distribution of returns in fishing, reinforced by consideration of the local variations briefly given in Appendix IV, has, perhaps, given the reader an impression of bewildering complexity. The principles followed in the different types of fishing seem at first sight to bear little relation to one another, the shares allotted to boats, nets and men apparently not being reducible to any kind of common formula.

Closer scrutiny, however, shows that certain general themes do run throughout the whole system. The proportionate returns to capital and labour, for instance, tend to correspond to the degree to which each contributes to the total yield. In fishing with the larger nets and boats the total share of the fixed capital amounts to somewhere round one-half of the yield, whereas with the smaller nets and boats, and in line fishing, it falls to very much below this, and the share of labour is correspondingly increased. Then, again, there is a tendency to equilibrium visible in the various types of return. This is manifest in the returns to labour. Whatever the fractional basis of calculation, these tend to lie broadly between $1.50 and $3 per week, except in fishing with a very short season such as that with the *jaring*, which may yield $3 and upwards per man per week. On the whole, the more regular the type of fishing, the lower the rate of general cash return to the individual fisherman. It is manifest also in the returns to boats and nets, the value of their shares in the distributive scheme being roughly proportional to their capital cost, the frequency with which they need replacement, and the frequency of their use. Small boats tend to earn between 50 cents and $1 a week, which represents roughly from 1 per cent. to 4 per cent. on their capital cost, and large boats from $1 to $5 a week, representing roughly from 1 per cent. to 5 per cent. on their capital cost. The higher figure in the latter case, however, is gained by those boats engaged in the less regular *jaring* and *pukat dalam* fishing. The larger nets, earning broadly from five to ten times as much as a boat, may recoup between 10 per cent. to 20 per cent. of their capital cost in a week, but their repairs are very heavy and their total life only between one-fifth and one-tenth as long. These figures, calculated on the basis of broad averages, cover a range of variations due to individual luck, skill and business enterprise. But they demonstrate that the complex schemes of distribution, sanctioned as they are by convention, are not merely customary structures of a haphazard kind ; they follow underlying economic principles.

Moreover, these economic principles do not work entirely unperceived and unrelated to their social context. Notions of equity, of a " fair return " to the factors of production, though not precisely formulated in any general rules, bring influence to bear on the distributive system. The schemes of allotment of shares by division and sub-division, with all their variations for the different types of fishing, serve as general nuclei or guides ; they obviate the need for argument on basic principles each time a fresh act of distribution is performed. Nevertheless they are flexible in that the organizer of production, normally also the owner of the major capital, is commonly the person responsible for the distribution of the returns, and can use his discretion without evoking resentment. The modifications he introduces usually arouse no opposition ; they are accepted as ethical, on one of two general grounds. As the owner of the major capital and the primary organizer of the activity, he bears the basic costs and has the labour of management ; so long as he secures a " fair return " of fish and cash to his men they are prepared to allow him to interpret the conventional scheme to his own advantage. But should he push his own interests too far then he is " not straight ", and his men leave him. On the other hand, the flexibility of the scheme allows the distributor to make allowances for special services, or even for special social conditions, which are admitted by his crew on grounds of equity. He can give one man a bonus because of extra work done, or because of a labour shortage, or because of his age or poverty. He thus diminishes in effect either his own share or the shares of all the others, but no objection is raised. The flexibility of the system, while it does allow the organizer of production some advantages, is recognized as useful, since it allows recompense to be given in circumstances felt to be justifiable, though not covered by the strict rules.

CHAPTER IX

OUTPUT AND LEVELS OF INCOME

Calculations of output from fishing in such a community are most easily made in terms of values at market prices. Fish are not weighed when brought in, and the quantities recorded for official purposes by fishermen and fish dealers are merely estimates and in any case are incomplete. A general idea of the amount of the catch from individual boats can be obtained from the number of baskets—which is used as a rough measure by the fishermen—but these are so often only partially full that no conversion into exact weights is possible. (Some estimates of the physical volume of output, calculated by indirect methods, are given later, in footnotes.)

The absence of a census of production or other effective machinery of record makes it necessary for the investigator to build up the material from small units. The impossibility of covering every individual act of production makes it necessary to rely upon samples. In my eight months in the Perupok area I noted the results of fishing in well over 3,000 cases, with various types of units involved, covering a total output of more than $30,000 in bulk sales of fish alone. In terms of labour this corresponded to approximately 40,000 man-days, giving an average of 75 cents per man per day when fishing was actually carried on. About 10 per cent. of these cases were samples collected at random, especially during the early part of my study, when I was familiarizing myself with the general principles of the organization of the industry. In view of the many variations between units it is desirable to base the calculation of output primarily upon my more systematic records. These cover the half-year from mid-November 1939 to mid-May 1940. They are complete for bulk sales from lift-nets, nearly so for deep gill-nets, heavy drift-nets and seines, and cover from two-thirds to three-quarters of the estimated number of cases of fishing with light drift-nets, sprat nets, and some forms of line fishing. The danger of weighting individual or seasonal variations unduly during the period is thus small. A synopsis of the records of two important types of fishing during the half-year is given in Appendices VI and VII. I was not able to take such systematic records of most of the minor types of fishing, but the samples taken are, I believe, representative ; they are mentioned later in this chapter.

ESTIMATED VALUE OF ANNUAL OUTPUT

The calculation of annual output has many difficulties. Apart from the fact that a complete record for the six months under observation could not be obtained for all types of fishing, even where it has been taken a simple doubling of the result would not be an accurate figure of annual output, because of seasonal variations. The method I have followed is first to allot to each type of fishing an estimated full output for the six months' period on the basis of what was probably produced by those units not fully recorded. Such known factors as the state of the fishing each day, my counts of boats at sea and boats remaining on the beach, the prospects of individual units in terms of available crew and equipment and other available occupations have been taken into account. The estimated full output for the six months' period is then converted into output for the whole year from information available to me from the fishermen about conditions during these unrecorded months, and supplemented by sample observations I made before and after the period of systematic record. The results are approximate, but I think it probable that they represent the 1939–40 output for the area within about 10 per cent. margin of error (see Table 11).

It may be pointed out that this is the first time that a calculation of this detailed kind has been made for any Malayan fishing area. It is based upon records for all types of fishing, includes that portion of the output which goes into domestic consumption, and takes as its foundation the wholesale prices actually received by the fishermen. (Freshwater fish from the river and the rice fields have not been considered as they are unimportant in the coastal area, though in the agricultural regions inland they are of some value.)

This gross annual output of fishing must be modified by three factors, to which only approximate values can be assigned. The first is the custom of the fish dealers of " cutting " the price when they lose, which reduces the nominal figures of cash return. I estimate that for lift-net and deep gill-net, where this mostly occurs, a rebate of $1,000 is not too high a figure to allow for the ten months of major fishing. On the other hand, the value of

TABLE 11

ESTIMATED VALUE OF ANNUAL OUTPUT FROM FISHING, PERUPOK AREA,
1939–40

Type of Fishing.	Output for Six Months, Mid-Nov. 1939 to Mid-May, 1940.		Estimated Annual Output.
	Sample recorded.	Estimate of Full Output.	
	$	$	$
Lift-net (*Pukat Takur*)	18,851	18,851	42,500
Deep gill-net (*Pukat Dalam*) night.	4,848 [1]	5,000	10,000
,, ,, ,, ,, day .	308	350	500
Heavy drift-net (*Pukat Hanyut*) . .	545	600	1,000
Light drift-net (*Pukat Tĕgĕlang*) . .	948	1,250	1,750
Seine (*Pukat Tarek*)	92	100	200
Sprat-net (*Jaring*)	135 [1]	150	500
Line fishing (various)	817	1,000	1,500
Line fishing (dorab)	179	500	1,000
Trolling	63	300	500
Shrimp-net (*Takur Baring*) . . .	10	100 ⎫	
Small lift-net (*Takur Kĕchil*) . .	55	75 ⎪	
Fish-trap	51	60 ⎬	750
Scoop-net and Casting-net . . .	12	200 ⎪	
Shell-fish collecting	—	100 ⎭	
Totals . .	$26,914	$28,600	$60,200

[1] Figures for *pukat dalam* (night) and *jaring* begin from 20th October and end of October respectively.

the fish or cash obtained from secondary catches by the crews of lift-nets has to be added. I recorded about 70 cases of such extra yields, giving probably about $350 in all to the boats' crews. For the year's fishing the total from this source is probably between $750 and $1,000. Output from *makan lau'* is difficult to estimate in bulk, but the total value of the fish sold on the beach and taken home for food is probably in the region of 10 per cent. of the bulk sales of the general catch, that is, between $5,000 and $6,000.

The total annual output, valued at market prices, is thus probably in the region of $65,000 or $66,000 for the Perupok area.

In calculating average output per head, the number of fishermen concerned may be put at about 550, comprised of about 500 men primarily occupied with the lift-net and complementary work, and about 50 men who engage in fishing other than with lift-nets. This means that average output per annum is approximately $120 per head of producers.[1] Taking the total fishing population involved at about 1,750 persons, a general average of $37.70 per head is obtained.

Considering that the 1940 season was, on the whole, a poor one for fish it is probable that these figures are somewhat lower than a general average over a period of years would give. But I got the impression that the Malay fishermen, like other primary producers in analogous situations, tended to exaggerate the abnormality in the 1940 season. A long period average would probably be not much higher than the figures here given. It must be emphasized that these figures of output refer to fishing only, and that the production of rice, vegetables and copra, and craftwork, by many fishermen and others helps to give a higher level of output for the community as a whole (cf. Chapter X).

There are great differences in level of output in the various types of fishing, and a further analysis of these is required.

LEVELS OF OUTPUT FROM LIFT-NETS

Complete records of the sales of bulk fish from lift-nets from November 11th, 1939, to May 10th, 1940, inclusive, gave a total output in cash terms of 18,851 dollars 50 cents. With a total number of 1,027 net-days on which fishing was undertaken during this period, an average cash output of $18.35 per net per day at sea was thus obtained. For the whole period, covering the time during which the nets were not at sea, the general average is $5.40 per net-day, including Fridays, for a total of 3,489 net-days ; or $6.64 per net per day excluding Fridays, for a total of 2,990 net-days.[2]

Appendix VI shows the variation in weekly yields. There was a wide range of variation in the daily yields, in many cases no fish at all being obtained, in others the catch being too small to be worth selling in bulk. When a cash income was got it varied from under $10 to as high as $137. The results are given there in synoptic form. For the fleet as a whole the daily variation

---

[1] Taking the average price of fish at the 1939–40 figure of $5 per picul, the total annual Perupok output would be about 775 or 780 tons, and the average output per head of fishermen nearly 1½ tons.

[2] These figures of total net-days exclude 6 net-days at sea and one Friday spent by one net-group on an expedition to Trengganu fishing grounds ; no precise data of their catches were recorded.

in total output can be seen in Fig. 21, which shows the extreme
irregularity of production in this branch of fishing.   Comparison
with Fig. 22 shows, however, that there is a broad correlation
between this and the variations in the number of fishing units
at sea, though the curve of the latter tends to lag behind the curve
of output owing to the persistence of nets in going out in the hopes
of obtaining a catch, though they are often not successful.   The
frequency distribution of the daily output of lift-nets as a whole
can be seen from the following table.   In 62 per cent. of the fishing
a cash income was obtained, in 10 per cent. there was no cash
income for the net-groups, but only fish for the individual use of
the crews, and in 28 per cent. no catches at all were made.

TABLE 12

FREQUENCY DISTRIBUTION OF DAILY OUTPUT OF LIFT-NETS

Day's Casts Yielding.	Number.	Day's Casts Yielding.	Number.
Nil	285	$	
Fish for crew only . . .	104	61 to 70	30
$		71 ,, 80	12
1 to 10	129	81 ,, 90	5
11 ,, 20	144	91 ,, 100	4
21 ,, 30	116	101 ,, 110	2
31 ,, 40	86	111 ,, 120	1
41 ,, 50	64	121 ,, 130	1
51 ,, 60	43	131 ,, 140	1

From the table it will be seen that a figure of $20 to $30 may
be regarded as an average yield in cases where a cash income is
obtained—representing roughly about one-third of a ton of fish—
but that about $10 is a modal figure for results as a whole, taking
in times when no catch is made or no bulk fish sold.

It must be pointed out that this distinction between the types
of yield of the fishing—cash and kind—is one which corresponds
to Malay terminology, and is important from the economic
standpoint.   A cash yield means that rice and other articles can
be bought, repayments made on equipment, and possibly something put by for the future.   The net has " got ", in the fisherman's jargon.   A yield too small to be sold in bulk for the group
as a whole, and therefore distributed among the crew, means
that they have fish to eat, and probably some for retail selling,
giving them a few cents for coffee, snacks or cigarettes.   But it
is essentially a subsistence yield ; they can live on it for a short
period, but not buy rice or meet any obligations.   Nevertheless,
on occasions crews are willing to fish simply for *makan lau'* alone.
To get nothing at all, to come in " empty " is a waste of time and

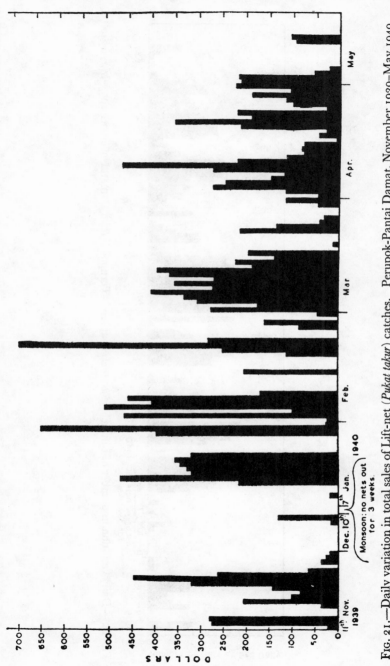

FIG. 21.—Daily variation in total sales of Lift-net (*Pukat takur*) catches. Perupok-Pantai Damat, November 1939–May 1940. (Fridays omitted, since no lift-nets put to sea.)

263

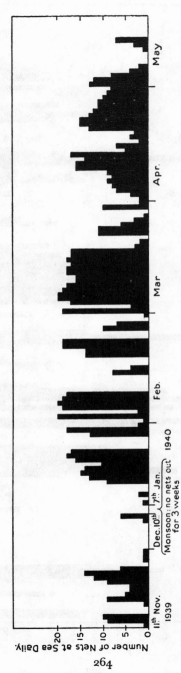

FIG. 22.—Daily variation in number of Lift-nets (*Pukat takur*) at sea. Perupok-Pantai Damat, November 1939–May 1940.

labour. But at times a crew may prefer not to cast the net for probable *makan lau'* but to turn to line fishing instead, so getting nil with the lift-net, but saving themselves some hard work, and relying on their lines to bring them in a cash yield.

A question of interest is what proportion of these types of yield is to be seen in the daily results of the individual net-groups.

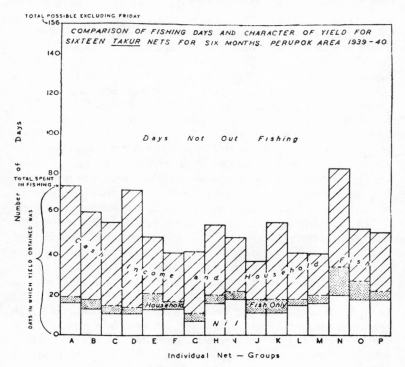

FIG. 23.—Comparison of fishing days and character of yield from sixteen Lift-nets for six months, Perupok area, 1939–40.

For convenience I have shown this in the form of graphs (Figs. 23 and 24). The first illustrates the position for the 16 net-groups which were in being during the whole of the six months' period of observation, and which may therefore be regarded as stable. The graph shows that comparing output of each net-group from this point of view of cash, kind or nothing, in every group the number of days on which a cash income was obtained is highest, those with nothing come next, and those on which fish for the crew (*makan lau'*) only was obtained is the smallest proportion. This agreement of the broad distribution of yields allows one to infer that assuming an approximate equality of chances for each group, there is a general standard of skill prevailing throughout them all—to put an extreme case, there is no group which shows three times as many failures as successful catches. This is what may be expected from the fact that the experts who are primarily responsible for the fishing are all trained men, and that the groups of men who are unsuccessful for a long time break up. But it is apparent also that within this broad agreement there is considerable variation in the proportions of types of yield from one group to another. There is only one case, group *O*, in which the ratio of days on which a cash income was secured is less than 50 per cent. of the total days out. But in groups *J* and *M* this is only 51 per cent., whereas in groups *A* and *C* it is 74 per cent., and for group *D* it is as high as 80 per cent. This is the group of *juru sělam* Japar, who in this as in other respects leads the van. Group *N* is an interesting case of an expert whose energy drives his crew out in all but impossible weather, with the result that his output is characterized by a fairly low proportion (59 per cent.) of days with a cash income, though his aggregate output is high.

In the second graph, to give more detail, I have plotted the frequency of the values of the catches, in rising units of ten dollars, obtained by five net-groups with aggregate outputs ranging fairly evenly from highest to lowest, among the 16 stable groups examined. From the curves it is evident that the same general type of distribution of daily yields is characteristic of all the groups. In each case the greatest frequency of cash yield is in the lower range of figures, below $50, followed by a rapid decline, with irregular low frequencies thereafter. The outstanding position of group *D* (Japar), however, is shown by the tendency to a more even concentration in the $20–50 range.

The differences in the aggregate output of the various net-groups are shown by the totals given in Appendix VI. Group *D* easily heads the list, with a total of more than $2,500, followed by *A*, *B*, *C* and *N*, in that order, all between $2,000 and $1,500. None of the remainder gain more than $1,000, but, on the other hand, only one of the groups operating during the whole of the six months (group *J*) gets less than $500. The four leading groups, *D* (Japar), *A* (Sa'e), *B* (Jakob) and *C* (Ma' San) are all from Perupok village, which bears out the general opinion of the fishermen that the men at this end of the beach are the most

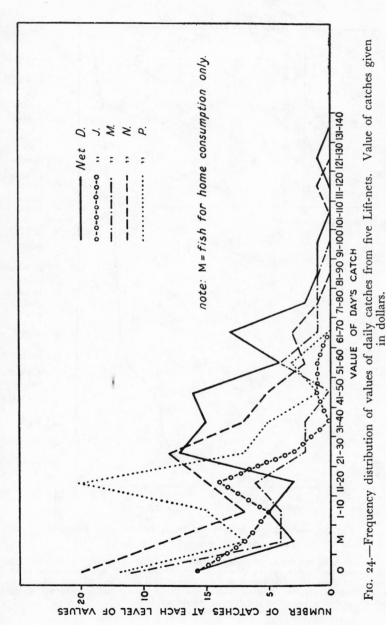

FIG. 24.—Frequency distribution of values of daily catches from five Lift-nets. Value of catches given in dollars.

note: M = fish for home consumption only.

Net D.
" J.
" M.
" N.
" P.

NUMBER OF CATCHES AT EACH LEVEL OF VALUES

VALUE OF DAY'S CATCH

267

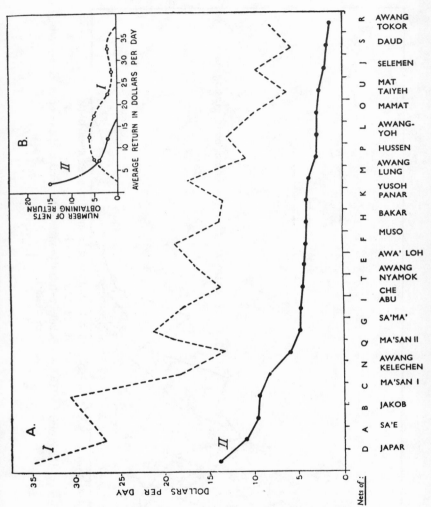

Fig. 25.—Average daily yield from twenty-one lift-nets.
Letters denoting nets as in Appendix VI.

Curve A II sets the net-groups in descending order of average daily yield (in cash) over the whole period, November 1939–May 1940 ; A I follows the same order, but gives the average yield per number of days fished in each case.   Curve B II indicates the number of nets obtaining the given average return over the whole period, and B I the same per number of days fished.

268

efficient. The outstanding position of group *D*, which has an aggregate of more than $500 greater than that of any other group, also corroborates the popular view that Japar is the best lift-net expert in all the area. While Japar was among the most energetic in taking his net out, his superiority in output did not simply depend on this ; he obtained about 25 per cent. more during the period than did his nearest rival, Yusoh Sa'e of group *A*, who was actually at sea two more days than Japar.

In comparing levels of output among the various net-groups it is necessary to consider the different amounts of time that they put in at sea, as well as the average levels over the whole period under consideration. Fig. 25A allows this comparison to be made by showing in curve *II* the average daily cash yield for each net-group for the whole of the period under review, and in curve *I* the average daily yield in terms of the number of days for which each group actually went out fishing. Curve *II* is plotted with the groups arranged in decreasing order of averages ; curve *I* follows the same order of groups as *II*. Set thus, side by side, the two curves allow an opinion to be formed on the comparative energy and fishing skill of the various experts. The net-group of Japar has the highest average of yield over the whole period, and also for the total number of days actually fished. But Sa'e, who comes next to him as regards average daily yield for the whole period, is below both Ma' San and Jakob as regards average per number of days fished. The inference is that, while the energy of Yusoh Sa'e in fishing is greater than that of his two rivals, his skill is rather less. The advantage to be gained by persistence and energy is shown also by Awang Kelechen and by Ma' San II. The latter, in particular, has a very medium average of yield per number of days actually fished. But though he started the season a little late, he occupies sixth place in the list of general averages by reason of the frequency of his fishing. Awang Lung, on the other hand, is an example of an expert of some skill who did not push his advantage. His average yield when he was out fishing was comparatively high, but his frequent stoppages of work reduced his general average over the whole period.

LEVELS OF OUTPUT IN OTHER TYPES OF FISHING

The detailed analysis of output given for lift-net fishing is not necessary for the other types of fishing. The principles are similar and the fishing methods are of much less importance in the area.

But a few summary observations are desirable.   The output from
the deep gill-nets, which ranks next to that of lift-nets in value, is
also highly variable, as the data of Appendix VII show.   In
1939–40 there were approximately 25 complete nets of this type
in use in the Perupok area, and during a little over six months I
recorded the catches of 24 of them on nearly every occasion.   In
fishing by night for mackerel, which is their principal use, an
output of a total value of $4,797 was obtained in the 253 cases
given in the Appendix, and an additional output of a value of
$51 in 9 other cases.   This gives an average output per net per
night of $18.50, which is slightly greater than the average lift-net
output per net-day at sea.   When it is remembered that the crew
of a gill-net is less than half of that for a lift-net, being only about
10 men on the average, the output per man is rather more than
double.   This explains the enthusiasm for this form of fishing.

The aggregate yields of the various nets show great variation
for the same period under review.   This is partly due to the run
of the luck in meeting shoals of the fish, but skill also plays an
important part.   The net of Japar (case *e* in the Appendix) has
one of the highest aggregates, $410, and his average yield per
night of $34.2 is the highest of all.   The only other two aggregates
above $300, those of Bakar (*r*) and Awaʻ Loh (*s*) of Paya Meng-
kuang, are of men who specialize in mackerel work and do not
normally go out with the lift-net fleet.   Their average output per
night is $30.9 and $18.2 respectively.

The value of an individual catch is sometimes very high,
three instances during the period recorded having figures in
excess of the highest recorded lift-net catch.   Two of these figures,
$172.50 by Bakar (net *r*), and $138 by Selemen (net *w*), were
both obtained on the same night.   The actual volume of the fish
taken was large in each case, 7,500 in the one and 6,000 in the
other.   But the outstanding cash yield was due to the fact that
these catches were made at the beginning of a mackerel season,
on a night when hardly any other boats were out, and the price
per thousand given in the morning was correspondingly high.
It will be noted that in the middle of the May season the catches
of fish were in some cases even heavier, but supplies were so
abundant that the price per thousand was only about half that
obtained earlier.   The catch of Japar (net *e*), sold on the morning
of May 5th for $115, was said to have comprised over 9,000 fish,
which was almost certainly the case, since the price per thousand
was $12 then at the Perupok end of the beach.   When catches

are very heavy a sale is often made not on the basis of counted
thousands of the fish, but in bulk on estimate of numbers.

In addition to night fishing for mackerel the gill-net is used at
times for fishing by day for jewfish and other fish. I recorded
practically all the cases of such fishing that occurred during the
period of investigation—nearly all were in November, 1939. For
a total of 34 days' fishing by different nets or pairs of nets the
total output in cash values was $315, giving thus an average of
$9.3 per unit, the catches of highest value being $30 by a single
net and $44 by a pair of nets. In 7 of the cases *makan lau'* only
was obtained, and in one no catch at all was made. The average
value of a day's fishing, less than half of that from a night's fishing,
compares still less favourably when the fact is taken into consideration that many of the day-fishing nets work in pairs, and divide
the catch. In the 34 cases mentioned, 15 were of nets working
singly, 18 of nets working in pairs, while in one case there was
a triple combination.

The output from heavy drift-nets (*pukat hanyut*) does not nearly
match that from lift-nets or deep gill-nets in either gross values
or value of individual catches. There were in the Perupok area
9 *pukat hanyut* in 1939–40, and the total value of the output
I recorded covering most of the cases of fishing, was $545.50,
giving an average yield of $6.57 for the 83 nights of fishing.
Most of the output was produced by 5 nets, all of them run by
men who normally did not participate in lift-net work, though
one was a prominent gill-net fisherman. With an average crew
of 6 men, average output per head is approximately $1 per night.
Of the 5 regular nets in operation the average nightly yield of
4 was very similar—between $5.80 and $6.80—while that of the
fifth was over $8. This last was the net of Ma'e, who was also
pre-eminent in his other major occupation of dorab fishing.

For light drift-nets (*pukat těgělang*), an important subsidiary
form of fishing especially in November, January and March,
I recorded the yield for a total of 464 cases, comprising 1,613
man-days, of $948, giving an average output of practically $2
per boat per fishing day, or 59 cents per man per day. The
distribution of individual boat yields is shown in Fig. 26. In
January, larger boats tend to be used, with a crew of from
3 to 7 men, and the output is often as much as $1 per head ; in
March boats and crews are apt to be smaller, two or three men
being the rule, and output per head is usually smaller also.

The 1939-40 season for seines and sprat-nets was a poor one, so the records of output for these two types cannot be taken as a fair index of normal yields. They do indicate, however, the kind of conditions with which these fishermen are sometimes faced. There were only two seines (*pukat tarek*) in the Perupok area, and their total output for 13 days' fishing reached a value of only $92, or an average of about $7 per net per day ; the value

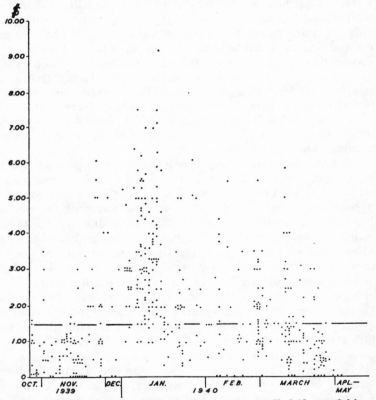

FIG. 26.—Daily yields to individual boats in small drift-net fishing.
Each dot indicates the cash value of one boat's catch. The line indicates the general average over the period.

of the largest catch was $20. In addition, about $260 worth of fish was brought in by Perupok boats from " foreign " seines met at sea, as well as $15 worth from *pukat payang*, making a total of about $350 of total output of fish of this type reaching the community for sale and distribution. The yields from sprat-nets (*jaring tamban*) were, on the whole, very low. There are about

25 such nets altogether in the area, and the total output of 23 of them for 51 days' fishing, $135, represented an average output of $2.66 per net per day, or about 65 cents per man per day. The catch of highest value was $20, taken by 4 men, but many

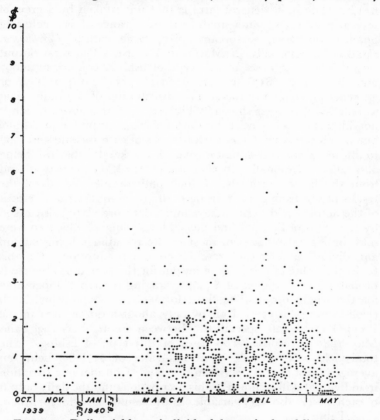

FIG. 27.—Daily yields to individual boats in hand-line fishing.

Each dot indicates the cash value of one boat's catch. The line indicates the general average over the period. It will be noted that hand-lining is most concentrated between March and May, while drift-netting (Fig. 26) is more widely distributed.

of the casts yielded nothing at all. A subsidiary form of fishing, resorted to at the time when ordinary lift-net fishing is dying away in April and May owing to the scarcity and small size of the fish, is the " small *takur* " netting. There were only 2 of these nets used in the Perupok area in 1940, their total yield for

9 fishing days being $55. With a crew of a dozen to 15 men this gives an average output per man of rather less than 50 cents per day.

The output from line fishing must be considered under three heads. The most general form is that undertaken by a crew of several men, usually in a small boat, for a range of fish including Spanish mackerel, sea-bream, pike, large catfish, *Coryphaena*, squid, etc., especially in March and April. The total output recorded for 739 cases of this type of fishing, comprising 2,097 man-days, was $817, averaging $1.11 per boat per day, or 39 cents per man per day. The distribution of the daily yield to individual boats is shown in Fig. 27. Another form, which is not important as a contribution to fishing output as a whole, but which serves as a very useful source of cash income and food to lift-net crews, is trolling for Spanish mackerel (other fish being also taken occasionally) in March and April, on the way to and from the fishing grounds. I took only sample records of the results of this fishing, comprising perhaps one-quarter or one-fifth of the actual yield. In these samples I recorded 105 fish taken by the crews of Perupok and Paya Mengkuang, of which 40 were sold for a total of $24.20, most of the remainder being cut up and divided among the crew for home consumption. On this basis, the value of this item of output for the period as a whole is probably in the region of $300, averaging perhaps $1 per man for the whole time. The third form of line fishing, for dorab (wolf-herring), is carried on by only about a dozen men in the Perupok area, all specialists. Between September 1939 and May 1940 I recorded 166 cases of this type of fishing. The total yield was 652 fish, of a total value of $178.80, being an average of just over $1 per man per day. Considerable variation in individual skill is shown in this occupation, and the daily output of each man is also very variable, as the results show for four men. (See Table 13.)

These sample records are not exactly comparable as a measure of comparative skill, since all the four men were not always out on the same days in each case. But an average for the days on which all were at sea and their catches were recorded gives the same relative levels of output. *A* (Ma'e) was acknowledged to be the best of them all in the area. On the whole I recorded about one-third of the cases of this fishing in the area during my stay. The total value of the annual output from this form of fishing may therefore be in the region of $1,000.

TABLE 13

INDIVIDUAL RETURNS IN DORAB FISHING

Fishermen.	Days Out.	Total Fish.	Total Value.	Average Output per Day.		Highest Catch in one Day.	
			$	Fish.	$	Fish.	$
A . . .	32	172	46.25	5·4	1.44	17	5.20
B . . .	22	85	21.75	4	1.00	10	2.80
C . . .	13	37	10.40	3	.80	10	2.75
D . . .	10	18	4.80	1·8	.48	4	1.20

A form of fishing of great importance for small-scale cash requirements and domestic consumption during the monsoon is the taking of grey mullet with the scoop-net in the surf. Since the fishermen were in and out of the water at almost any time of the day, all up and down the coast, it was impossible to get anything like a complete record of their output. All that could be done was to take a random sample of fishermen each day, and supplement this by following the fortunes of several men consistently for a period. Altogether I recorded 93 cases, in which a total of about 1,270 fish were taken. Most of these were taken home for food, but sales were made at rates varying between 7 for 10 cents and 12 for 10 cents, depending on local abundance of the fish. A general average was from 15 to 20 fish per man per day, for about 20 days out on the average during the month in which the fishing took place. At a rough figure, the total cash value of the output for the month may be put at $200 for the area, there being, on an average, about 15 men out each day. (Many men do not go out with this net.) Individual output varied from half a dozen fish per day, taken home for food, to $1 in cash, by a young man who was especially skilled.

A complementary type of work to this scoop-net fishing is the collection of cockles or similar small shell-fish known as rĕmi. The work is done mainly by women and children on the beach after the heavy weather of the monsoon. At times there are as many as 40 or 50 people engaged in this, digging the shell-fish out from the sand with their toes, and putting them in wire baskets. The product fetches about 5 cents a quart, and a day's hard work can yield about 50 cents. But most women are content with a quart or so for home consumption. The total value of the output is small, perhaps a figure of $100 being somewhere near the mark.

Two other subsidiary methods of fishing, the shrimp-net (*takur baring*) and the fish-trap (*bubu*) also contribute a small aggregate output. Each has a fairly short season. There were about 8 shrimp-nets in the area, and towards the end of April some of them were taking catches of 100 gantangs or so of shrimps per day each, worth between $3 and $4. I would estimate the total output for the season at less than $100. There were only two owners of fish-traps in the area, and one man, who had two old traps, used them once only during my stay and got a negligible catch. The other, a former lift-net expert of Pantai Damat, who had sold his net at the opening of the 1940 season, had 4 or 5 new traps. I recorded most of the times he took these up, in April ; for a total of 7 settings the output reached a value of $51.30, mostly from snapper. This gives an average yield of $7.30 ; or, for a total crew of 4 men, an average output per head of $1.84 per day.

IMPORTANCE OF THE OUTPUT FIGURES

This analysis of empirical records of output from the various types of fishing in the Perupok area gives a quantitative basis for the general descriptive material. It shows how considerable is the range of average output per head in the different forms, from about 40 cents per day for line fishing to about $2 per day for night-fishing with deep gill-nets.[1] It shows also the variation in output between specific net-groups or other units in each type of fishing, and for the same unit at different times. In this way it gives some idea of why men come to prefer one kind of fishing to another, or one group to another.

In the second place this material provides a basis for the estimation of levels of fishing income. Here knowledge of the variations is important. Any estimates of income based upon a few random inquiries would almost certainly be inadequate, since they fail to take into consideration the effects of seasonal change, individual differences in skill, and the lack of continuity

[1] The figures also allow some rough inferences to be drawn regarding comparative physical output in different types of fishing. Thus the average output per lift-net for over 1,000 net-days at sea was nearly 4½ cwt. of fish per day (calculated from an average price of $5 per picul of fish), and the average output per man was about ⅓ cwt. per day. In fishing for mackerel with the deep gill-net approximately 355,000 fish were taken on 262 net-nights (see Appendix VII). Calculating the average weight of mackerel as 6 oz. per fish (from samples of fish weighed) the average output per net was rather more than 4¼ cwt. per night, and the average output per man was nearly ½ cwt. per night. In hand-line fishing average output for over 700 cases was about ⅓ cwt. of fish per boat, and 1/16 cwt. per man per day.

in work. I am inclined to think that one of the most useful features in my records of the major types of fishing in this area is their systematic nature, the way in which they follow out the results obtained day after day by all or by the majority of the units engaged over a long period. The conclusions drawn from them, approximate though they are, are more objective and more valid than any formed from haphazard inquiry.

LEVELS OF INCOME

In the preceding analysis I have taken output as the aggregate of the values of the product of all units engaged, at current wholesale prices—that is, what the catches of fish actually fetched on the beach according to my records, or what they would have fetched if they had been sold and not carried off for home consumption. In assessing income levels I take gross income as the aggregate of the amounts received by individuals from the output of the units in which they participated, taking into consideration the principles of distribution of the earnings from each type of unit. Net income I take as this aggregate *less* the costs of obtaining it—i.e., *less* taxation on equipment and the costs of maintaining or repairing equipment. This treatment is, in essence, the same as that used in one method of calculating income in an ordinary economic analysis. But it has justification not only in theory but also in practice. In a Western economy, where incomes are usually studied, the relation between output and income is normally less direct than in the present case. In the fishing communities of Malaya the incomes of individuals are received in nearly every case by direct division of the *actual cash* obtained for the fish produced by the net-group or other unit concerned. Production is still to a large extent of the " round-about " kind, that is, the largest section of the goods produced is not itself shared out among the agents of production, but is sold to outside buyers. But the agents of production are recouped primarily from the immediate proceeds of the sale. They are not maintained—as in an industrial community—by a flow of cash from banks or from other accumulated reserves to which the funds received from the sale of the product percolate through. Moreover, in the fishing community a certain share of the product, namely, that reserved for domestic use and sale (the *makan lau‘*), is actually shared out among the agents of production on the spot and thus forms an immediate part of their income.

A broad calculation of the average net income per head from fishing in the sample area (Perupok) for 1939–40 gives the following result. Taking the value of the annual output at $65,000 to $66,000, the annual replacement and repair costs for boats, nets and other gear at $5,000 (estimated from differential rates of depreciation of each type of equipment), and the annual cost of taxation at $500, a total net income of $59,500 to $60,500 is obtained. Taking the number of fishermen concerned as 550, the individual net income on the average is $108 to $110, or approximately $9 per month. This may be compared with the figures given earlier (pp. 22, 40) for the whole of Malaya and for the Kelantan-Trengganu coast. On this basis the men of Perupok would seem to do rather better than the average of the east coast fishermen, though not so well as the average for all Malaya. But these general figures involve many estimates— sometimes only rough estimates—of quantities, and can be regarded as no more than likely approximations in the absence of more exact data.

The average annual income per head is a useful figure for comparative purposes—if allowance be made for differential purchasing power. But the bare citation of this average can be misleading. It tends to strengthen the impression commonly held that these Malay fishermen are all on the same fairly uniform income level. It conceals the extent of the variations that exist, and may perpetuate an idea that increase in the volume of output is all that is necessary to raise the general standard of living for all fishermen. To grasp the importance of the economic differences between individuals in such a community, as regards capacity to save and accumulate capital, or capacity to alter the direction of their consumption and possibly enlarge the effective satisfaction of their wants, it is necessary to compile a schedule of incomes from more detailed analysis. This schedule can be built up by adding together the distributive shares received by individuals from their participation in the fishing industry.

In a community of this kind, where, of course, no returns of income are made to the State, one must rely on empirical investigation. The difficulties in this will have appeared at various points in the earlier analysis, but they may be briefly recapitulated here. The income of each individual is derived from many sources. Each not only gives yields of different magnitude, but also distributes this yield among the agents of production on different principles. The yields are highly variable, not only seasonally but even daily, and variations due to weather and other physical factors are complicated by differential skill and efficiency in organization. Apart from variations in output, the number of persons in a given unit may vary from day to day, thus providing larger or smaller shares to the participants, and even though the numbers be stable the actual individuals may differ owing to re-alignment of groups or borrowing of crew and equipment. Again, the principles of distribution of output, though following a general form for each type of fishing, are not entirely constant ; they may vary according to the decision of the chief distributor, who is usually the principal owner of the main equipment. And, finally, variations in the ownership of the equipment used, even between different units in the same type of production, mean that the users of the equipment in different groups derive incomes of different magnitude from similar levels of output.

It is clear that no system of empirical records, however detailed, could cover in practice every single variation day by day in all these quantities—though I can claim that most of the main ones have been fairly closely measured. My estimates of income are therefore only approximate.

The steps in the analysis of each individual case need not be given here. But the principle followed has been to calculate from the figures of returns to the individual boat and net groups of different types the receipts of ordinary crew-members and of owners of the various forms of capital, and then to re-combine these receipts as they accrued to individual fishermen in the course of their changing activities. For instance, in the case of an ordinary lift-net crew-man during the period November 1939–May 1940, when the records were taken, the items to be computed are : his share, approximately 2 per cent., of the takings of his particular net ; his share from the trolling of Spanish mackerel and bonito by his particular boat ; his share of the fish for domestic consumption and sale ; his receipts from line fishing for sea-bream and other fish during the " off-season " for lift-netting in March and April ; his share, approximately 4 per cent., of the proceeds of mackerel netting at night, when he probably went out in the boat of the expert or carrier agent of his lift-net ; and what he received by the mullet caught by himself in his scoop-net during the monsoon. In the case of a lift-net expert, on the other hand, the items are : his share from the lift-net fishing, approximately 25 per cent., including the shares for his net, his boat, his " body " and his specialist skill ; his share of receipts from trolling from his boat, and from the personal sale of *makan lau'* ; his share from mackerel fishing by night, this amount depending on whether his own boat is used, and how many sections of net he owns ; and his receipts from line fishing for sea-bream, etc., during the off-season for lift-nets. Normally, the last is smaller than for a crew-man, since the *juru sělam* goes out line fishing less consistently. There is ordinarily no item for mullet in the monsoon, since hardly any *juru sělam* bothers with this method of fishing. In all cases, including those where other types of fishing than lift-netting are followed, individual earnings have been calculated as far as possible in terms of the actual groups to which individuals belonged, and not in terms of general averages. A considerable amount of approximation has necessarily been used, particularly as regards receipts from the minor items. But the general picture of the various levels of income does have the merit of being composed from empirical records, and does show fairly accurately the distribution and range of incomes.

The broad results of the analysis, rendering the figures into terms of weekly income, is as follows.

Out of the total of approximately 550 fishermen in the area, about 90, or roughly one-sixth, get less than \$1.50 per week from fishing alone, on the average. These are mostly men who are also cultivators, and for whom their fishing is a part-time occupation. About 200 of the men, or a little more than one-third, get from \$1.50 to \$2 per week on the average. These are mainly ordinary crew-men, supplying their labour primarily in lift-net and mackerel fishing, with no capital of their own in nets or boats. About 180 men, or one-third of the whole, get from \$2 to \$3 per week on the average, and about another 50 men, about 9 per cent. of the whole, get from \$3 to \$5 per week. These are mainly men who participate as crew in the lift-net fishing, but as boat owners, owners of sections of mackerel net, or other forms of capital have a subsidiary source of income beyond their labour. In the higher ranges, about 25 men, or nearly 5 per cent. of the whole, get between \$5 and \$10 per week ; 4 men get from \$10 to \$20 per week, and 1 man from \$20 to \$30 per week. Of the men with \$5 to \$10 per week, 10 are lift-net experts, and the rest are specialists in other types of nets, and owners of considerable capital in nets and boats as well. All the 5 men getting more than \$10 per week are lift-net experts.

Putting the results from another point of view, while 20 per cent. of the total fishing income is absorbed by little more than 5 per cent. of the total men engaged, 85 per cent. of the total personnel divided about 65 per cent. of the total income.

An important problem is what proportion of these people live at or just above the subsistence level, which in 1940 was about $1.20 for a family of three persons, of husband, wife and one small child. An answer can be given only in very broad terms, because of differences in the size of families, individual earning capacity and subsidiary sources of income by husband and wife. But on a very rough estimate, perhaps about one-fifth of the population of the area have little if any income surplus above the requirements of their daily domestic needs. If the standard be reckoned as $1.50 to take in items regarded as part of the normal costs of community life, perhaps between one-quarter and one-third of the population are at this level, with no margin to put by for capital accumulation.

The amount of saving possible in such a community is difficult to calculate. It is clear from the income distribution in relation to the current cost of living that the surplus available, where it exists, is in most cases small. But there is some margin, amounting for the community as a whole, to something probably between $10,000 and $15,000 per annum, or between 15 per cent. and 25 per cent. of the total fishing income. Much of this potentiality for saving goes, of course, into current consumption, and much also into periodic items of special consumption such as more elaborate houses, ornaments, hospitality, feasts and ceremonies. But there is sufficient margin to maintain the fishing capital and even increase it. This is shown by the fact that unlike the situation in south Trengganu, there is hardly any capital from outside invested in fishing equipment. Moreover, in the higher range of income levels, there is a significant investment of savings from fishing in land, either directly by purchase or indirectly by loan.

In sum, this analysis of fishing incomes, brief and approximate as it has been, does show the differences in the position of these fishermen. It indicates how a considerable proportion of them are near the poverty line. But it also shows how incorrect is any view that regards them as all on the same level, having neither the possibility nor the capacity of improving their position.

CHAPTER X

FISHERMEN IN THE GENERAL PEASANT ECONOMY

Throughout most of this book I have been dealing with the internal organization of the fishing industry—its productivity, its planning and the returns it yields to those who take part in it. But to complete the picture a brief account is needed of its wider economic and social relationships. We must see how far the income of the fishermen is supplemented from other sources, what they do with this income when they get it, and how their life merges with that of the general Malay peasant community.

The Malay fisherman is not a race apart. He is a specialist in an exacting occupation, often brought up in a family which for generations has followed its calling at sea ; he spends most of his days among others of his kind, amid talk of boats and nets, wind and weather. But he often has some other pursuit on shore to occupy his slack time and bring in a little extra income ; his wife may grow vegetables, make mats or sweetmeats for sale or go to trade in the markets. On Fridays he mingles at the mosque with rice-planters, craftsmen and shopkeepers ; in the evenings he attends marriage feasts and circumcision feasts, spirit-healing seances or shadow-play entertainments in his own village or in others near by. Through marriage or blood relationship he has kinsfolk inland, with whom he exchanges visits and gifts ; who may be bringing up some of his children or he some of theirs ; and with whom he often has complex ties of mutual assistance and monetary transactions. As an *orang ka laut*, a man who goes to sea, he is distinguished from the *orang darat*, the inland folk, but he is at one with them in being essentially a peasant with the same kind of economic and social outlook, the same type of social institutions and the same general standard of values.

SUPPLEMENTARY SOURCES OF FISHERMEN'S INCOME

The fishing communities along the Kelantan–Trengganu coast vary in the degree to which they supplement their income from other resources. The situation of each village is important : its nearness to good agricultural land, to a forest or a plantation, to a river or to a small town tends to open up opportunities for additional employment for some of its members. But initiative and energy, and the existence of a craft tradition also play their part.

At some villages, such as that of Ayer Tawar in Besut, the people live almost entirely from fishing. They do not cultivate rice, though there is some land near that might be so used. Though they are poor, their women do not take up agriculture, but sit at home till midday and then come down to the beach to collect the day's fish. No nets are made there and so the women do not have this work as a source of income ; a few of them engage in petty trade and a few others possibly earn a little by working for a Chinese who runs a small shrimp-paste factory. At the village on the Tanjong on the south side of Kuala Trengganu, no rice is cultivated either, since there is no available land. Not only rice but nearly all coco-nuts and firewood must be bought with money earned by fishing. " We all eat from this," said a fisherman, pointing to a net. But a few fishermen have fruit orchards inland ; others take some part in net-making and boat-building, while some of their wives engage in weaving cloth. At Saberang, on the north side of the estuary, there is no rice cultivation ; the soil is almost pure sand there, and the fields are too far away. But some of the men work as labourers at 50 cents to a dollar a day when a steamer is in, handling copra and other cargo ; some fishermen who have boats ply as ferrymen across the estuary when fishing is poor ; and some of their womenfolk make bamboo trays for drying anchovies or mat-awnings from pandanus leaf. These awnings, which are sold locally, serve a variety of purposes, from covering boats and nets to acting as a temporary house-wall. At Penarik and other villages along the sandspit on the seaward side of the lower reaches of the Setiu River there is also little opportunity for agriculture, the soil for some distance inland being sandy, with the gĕlam, a " paper-bark " (Melaleuca leucadendron) as predominant cover. But the bark of this tree is used for caulking boats, mat-awnings are made and sold in the larger centres, and rattan is collected inland and exported to Siam. In the monsoon season, when the men cannot go to sea, a considerable amount of river fish is taken for food. By the people of Kuala Marang, Batu Rakit and Kuala Besut (all important fishing villages in Trengganu) no rice is cultivated, though in each case there are other occupations. At Kuala Besut, for instance, there is some boat-building and net-making. Women prepare and sell dried fish, make containers for the export of this and of anchovies, and make sweetmeats for sale. A few people get some income from small coco-nut orchards, a few rubber trees, or a little rice land share-cropped by planters inland. A small amount of rice is obtained from kinsfolk, cultivators, in exchange for cured fish. At Paka, Chenering and Merang a little rice is grown by some fishermen. At the last-named village cultivation is done on land newly opened up by the government ; other sources of income here are wood-cutting, making mat-awnings (at piece-rates for Chinese) and working for wages for Chinese at cooking anchovies.

In the Kelantan fishing villages things are very similar, though their nearness to the great agricultural plain allows of rather more rice-growing and vegetable-growing by fishermen and their womenfolk than in Trengganu. I have shown in Chapter III that in the coastal villages of the Perupok area taken for statistical inquiry three-quarters of the men followed fishing as their primary occupation ; and while four-fifths of these had no definite secondary occupations the other fifth did. Nearly a score of occupations were involved, the most important being rice-growing and other forms of agriculture, fish dealing, and making fishing equipment. Moreover, at least one-quarter of the womenfolk of the community had fairly regular occupation, while many others did casual work. From supplementary investigations further inland it was clear that many of the men living upwards of a mile away from the coast divided their time fairly evenly between fishing and agriculture, and some of them had subsidiary sources of income by preparing coco-nut sennit, collecting coco-nut sap for the manufacture of sugar, tile-making for a small Chinese factory a few miles away, and so on. Their womenfolk were busy with agriculture, net-making, tile-making and petty trading.

It is difficult to estimate at all accurately the total contribution to income made by these subsidiary employments, and by owner-ship of resources such as rice land. Out of the 331 households in the census area, only 130 had resources in rice, bringing them in altogether about 25,000 gantang of padi per annum, worth about $2,500. Various costs would amount to about half this sum, leaving the other as net income. The proportion of padi-growers among the fishermen outside the census area was much higher, and the total net income from this source among the whole set of fishermen concerned in our inquiry was probably in the region of $4,000. From coco-nuts, 160 households in the census area had resources giving an annual net yield of about $1,000 ; for the whole set of fishermen the total net income from this source would probably be about $3,000. From vegetable growing and the rearing of livestock the income would probably be in the region of $2,000, while from fish dealing, craftwork and other special services (impossible to estimate closely in the absence of detailed inquiry) the income might be $5,000. All this gives a total supplementary income of about $14,000, an average of $25 per man per annum. This, added to the income from fishing already calculated earlier, raises the total to an average of about $11 per man per month. As already noted, however, a considerable proportion of the fishermen and their families do not have such subsidiary income sources.

SAMPLES OF HOUSEHOLD INCOME

The results, in terms of household income, of this combination of fishing with other activities may be illustrated by four examples, representative of some fifty cases studied in detail.

1. A fisherman of middle age, with wife, three young sons and a daughter, living in a single-roomed thatched house a few hundred yards inland from the beach. He owns a small boat bought for $39, and a small drift-net. He has rice land, which he cultivates himself with a hoe (owning no cattle for ploughing). The land yields him 50 gantang of rice per annum, but it is far too little for the family needs, and he buys about $45 worth of rice a year. He has also some coco-nut palms, which yield about 50 nuts at a climbing (about 250 per annum), enough for food ; and a small vegetable garden (typical contents being about 20 taro plants, 20 cucumber plants and about a dozen yam vines). As spare-time work he makes a little sinnet cord, selling from $4–5 worth in a season, and also some hooks for taking Spanish mackerel, selling perhaps $2 worth. His recorded income from fishing was $60 for 167 days, including the monsoon, and his annual fishing income would be in the region of $150. With a total income of about $160 per annum, his expenditure on routine needs for himself and his family would be in the region of $115, leaving some $45 for expenditure on social affairs, etc.

2. A young fisherman with a wife but no children, living in a single-roomed thatched house near the beach. He goes out as a regular occupation with the lift-net and mackerel net of Japar, the most successful fishing expert in the area. In addition he owns a small drift-net, but no boat. His income from fishing

alone is about $100 a year from lift-net and drift-net, and another
$20 from the mackerel fishing. He owns some coco-nut palms,
which yield about 500 nuts a year, and he sells $7–8 worth of
these per annum. He also owns a small amount of rice land,
yielding 40–50 gantang a year. As an extra occupation he goes
out on occasion with a friend as a member of a shadow-play
team, which brings him in a few dollars every year. His wife is
industrious. Not only does she attend to the cultivation of her
husband's land but in the harvest season she helps in other fields
and so earns extra rice. She also goes inland about twice a week
to sell fish, making about 40 cents a time by this. The total
income of the pair, excluding their rice resources, is in the region
of $160 a year, while their routine household expenditure is
probably only about $60.

3. A fisherman, his wife, four sons and a daughter, living in
a two-roomed thatched house near the rice-fields. He owns a
large boat and several small nets, but spends most of his time
out with a lift-net, with occasional sprat-netting. His eldest son,
though still young, also goes to sea as a crew-man. Their
combined income from fishing for a period of six months' observation was between $40 and $50 only, but this was a period of poor
catches and they were hampered by being in an unsuccessful
net-group. The man owns land which yields 200 gantang of
padi ; in some seasons he works it himself, but usually lets it out
for share-cropping because he finds his fishing leaves him with
too little time. He also owns coco-nut palms which yield 130 nuts
at a climbing (650 a year), and he sells about half the produce.
His wife adds to the family income by making sweetmeats, by
which she earns 20–40 cents per day—" she provides food for
the children ". The total income is probably in the region of
$150 per annum on the average ; their total routine expenses
about $120 (excluding the value of their rice, which lasts them
only about 2 months, owing to the calls of kinsfolk).

4. A man, his wife and one small child, living in a single-
roomed thatched house among the rice-fields. The man divides
his time mainly between agriculture and fishing, with rather
more emphasis on the former. From his father he has inherited
about 10 padi fields, of which he works 5 himself, yielding him
about 150 gantang a year, and lets his brother work the others,
getting from them about another 75 gantang. This gives the
household rice for a little more than half the year (making
allowance for the calls of kinsfolk, ceremonial and hospitality

upon it). After the rice harvest he spends much of his time growing vegetables, having several hundred pumpkins, melons and sweet potato plants under cultivation, and getting from them an income of perhaps $25 or so in a normal year. Most of his fishing is done in March and April, for Spanish mackerel and squid ; he averages 75 cents a day at this over the period, and his total fishing earnings during the year are probably $30–40. In addition, he goes out in the evenings as a *bomor*, a medicine-man, earning thereby 10 cents or so, and some betel. His total annual income is probably in the region of $75 in cash; the routine expenses of the family, as studied by my wife, were almost exactly $3.30 a month for two months, or approximately $40 a year.

Each of these samples shows a balance of income over ordinary routine expenditure. Some of this balance would be absorbed by items such as clothing, contributions to feasts, travel expenses in visiting relatives, etc. But there is some margin for saving, and case 2 is one of a man who was likely to invest his savings in a boat and so enter the ranks of the more substantial fishermen. In general, these fishermen with subsidiary occupations are the more energetic men, aiming to build up their resources.

PEASANT STANDARDS OF LIVING

A detailed study of consumption in the fishing communities of Kelantan and Trengganu, with examination of sample household budgets, has been made by my wife.[1] My treatment of the subject in this section is therefore brief and general.

In this region the primary unit concerned in consumption is the household, of which the personnel in the census area studied ranged from one to thirteen persons, with an average of about 4 persons per household. In most cases the household income is pooled, the wife acting as holder of the purse, undertaking the ordinary domestic expenditure and even controlling the outlay of the husband on his personal wants and the finance of his productive activities. The problems of consumption, however, cannot be considered simply on this basis. Where the household includes others than parents and children, as a married child and his or her spouse, a widowed mother, or some other dependent, some economic autonomy is often observed. The dependents, while sharing the same dwelling, usually " calculate separately " (*kira suku*). While fish is commonly treated as joint income, rice

[1] Rosemary Firth, *op. cit.* The summary here given draws largely from her work.

is bought separately and used as part of an individual budget, as also vegetables and snacks. Cash incomes are usually not shared. Since the dependents often do not gain their living from fishing, but from other occupations, an improvement in fishing incomes alone does not necessarily benefit the whole household, as one might think at first glance.

While a primary interest of every household is in obtaining enough food, even at the lowest income levels mere subsistence is not the sole aim of spending. Rice, the staple food, is not wanted simply as the easiest available means of satisfying hunger ; it is prescribed by convention and by taste as an essential item of a meal, and the quality and flavour of the different varieties used for different social purposes are an important element in its purchase. Similar canons guide the buying of other food items, and find expression in the cooking and eating habits of the people and in the kind of food they set before guests. It is not surprising therefore that the amount of money spent on food, even among the poorest households, varies considerably, even after allowance has been made for differences in the number of individuals catered for, in their age differences and in the subsidiary food resources upon which they may be able to draw.

In general, a figure of 20 cents a day spent on routine items was regarded in 1940 as a normal standard by the people themselves, for a household consisting of two adults and one child. This, representing just under $1.50 per week, would absorb practically all the income of some of the poorest families in the community. In practice, however, as seen by analysis of a series of household budgets, the ordinary weekly figures ranged round a mode of about $1.20 per week. The individual variation was considerable, the cash expenditure being as low as 40 cents in a week with a household consuming its home-produced rice (equal to a total cash outlay of about 65 cents if the rice had been bought) and as high as $1.89 with another household. Variation from week to week for the same household was also fairly marked, and though partly due to the lack of correlation between rice purchases and weekly periods, was also the result of differences in the amounts of fish, snacks and other more marginal items bought.

The major items of routine household expenditure are rice, snacks, tobacco and betel-chewing materials, spices and sugar for flavouring, oil and matches for lamps, vegetables and fruit, and fish. For all but rice cultivators the first three items are of most importance in the budget. Rice, taking from one-third to one-half of the budget in cases where there is no private source of supply, is consumed at a rate expressed fairly accurately by the peasants themselves as " a *chentong* a person for one cooking ", that is, about a pint of uncooked rice a meal, or a pound a day. Several qualities and prices are available in the market, and very definite preferences exist. Snacks, which apart from coffee are mainly sweetmeats with rice as a basic component, are an important element in the schedule of wants, and take up from one-eighth to one-fifth of the budget. This would hardly be inferred from casual observation, and indicates the extent to which other than purely nutritional factors enter into the purchase of foodstuffs. So also with smoking and betel materials, the demand for which is widespread and regular, and which are regarded as essential items in the satisfaction of wants. The other items named do not call for comment here, save that of fish. It may seem surprising that fish should be bought in such a community, where so much fish is distributed free among the families of the producers. The demand comes partly from households which have no direct source of supply, and partly also from those whose members have been unsuccessful on any given day.

To discuss the standard of living or level of consumption of these Malay peasants simply in terms of their routine household expenditure would be unjustifiable. Allowance has to be made for their outlay on clothing, housing, household furniture, taxes, recreation and other social items. Here, though social standards dictate their requirements, there is much variation.

In clothing the range of expenditure is not very wide. Rich and poor in everyday life go about in much the same garb ; it is only on holidays and festivals that wealth becomes apparent, and even here it is among the women rather than among the men. For most of these peasants a few dollars a year would cover their clothing budget. With the women differences are noticeable in their ornaments. But in the intricate system of loans, often between kinsfolk, a piece of jewellery may be pledged as security and may indicate its wearer as no more than a temporary creditor on the one hand or a potential debtor on the other.

Standards of housing vary greatly, from the simplest shed of thatch roof, split-bamboo walls and palm-trunk posts and floor to the elaborate dwelling of tiled roof, dressed timber gables, walls and floor, and concrete posts and entrance steps. A simple house may cost $25 to $50 ; an elaborate one $500 to $1,000. On the whole, perhaps a mean figure for this area might be in the region of about $100. Furniture in most cases is simple even in the more elaborate houses. It consists mainly of pandanus mats, some pillows, some pots, jars and other simple domestic utensils with some imported china in the more wealthy households. Tables, chairs and other furniture are practically unknown.

Taxation is an item of some account in the budget of the Malay peasant. In Kelantan there is the annual quit-rent on land, on which the ordinary small-holder expends, as a rule, one or two dollars, and the annual tax on boats, which may amount to $5 or so for a large boat. Apart from these, the only items of regular incidence are the *zakat* and the *pětěroh* (*fitrah*). The former, the tithe on rice production, is levied only on those cultivators whose annual production is more than 400 gantang ; as such it is of no account for the coastal community under consideration. The yield from this tax goes to the Majlis Ugama, the governing religious body of the State. Part is returned to the local Imam (chief Muslim official) of each parish to go towards his income, and part is retained for general religious purposes. The *fitrah*, paid in cash or rice, is levied at the rate of 20 cents per head per annum (in 1939–40 ; the rate varies in different years according to the price of rice) on all the community, fishermen and rice cultivators alike. According to the Imam of Perupok the *fitrah* is paid on a household basis, two persons paying at the rate of one, four at the rate of two, and so on. This, however, may be a concession, since if a person has no property, he said, the tax is not taken. Though in theory only rice and cash are permissible, other types of property may apparently be distrained upon by the Imam in lieu of these. The Imam in Perupok derives no revenue from any *zakat*, but gets one-fifth of the *fitrah*, which amounts in this parish to about $150 per annum. Apart from the *fitrah*, which is a religious due, there is no poll-tax. A number of dues of various kinds levied on commercial and other activities, however, are somewhat of a burden on the peasant. Transfer fees on the sale of boats and of land are small, and constitute a reasonable administrative charge, but there is no doubt that they lead to evasion and to the sale of property by verbal agreement alone, without registration. Market fees, payable by all stall-holders, and even by people who come to the market with a basket or two of produce, though also light (one cent per basket, for instance), are felt as burdensome, and also are often evaded. There is some justification for this. The turnover of many market sellers is so small, and their margin of profit also, that even a fee of one or two cents may make a substantial difference to their position. Irritating also are the dues that have to be paid when cattle are killed, and when a shadow-play or a spirit performance with music is held. The object of the government here is possibly rather to exercise some control over these activities than to derive a revenue. Both objects, however, would seem to be only imperfectly realized, since the revenue is small, and it is the cost of the affair to himself rather than the payment of the fee that restrains the peasant from greater indulgence.

Malay community life in the north-east States is still very strong, and the social values based upon it are of great importance in dictating stands of consumption. This is true not only in such relatively minor spheres as the need to have supplies of smoking and betel materials on hand in every household with any pretensions to status, for the entertainment of any visitors who may come ; or in the calls upon a rice-cultivating household to make gifts of rice, especially new rice, to less fortunate kinsfolk ; or in the obligation to provide food for kinsfolk who may arrive from other villages. It operates also in the sphere where recreation and more formal social institutions combine, particularly under the influence of Muslim religious values.

Charity, as in all Islamic communities, is enjoined upon all who can afford it, and is practised to a considerable degree by the wealthier peasants. One of the commonest forms is payment for the erection or repair of a *rokaf* (*wakaf*), the open-sided wayside shelter which is such a boon to traveller or lounger along the paths of the hot countryside. So much is this form of charity regarded as a duty that I heard a wealthy fisherman criticized for not having yet built a shelter. School-houses for teaching the Koran, praying-places, and wells, are also provided by men of wealth.

Basically religious too, though overlaid by much Malay social practice, are such customs as the invitation feast and the presentation of sweetmeats at the breaking of the fast of Ramadan ; the ritual sacrifice and accompanying presentations of meat at the feast of Hari Raya Haji, in commemoration of the Mecca pilgrimage ; and the celebration of the *suro* (*Ashura*) about the tenth day of Muhurram. Each of these at one time or another affects the economy of all but the very poorest households, raises their standards of consumption for the time being, and involves them in considerable extra expenditure.

Apart from these regular annual events, from which a household can abstain, or in which it may play a minor part, according to financial circumstances, there are other social institutions of less regular occurrence but in which it is difficult or even impossible not to participate and which demand some outlay. There are funerals, circumcision rites and marriages within the circle of kinsfolk and immediate neighbours, and, most important of all, within the household itself. There is no space here to give a detailed description of these institutions, even on their economic side alone. But from the point of view of their relation to standards of consumption they have one important feature—they all demand contributions either in cash or in kind, or in both. The system of *měngelen*, of taking contributions to a feast, is deeply rooted in the local Malay social and economic system, and while some recompense is normally gained by a meal, the immediate kin of the host are forced to incur expenses which, for the time being, are well above their ordinary routine standard of living. On a long-term basis, as shown in Chapter VI, reciprocity is commonly obtained, and even capital resources built up or mobilized, but the short-term effect is often to put a strain on the household finances through the need to save or borrow.

In connection with the feasts or other social events, or standing alone as entertainments, are performances of shadow-plays, mediumistic seances, religious chanting, formalized wrestling, organized drumming, etc. From the economic point of view these activities may involve a certain amount of communal contribution, but in practically every case the main burden of the expenditure is shouldered by one individual who thereby obtains satisfaction and prestige as a public benefactor. The sums disbursed by him may vary considerably but a figure of from $15 to $30 is frequent for such entertainments. Naturally it is most commonly the more wealthy members of the community who undertake such entertainments, but persons of even comparatively moderate means sometimes incur the outlay, either in fulfilment of a vow, or to obtain social kudos.

In sum, the standards of consumption are governed by these social standards of ceremonial and recreation to a significant degree.

It is difficult to estimate the weight of all these social items in quantitative terms, and no scheme of expenditure schedules can be given here, especially since the incidence of these items is so irregular. But, on the average, allowance must be made for an annual expenditure on them by households of the community studied of amounts ranging from a few dollars to fifty dollars or so.

During the normal fishing season, which with its various types of major production extends throughout the greater part of the year, except during the period of the monsoon (roughly between the beginning of December and the end of January) the income of these fishermen is adequate to meet their current needs for routine household expenditure. The monsoon, however, introduces a complication. The only type of fishing available, that for mullet with the scoop-net, does offer a cash income, but it is irregular and small, save for the few most skilled and energetic or hardy fishermen. The scheme of consumption has therefore to take account of this reduction in resources. The situation is met in one of three ways, or by a combination of them. In the first place, the level of consumption during the monsoon tends to be reduced. Except perhaps for households absolutely on the margin of income, this does not mean that the amount of food is lessened, but that expenditure on snacks, and tobacco and betel materials is cut down. This reduction, primarily a response to a lower amount of cash at command, can be also linked with a change in the habits of the men, who, having no longer the same arduous labours at sea, may perhaps be satisfied with rather less. The second method of meeting the situation is by saving for the monsoon during the full fishing season, which means in some cases a restriction of consumption for three or four months beforehand. The amount so saved depends upon the prosperity of the fishing season. But a small household depending on one man for its major provision normally hopes to save about $10 for the monsoon, and so be able to carry on during this time at much the usual level of consumption. The third method, adopted by the poorer households, or those which have been unfortunate during the fishing season, is to borrow food or money or both during the monsoon and repay the loan after the new fishing season has started again. This involves a restriction of consumption retrospectively. Shopkeepers and wealthy *juru sělam* are the most frequent lenders, and rice is a common item so lent. The aim of good housekeeping, however, is to get through the monsoon period without borrowing, and the level of consumption is usually adjusted with this in view.   Households which have rice resources of their own apply this principle in a special way.   If, as is commonly the case, they have not sufficient rice to last them the whole year round, normally they do not begin by eating up their stores and then proceeding to buy rice.   They first buy rice for a number of months, with the cash they get from fishing, and then, when the fishing income declines during the monsoon, they draw on the accumulated stocks.   The procedure is a sensible one, but it is striking to the observer when he first encounters it, to find people buying rice when they already have quantities in store. Incidentally, this shows that in appropriate circumstances a Malay peasant does exercise foresight in economic affairs.   He acts, in effect, on the principle that his rice demands are comparatively inelastic, while his demands for cash are elastic ;   the former can be easily estimated in advance, the latter not.

POVERTY AND WEALTH IN THE PEASANT SCHEME

It will have been clear from the analysis given in this book that we are not dealing with a community in which wealth is fairly evenly distributed.   Even in this comparatively simple peasant organization resources and income are sufficiently varied to allow one to speak of poverty and wealth, to cause some individuals to be constantly conscious of the pressure of having to find enough to eat for themselves and their families, and others to be free from worry about their subsistence and to have a surplus for investment.

The Malays express these economic differences in a number of ways.   The commonest expression for a poor man is that he is " hard up " (sĕsok).   Amplified, the position is put as follows : " He is hard-up, he is living in some one else's house " ;—" he hasn't a thing, the house is that of another man " ;—" except for his food box and betel box (taken to sea as the equivalent of a lunch-basket) he has nothing " ;—" he has a paddle, a food box and a betel box, and nothing else " ;—" he who has no boat, no net, no padi land, no orchard, and who dwells in the house of another person, that is a poor man."   These statements crystallize the position of a fisherman of the poorest type, who has no capital, who gains his living by his labour as a crew-man alone, and who probably has to borrow to keep his family going during the monsoon.   At a higher level are people whose income is sufficient for their needs, but who have not the capital to launch out into productive enterprises of any size. One such man described his household thus : " We have enough to eat ; we don't borrow ; we can't put anything by—just enough to eat." This was an old man and his wife, with a child ; he possessed an old small boat, an old small net, one small field of rice land giving about two months' food for the household, no coco-nut orchard, but a small vegetable patch. He got the major part of his income by line fishing and going out with his net. Another man, without a boat, but with a small drift-net and two sections of mackerel net, a small field of padi land and a small coco-nut orchard described himself as " hard-up " because he could not afford to buy a lift-net as he would have liked to do. Others in similar positions complained of their inability to buy boats. At the other end of the scale are the men locally known as " Rich Awang ", etc., possessing, for instance, a large boat, several sections of mackerel net, a drift-net, making and selling several lift-nets a year, having enough rice lands to yield food for their households all the year round with a surplus for sale, and large coco-nut orchards from which they draw a substantial cash income.

Some of this wealth is inherited, but the rules of the Hukum Shara', insisting on division of property among all the heirs, tend to operate against continued concentration of wealth in a few hands. On the whole, in this area men of wealth have accumulated their property by industry and saving. These two features, combined with the practice of charity enjoined on the rich, probably account to a considerable extent for the absence of any marked feeling of resentment towards the wealthy on the part of the poorer elements in the community. An allied factor also is the concept that wealth, while the product of individual exertion, has only come to a man because of the favour of Allah ; he has been " blessed ", and therefore, though envy may be natural, one is not entitled in the last resort to question the will of the Lord. Where resentment and criticism do enter is when the rich man does not show himself generous, when " his liver is thin ", when he does not practice charity to the poor, build wayside shelters or prayer houses, or entertain liberally. In brief, the control over wealth still exercised by religious precept and practice, and the cement of religious belief, tend to obviate to a considerable degree the possible causes of friction inherent in marked economic inequality. In this respect the Kelantan peasant community is comparable with the medieval European community.

TENDENCIES TO CHANGE (BY 1940)

There is little doubt that during the last thirty years the position of the peasant in Kelantan has tended to change, particularly in the direction of greater differentiation in levels of wealth.   Under the old system, with a large degree of arbitrary control by the ruler and the nobles and territorial chiefs, and much closer dependence of " their people " upon them, the wealth of the peasant was always subject to distraint on the part of his superior.   With the introduction of effective British administration, with its system of land registration, moderately impartial Courts, regular salaries to officials paid out of State revenue, and substitution of fixed annual dues with centralized exchequer control for the more local, variable and arbitrary produce taxes, " farms " and other levies, the personal, social and economic ties between peasant and his feudal lord have tended to disappear. While this has meant greater freedom, security and individuality for the peasants as a whole, it has also facilitated economic differentiation between classes of peasants.

In agriculture the security of land tenure, the comparatively low quit-rents for non-cultivating as well as for cultivating owners, the entry of rubber as a commercial crop and probably the growth of population also, have promoted the existence of *petits rentiers* living largely on the shares of the produce of their lands worked by others.   Definite information on the situation is lacking, but I am inclined to think that the practices of share-cropping rice lands and of leasing the produce of orchards have increased considerably in recent years, perhaps to an extent hardly realized by the Government.

In fishing, the same factors, linked with the extension of communications and the entry of wheeled transport have operated to develop newer and larger markets.   With this development has come the introduction or at least the extension of more elaborate technical equipment, wider opportunities for the investment of capital, and a tendency for the rise of a class of capitalist-entrepreneur fishermen.   These, while they nowadays work themselves and supply a great deal of the specialist skill and organizing functions, may well tend in the future to restrict themselves more and more to capital control, leaving the supply of labour and technical skill to others.   This tendency to the rise of a petty capitalist class is still rather a prediction than a demonstrable fact.   But the analogy presented by the entry of

the Chinese fish-dealers as capitalists for the Malay fishermen of
south Trengganu and Pahang strengthens the inference.

The gradual diversification of the peasant economic structure
has also been promoted by the increased opportunities of spending
money.   Much expenditure has been diverted from local goods
and services to investment in distant property, and to the purchase
of a new range of consumer's goods such as European furniture,
clocks, china, sewing machines and even to some tinned foods
such as fish and milk.   The unequal spread of education and
the chances of foreign travel also allow some individuals to widen
the gap between themselves and the great mass of the peasant
communities.

Such changes, similar in many respects to those which
have taken place during the last century in the economic and
social structure of many other peasant communities in the Orient
and elsewhere, are of great significance.   They raise fundamental
problems of how these communities can adapt themselves to the
new wants, opportunities, values and requirements presented
to them by the modern world.   These processes of change, and
the problems arising from them, have been the object of much
study by anthropologists, particularly in reference to the ' culture-
contact ' or ' acculturation ' produced by the impact of Western
civilization on hitherto primitive societies.   But it should be
remembered that over much of the Orient such changes represent
but one phase—albeit a very pronounced one—of a historical
process that has been at work for centuries.   The Kelantan
peasantry, for instance, are not a primitive people who have
remained until recently out of touch with an outside civilized
world.   Directly or indirectly they have had long-standing
trading relations with foreign lands ; the ravages of war were
not unknown to their ancestors ; their country formed part in
turn of a Buddhist empire and of a Hindu empire, and even
today Hinduism has left its mark ; about the 15th century Islam
came to give a new faith, a new law and a new orientation to
many peasant institutions ; later again the Siamese claim to
suzerainty had some political effects.   All this is relevant in forming generalizations about culture change in such communities.

CHAPTER XI

MODERN DEVELOPMENTS—A RE-STUDY AFTER TWENTY-THREE YEARS

The main part of this book has been a study from the period 1939–40 of a fishing community in a comparatively small area in Kelantan. The aim of that study was to unravel the complexities of organization in the fishing industry at that particular historical period, and to show analytically the relation between technology, economy and society at such a peasant level of living. In that account I indicated what appeared to be some tendencies to change in the fishing industry and in the local society. Since I was able to make further observations in this area at two subsequent periods, in 1947 and in 1963, this Chapter examines the continuity maintained and the changes that have taken place. This also gives some test of the accuracy of certain predictions made at the earlier stage.

THE WAR AND IMMEDIATELY AFTER

For the Malay fisherman of the East Coast the period of the war was a traumatic one, with considerable hardship. Not only was their country occupied by Japanese troops, but they also had to suffer both indignity and privation. No landings of troops took place in the Bachok–Perupok area, and there was no local fighting, although some foreshore houses were destroyed earlier, presumably to open up a field of fire. According to local reports, on the day when the Japanese landed on the coast the people who had not gone to sea were afraid and fled inland, deserting their boats on the foreshore and their nets in their houses, letting their cattle run loose. The next day they returned and settled down. Conditions grew harsh. Men were put to shore-watching, though not told specifically what to look for. People were slapped by the Japanese at the least provocation. Some folk were killed, others were taken away—the men with promise of work and the women as prostitutes—and did not return.

Fish were bought at prices set by individual Japanese ; if the owner haggled or refused, the fish were simply taken. There was a great shortage of food. The Japanese provided sacks and took away the rice, leaving the people with tapioca as their staple. There was almost a total lack of cloth and the people mainly wore clothing of gunny sacks. Some rural folk made clothing of thin rubber sheeting and sold it. This was all right in the rain, but stuck to a man at sea in the hot sun so that, they said, the skin came away with it. The Japanese seem to have interfered little with the fishing industry as such, but they made some attempt to induce the fish dealers to enter a combine, though this was abortive.

Among the most marked effects were those of inflation. Currency lost value greatly, so that towards the end $100 of Japanese-produced money were worth one ordinary Malay dollar.[1] A man who bought a boat for $11,500 Japanese money wanted $140 of ordinary cash (to give him a profit) and people were offering him $100. Some Perupok men told me in 1947 that the course of prices for, for example, rice and fresh mackerel during the war was as follows :

1940	Quart of rice	6 to 7 c.	Mackerel	3 to 4 c. each
1942	,,	,,	$1	,,	6 c. each
1943	,,	,,	$2.50	,,	10 c. ,,
1944	,,	,,	$10	,,	$1 ,,
1945	,,	,,	$150	,,	" Whatever is asked "

Towards the end of the war the price for a fish was " to follow what a person said ", that is, to pay whatever the seller asked. Inflation had proceeded so far that the stability of prices had been very badly affected.

SOCIAL AND ECONOMIC CONDITIONS IN 1947

By August and September 1947, though the memory of the war was very vivid, there had been a substantial measure of recovery. Poverty was still considerable, but food and clothing were once more fairly abundant, and a wide range of other consumers' goods, such as soap, matches, face powder, combs, buttons, spoons, pencils, exercise books, cigarettes, lamps and other hardware, were available in the market and shops. There were large

[1] The ratio had probably degenerated much more by the time of the Japanese surrender. With the reoccupation no value was given to the Japanese issue of currency, and a new currency was issued by the British Military Administration. Taking into account the pre-war currency brought out of hiding and the new issues, by the end of 1946 it was thought that the total currency in circulation was probably in the region of $400 million. (Pre-war volume of currency was about $220 million.) With the re-introduction of sound currency system prices dropped, but not proportionately ; the pre-war price structure had been a casualty of the war. For more details see, e.g., Raymond Firth (1948), pp. 10–12. (The active circulation of notes and coins in Malaya–Borneo in 1962 was over $1,000 million.)

wooden houses with tiled roofs, the result of investment of trading profits and inflated currency in Japanese times. More generally, life seemed much as before. Religious performances, marriage festivals and other evidences of community activity were frequent. Local traditional recreations of a dramatic order such as Menora and Ma' Yong were available for public occasions and, as an indication of freedom of movement, a Bangsawan theatre troupe from Selangor was just finishing a fortnight's season in the Bachok district.[1]

Yet the evidences of prosperity were very uneven, and there were in the Perupok section of Bachok district alone some 300 men classified officially as " poor " fishermen. Of these quite a number were in receipt of relief food and clothing—for instance, in February 1947 about 500 persons in the Perupok area had issues of rice, flour and clothing made to them by the State Rural Welfare Organization.

By 1947, though Kelantan in common with the rest of Malaya was on the threshold of radical political developments, despite the combination of novelty and privation of the war period, little technological change had occurred in the life of the peasantry. One of the more significant features, a presage of radical changes to come, had been the opening up by the Japanese of a road from Kuantan in Pahang through to Kuala Trengganu, thus linking up Kelantan by motor and truck road with the major centres of population on the West Coast. But the East Coast railway, which formerly ran through the centre of the peninsula through Kelantan to Thailand, had been torn up during the war and not re-established, and communication between east and west was still difficult.

The general fishing technology and the mode of organiza-tion of the fisheries were very much as in the days before the war. Travelling up the coast one could note that off the estuaries of the Kemaman, Trengganu and Besut rivers the large seine, known as *pukat payang*, was still very important. On the open Trengganu beaches the ordinary inshore seine was the principal net, while at Perupok and on the Bachok coast generally lift-netting and gill-netting were still most prominent. The fishing

---

[1] I attended a performance of a compressed dramatized version of the classical story of the hero Hang Tuah. This lasted from 9 p.m. to 2.30 a.m. Though the performance was fairly sophisticated, there were certain evidences of traditional Malay belief. Since there was to be *kĕris* fighting on the stage, incense was burned before the performance, and the actor who was to play the principal rôle " washed " his hands and arms in the smoke while a ritual formula was recited to avert any harm through accident.

boats were still sailing craft. The fishing crews complained much of poverty. One man said to me, " Nowadays expenses are high, there is little food and there are no fish—things are hard indeed." Another said, " Fish is cheap and rice is dear." Still another said, " We eat only wheaten flour "—meaning that they did not make enough income to buy their ration of rice. But even this was much better than the war-time tapioca, and on the whole the net fishermen seemed to be making a fair living.

What of the entrepreneurs and level of capitalization? In 1947 many of the fish experts and fish dealers who had been operating in 1940 were still active. In the Perupok area where lift-net fishing was still the major industry, against a total of about twenty large lift-nets before the war, there were in 1947 still about fifteen. Of these, nine were new, while another new net was being made and another was being purchased in Trengganu. The most marked difference from the pre-war situation was in price. Whereas in 1939 the cost of a good new lift-net was in the region of $150, and in 1940 was still less than $200, by 1947 it had become from $1,000 to $2,000, though prices had steadied recently. One fishing expert (F on p. 268), who himself had put up the capital to have a new net made, said that it had cost him $2,100 in all. Another expert, my old friend Awang Lung (M on p. 265, etc.), said that the total cost of his new net had been $2,000, including ropes, etc. (though a year before it would have been $3,000). The captain of his net boat had some investment in this and they divided the proceeds, two-thirds going to Awang Lung and one-third to the boat captain. Other experts had bought new nets for $2,000, $3,000, even $3,500.

Similar inflated price levels relative to 1940 operated with some other equipment. A woman dealer, Me' Besar, widow of Pa' Che Su, the fish dealer (p. 171), bought a seine for $3,000. If it had been before the war, it was said, she would have paid $250. But by 1947 the price of boats did not seem to be pro-portionately so high. A *kolek buatanbarat*, for example, bought for $350, would have cost about half this sum in 1940. A small to medium *kueh* fetching in good condition $50 to $100 if new, would have cost $25 to $30, or about $40 respectively in 1940.

By pre-war standards the prices fetched for daily catches of fish from the lift-nets were very high. I was able to obtain in-formation about the sales of catches in 54 cases over about a fortnight in early September. Takings for the fish in all cases ranged from $10 to $400, with a general average of about $100 (as against about $20 in 1940). The cases were drawn from most of the experts at work and their individual average takings per net per day when they were successful seemed to range from about $40 to $270. The sample was only a very small one, but it was clear that for catches of about the same volume and quality as in 1940 the prices paid were four or five times as high. Even a moderate catch was fetching $100 or $150 where before the war it would have fetched $25 or $30. Much the same situation seemed to obtain with those engaged in other kinds of fishing, such as seineing and hand-lining. But it was my impression from sales I recorded that the price of small bulk fish had not risen in proportion to that of large, fine fish—Spanish mackerel, pomfrey, etc. Moreover, the prices of other foods and of raw materials had risen at least as much. The fishermen said more, and an official opinion was that whereas the cost of cotton yarn was about ten times as high as pre-war the price of fish generally was only two or three times as great. Hence the relative economic position of the fishermen in comparison with the pre-war period had probably degenerated.

What seemed to me in 1947 to be a significant trend and what clearly foreshadowed later developments was a modification which had begun to creep into the system of distribution of fishing income from lift-nets. Whereas seine and gilling-nets still divided the proceeds as before, entrepreneurs of some lift-nets were beginning to adopt slightly different schemes. The principles for division of the weekly yield of cash have been described in Chapter VIII. Briefly, in 1940 in lift-net distribution, after setting aside one-tenth of the week's cash as share for providers of fish lures, the remainder was split into two equal portions, one of which was divided by the fish carrier among the crew (after taking one-tenth for his boat) and the other divided by the *juruselam* among boats, special task operators and crew after taking one-third for the net's share. This was known as a *bagi-tiga* system in regard to the net (p. 236). But a few *juruselam* had adopted a *bagi-dua* system, allocating half of their share, or some approximate variant, to the net. Now by 1947 this appeared to have become general.

The principles of division of two *juruselam* (Awang Lung [1] and Talib—operating the net of the woman entrepreneur Me' Besar)

---

[1] Meaning " eldest " of a set of siblings, this word is a colloquial form of *sulong* and is ordinarily spelt *long*. The vowel seems intermediate and I spelt it Lung in this book.

set side by side will illustrate this. For simplicity the division is represented as Talib gave it to me, in terms of fractions of takings of $200. Corresponding figures for 1940 are given for comparison.

TABLE 14

DIVISION OF EARNINGS FROM LIFT-NETS – 1947

Awang Lung		1940	Talib	
	$	$		$
Share of fish lures, 10% .	20	20	Share of fish lures, 10%	20
—			Entrepreneur (woman),	
			5%	10
Bagi tiga : praih . . .	60	90	Bagi dua : praih . . .	85
net . . .	60	30	net . . .	42.50
boats and			boats and	
crew . .	60	60	crew . .	42.50
	$200			$200

In the 1947 scheme the *praih* got less, in some cases distinctly less, to divide. But the essential difference was the size of the net's share. Whereas in 1940 the net received three-twentieths of the total takings, by 1947 it received on one scheme just over four-twentieths or on another as much as six-twentieths. This revised method of distribution was justified by Awang Lung by reference to the increased cost of the net, especially by the high cost of thread. Correspondingly, there was by 1947 a lower proportion of takings to divide among the crew, each member getting from a share-out such as that shown above, between about $3.30 or $3.60, instead of about $4 as he would have done on the old scheme (v. p. 247).[1] But the average day's takings, as shown earlier, was probably nearer $100 than $200. If, as in 1940, only about one-third of the days in the year were utilized for lift-net fishing, an ordinary crew man's total income for this primary employment would have been in the region of $200 to $220 p.a. in 1947.[2]

[1] Another modification (at least in the one instance I saw, that of the distribution by Awang Lung), the division was made daily, in the evening and by the *juruselam* to the *praih* and boat captain only—they were then responsible for dividing with their crews later. Compare with the scene described on pp. 237–8.

[2] My impression was that other incomes for other types of fishing were of the same general order. In line fishing an average of 43 man days gave nearly $4 per day per man, but this was swollen by a fortunate catch of four large rays on two days which sold for $98. Omitting these " windfall " gains, an average of 36 man days gave approximately $2 per man over about ten days ; in a week the pair of men in one particular boat only averaged $1.50 a day each.

Two points are noteworthy about this situation.   One is the
flexibility of the Malay entrepreneurs in taking advantage of a
rise in the proportion of capital investment in the undertaking,
to increase not merely the gross takings of capital but also the
proportionate takings.   The second point is that relations of
" labour "—the ordinary crew men—to the entrepreneurs were
essentially of the same social and economic order as before,
manifested by grumbling about the smallness of their earnings
but by no more effective protest.

RADICAL TECHNICAL CHANGE BY 1963

The foregoing accounts set the situation as seen in 1963 in
some historical perspective.   By 1963 a radical alteration had
taken place in the Kelantan situation.   Demographically the
population had increased considerably—from about 400,000 in
1940 to well over half a million (506,000 in 1957), though the
increase was rather less than in most other parts of the peninsula.
Technologically, the re-establishment of the railway link with
Kota Bharu, the continued improvement of the road system and
the development of internal air services, meant that the relative
isolation of Kelantan was broken down to a great degree.   In
1940 one could reach Kota Bharu only by sea or by rail, the
latter involving a ferry journey across the Kelantan river.   By
1963, though the sea route was little used, one could travel not
only by rail with relative ease but also by air or by road.   From
the capital of Malaya, Kuala Lumpur, to Kota Bharu (about
400 miles) hired car or taxi went at great speed, though there
were still several rivers to be crossed by ferry.   (These were
steadily being bridged.)   With great improvements in the local
roads and a general rise in incomes and standards of consumption
of the rural folk, not only had automobile taxis become common
on the Kelantan plain, competing with the buses, but also *teksi*
of another kind—bicycle trishaws, pedal-propelled—plying for
hire even in the small villages.

The apparatus of government had also changed fundamentally.   Malaya had been independent for a number of years, and
in common with the other States was being administered by
active, forward-looking Malayan officials with a direct and permanent stake in the country, and much closer local contacts than
had even the best-equipped British officials.   An elaborate system

of schools had been set up, education had come to the villages and universal literacy was in a fair way to being established.

Yet superficially in 1963 much of Perupok community life in the *kampong* was being carried on in traditional style. Technical processes in agriculture, such as ploughing, harrowing and transplanting of rice seedlings, were carried out as before. Some significant shifts in the application of the peasant labour force had occurred, to include road and other constructional work and seasonal migration to the islands off the coast or to Kedah during the monsoon. But the housing and domestic arrangements of the people, the general character of their clothing, their celebrations of weddings and religious festivals followed in major form the ways of a generation before. Some details varied. There were more wooden houses and tiled roofs ; an invitation to a wedding now came as a printed card instead of as a verbal message and a stick of toffee (*dodol*).[1] The villagers now were exposed to public advertisement by commercial loudspeaker from travelling vans. The attractions of the local open-air cinema had now been added to those of shadow play and Ma' Yong or Menora, and there was a radio as well as a sewing machine on the average for about every house in ten. But a growing reliance upon Western medicine was still underpinned by a faith in the powers of the spirit-healer. And by older men at least I was still told of the belief that a fisherman could adopt a *hantu*, fostering this spirit by occasional offerings of white fowls, so that when caught by storms at sea he might still reach the shore. (In a bad storm a motor boat may get one to shore, but without aid from a *hantu* no ordinary sailing boat can make it.)

Though no radical re-orientation of behaviour patterns had taken place over most of the social field, it was otherwise in the technological field of many of the fishermen. Here a dramatic change had occurred. What was in effect a local technological revolution had taken place : the introduction of motor propulsion for fishing boats had given much greater productive efficiency and the use of ice for the better preservation of the fish had greatly improved the market situation. These improvements were not novel in Malaya ; they had been foreseen by Fisheries Officers before 1940 and I myself had indicated them in my consideration of probable developments in the Kelantan fishing industry (1st ed., pp. 17, 19, 301, 302–3). They did not occur immediately after the war nor for some years later, but by about

[1] See Rosemary Firth (1966), p. 152.

1962 the combination of economic conditions and entrepreneurial initiative was such that they were acceptable. Moreover, the introduction of motor boats made possible the adoption of a new technique of net fishing—which had not been specifically anticipated. Furthermore, some nets were now being made of nylon, not cotton or ramie fibre.

The advantages of motor propulsion in fishing were several. With a motor boat to tow the fleet the fishermen could go much further out to sea—40 miles or so, twice as far as with sail—and so utilize more distant fishing grounds. They could also counter the effects of adverse wind and render the fleet independent of the alternation of land and sea breezes. With a motor boat the fish could also be got back to market more rapidly. The good judgement of the Perupok fishermen was shown by the fact that they did not adapt the less powerful and less efficient though cheaper outboard motors (*enjin sangkut*, the " hooked-on " type). They hung back until they were convinced of the superiority of the inboard diesel-fuelled motors, to which they converted very rapidly. They were able to observe the motorized craft in areas of the south. In a remarkably short space of time, about eighteen months apparently, all the leading *juruselam* of Perupok had invested in these motor boats. The advantages of ice in the fish trade are obvious. Since with an adequate supply of ice fish, even delicate mackerel, can be kept fresh much longer, the catch can be brought in good condition from much more distant markets. From the point of view of return on investment of capital and labour, as shown in the price of fish, ice is a valuable asset.

These new technical developments could have been applied to the traditional mode of fishing with the lift-net—they have been so used in fact by fishermen of Melawi, Sabah and some other coastal areas in Kelantan. But they facilitated the adoption of a type of net comparatively new to the east coast—the purse seine (*pukat jerut*). I was told that this type of net was introduced into the Perupok area only in 1962.[1]

[1] This had been in use for many years by the Chinese of Pangkor on the west coast of Malaya, as reported by Inche' Ishak bin Ahmad (*Annual Report of the Fisheries Department* (1931), pp. 6–7) ; cf. also Gopinath (1950), who gives a detailed description of the purse seine and its different operation by Chinese and Malays. According to Gopinath and Tham Akow (in Kesteven (1949), p. 51) the purse seine was introduced by Chinese from Hainan and Pakhoi in South China. In 1952 it was reported that the use of the *pukat jerut* or purse seine was confined to the fisheries of Kuala Kedah, Pangkor and Mersing and that a Singapore man who began to operate such a gear off the east coast of Johore had difficulty in obtaining skilled local fishermen (T. W. Burdon (1953), p. 87). Its novelty on the north-east coast is indicated

Pukat jerut are of two types, a smaller of fine mesh for taking anchovy,[1] and a larger of coarser mesh for taking fish, principally *kembong* (mackerel) *selar gilek*, *selar kuning*, *lolong*, *selayang* (types of horse mackerel), *tamban* (herring, sprats). The large net for taking fish is said by Perupok fishermen to be about 240 fathoms long by 40 fathoms deep. It is made of nylon and weighs about half a ton. It is furnished with floats about a foot long, made of plastic, lead sinkers, and a row of heavy brass rings at the bottom through which a rope is reeved. The fish are taken by encircling a shoal and drawing tight the bottom rope to form a kind of bag, hence the name purse seine. (Malay *pukat*—net ; *jěrut*—to tighten a slip-cord.) To attract the fish initially, a fixed lure (*unjang ibu*) of coconut front is used (cf. p. 99), but no movable lure (*unjang ano'*). The purse seine operates on dark nights, when the moon is not up. In early years, on the West Coast, the fishermen depended on the luminescent glow of the mackerel shoals just below the surface, to detect the fish. Now the modern practice on the East Coast is to have four or so frames of light wood which will float, and on each of these to tie a set of gasoline pressure lamps—up to a score in all. Each " house of lamps " (*rumah lampu*) is put over the side and gives out a fierce glare which attracts the fish in vast numbers if they are present, so that they come up as a boiling mass under the light. If conditions are very good a great weight of fish, up to 100 boxes or about 6 tons in all, may be obtained. Depending of course on the types of fish caught, $1,000 or more may be obtained from a night's fishing. This is a much better yield than from any other net, including the lift-net (see later, p. 331). The *pukat payang*, perhaps the nearest rival to the *pukat jerut*, cannot operate so effectively with a motor boat, which the pelagic shoals apparently ignore but which frightens off the jewfish, scabbard fish and other ground-feeding types taken by this net.[2]

To manipulate the purse seine in its modern East Coast form at least two boats are needed, one to carry the net, and another, a motor boat, to tow the equipment out against the evening sea breeze and in again against the morning land breeze, and to shoot the net around the shoal of fish. Sails may still be used,

---

[1] The smaller type of *pukat jerut*, for taking *bilis* (anchovy) has the advantage over its competitor, the *pukat tarek bilis*, that it can take the anchovy far out at sea.

[2] It may be for this reason that by 1963 *pukat payang* had decreased very markedly on the East Coast.

---

by the fact that Parry (1954, pp. 83, 104), using data from 1947 to 1952, lists only five *pukat jerut*, none of them in Kelantan, and gives no description of this type.

but as ancillary to engines.   Paddlers are needed much less than
before, but a great amount of power is needed in hauling the
net, so lacking mechanization human muscle is necessary.   (No
East Coast Malay fishing boat was as yet fitted with winches by
1963.)   Hence the purse seine crew is large, from about 25 to
40 men (the most successful net in Perupok had 48), and even
then the crew are inclined to grumble at the severity of the work.
Because of its laborious character the net can be shot only once
or twice per night ; three times would be too much.   I did not
go to sea with the purse seine, but I had vivid accounts from
crew men of their hauling on the net ropes from 8 p.m. to 8 a.m.
with hardly a break for food or sleep, and the *juruselam* shouting
Kuat ! Kuat ! (Harder ! Harder !) at them till they were nearly
dropping with fatigue.   On the other hand, their labour is easier
when they reach shore.   The former practice, still in vogue with
lift-nets, of hauling up the fishing boats from the beach at the
end of each day's work is no longer followed.   The motor boats
are too heavy for daily hauling up.   This is done only when
absolutely necessary for repairs, so they tend to be all left to ride
at anchor.   One result of all this has been a great reduction in
the picturesque sailing fleet which formerly made the East Coast
so attractive.

This alteration in their fishing technology has brought about
fairly radical changes in the habits of a considerable number of
Perupok fishermen.   The lift-net operates by day, and consequently its boats leave before dawn (p. 99), to take advantage
of the outgoing land breeze and to arrive on the fishing ground
about sunrise.   The purse seine operates by night, and hence
their crews leave in the afternoon, prepared to spend the night at
sea and return in the early morning.   A characteristic sight in
the village at about two o'clock in the afternoon is little sets of
young men walking in single file along the path, clad for heavy
sea work, and carrying their paddles, their supper bowls and
their boxes of trolling hooks and lines and tobacco, going off to
join one or other of the various purse seine groups.   At times
the purse seine remains out at sea for three nights at a stretch,
returning home only twice a week.   This of course is made easier
by their control of engine power and ability to move in search
of fish.   Again, the mobility conferred upon them by their engines
has meant that when the monsoon comes (at which season they
cannot lie at anchor off the coast) they can move to more sheltered
waters such as the Perhentian Islands, and fish in the lee, thus

maintaining the same form of production instead of changing over to another type of fishing. From general description it appeared that the fishermen bore these changes of production schedule and of domestic habit with equanimity—rather surprising in view of their opinions twenty years or so before (cf. pp. 19–20 in the first edition of this book).

A radical change had also taken place in the marketing procedures when the purse seine became established. In 1940 the fish from the lift-nets were sold in bulk from the boats on the beach to independent wholesale dealers who combined to buy, then broke up the catch into smaller lots or took it off as a whole to sell in inland markets or to cure (see Chapter VII). For this purpose they used public transport, the motor buses which plied along the coastal and inland roads and found their way down to the beaches by sandy tracks between the coconut palms. But by 1963 the pattern had altered. Whether the capital involved was too great for the beach dealers or whether, more probably, it was found more profitable to combine net ownership with fish dealing, the bulk of the purse seine fish was not sold on the beach at all. It was sorted roughly at sea and handled for economic purposes either by a few fish dealers working individually on a large scale or by the net owner—for whom an emissary often did the actual work. The fish was brought on shore in boxes from the boats and carried directly to light trucks (*beng*, i.e. van). The dealers, practically all Malay, usually bought fish in five or ten box lots. The private use of vans for transporting fish was possible because of the modern availability of large numbers of these vehicles in Malaya ; it made for speed and flexibility in market arrangements as compared with the use of public transport, which might not even be available. As compared with the situation in 1940, very little drying and curing of fish took place.

These developments seemed on the whole to be an indication not so much of the way in which fishermen became marketers of their product, but in which fish dealers entered more deeply into the scheme of production. In order to retain control of the sale of fish they tended, wherever they could, to acquire an interest in a net and motor boat ; thus, they cut out other beach dealers who had previously acted as middlemen. It seemed to be symptomatic of this emergence of a category of larger-scale capitalist fish dealers that in 1963 the term *tauke* was used freely by the local fishermen not only for Chinese dealers as formerly was mainly the case, but for Malay dealers and financiers as well.

On the beach the scene was superficially very similar in 1963 to that in 1940, but there were in fact notable differences. Purse seine boats normally came in from their night's fishing about 8 or 9 a.m., not as did the lift-nets from about midday onwards. It was only on the catch of the line fishermen (see later) that the afternoon beach marketing relied ; in the morning the main body of the catch was at once packed in ice and carried off to inland market. Some of the fish dealers who formerly operated on the beach were still there, but their numbers and their rôle were much reduced. Hardly any petty dealers with bicycles or carrying baskets (*praih kandar*, pp. 213-15) were left.[1] The others bought fish in much smaller quantities and from the crew individually, not in bulk from the boats. This was associated with a modification in the scheme of distribution. Whereas with the lift-net crew members would get only twenty or so fish for personal use and possible sale, under the purse seine scheme if a reasonable catch were made each man was given 100 or so fish—mackerel, horse mackerel, sprats, etc.—some of which he was expected to sell to help make up his earnings. On a reasonably good day a crew member could get at least 50 cents, and more likely a dollar or more by such means. In addition, he had the individual results of his own trolling, which sometimes yielded him a Spanish mackerel or a scabbard fish. These, though to some extent a reflection of individual skill, were primarily windfall gains.

Depending upon circumstances crew members sold their personal fish (*makan lau'*) which they did not wish to keep for domestic consumption, either wholesale to a single dealer or retail to individual purchasers. A simple record from the beach one morning will illustrate the procedure. A crew man was calling out " Come and buy " to prospective purchasers, both men and women. Another was quoting mackerel at 3 for 10 cents. As they were small ones the buyer wanted 5 for 10 cents, which the seller refused. As the buyer turned away, another fisherman called over " 5 for 10 cents ? Here ! " The pair squatted down on the sand, and the fisherman untied the cloth which wrapped up his bowl of personal fish—a mixture of sprats and small mackerel.

---

[1] The few that still operated lived inland. One " bicycle *praih* " living at Gunong some 10 miles away went off with a load of mackerel packed in ice (60 cents worth) ; he had bought the fish for $12.50. Carrier *praih* were poorer. One I saw who had come by ferry across the Kemassin river had bought a basket of fish for $1.50 ; another had brought glutinous rice in to sell for cash with which to buy fish for sale inland.

The buyer wanted 40 cents worth and the seller counted out 20 fish. The buyer proffered a 50 cent piece. The fisherman, who had no change, said " What am I to do with this ? " A woman came up, looked at the situation and said to him, " Give another 10 cents worth of fish, of course." He did so, and the buyer did not object ; but he said as the fisherman counted them out, " Give 6—one more," and the fisherman slowly did so. In this typical transaction is the same sense of a bargain, of the function of haggling, of the significance of margins, of the economics of reciprocity, as obtained twenty years before (cf. p. 217). In general, though the structure of the marketing procedures had radically altered, the principles, the essentials of the marketing process had not altered.

FINDING NEW CAPITAL

Utilization of the new technical facilities depended of course upon their availability in realizable economic terms. What is striking is the rapidity with which the Perupok fishermen abandoned their lift-net fishing of about half a century's pursuit, and took on the new form. As I have pointed out before, peasant " conservatism " is often an outsider's label for reaction to lack of demonstrable economic advantage or to lack of economic resources. In this case the major problem was not one of skill, organization or labour power, but of capital.

The problem of capital, in order to take advantage of a supply of ice, lies of course in the initial manufacturing installations, and this demanded an outlay far beyond local resources. Some ice was available from Kota Bharu. But Government plans for the establishment of ice factories for the fishing industry on the East Coast had been long drawn up, and advantage was taken of facilities offered under the Colombo Plan. In the Bachok area a Cold Storage plant was set up with aid from Canada, and came into operation in January 1963. The plant was used by fish dealers to some extent to store boxes of fine fish overnight (at a cost of $1 per box—several sometimes shared a box), and they also bought ice from it (at $4.80 per box or proportionate figure per half-box) to pack fish in for preservation and transfer to inland markets. The prices charged were for cash, but were within the means of even fairly small-scale fish-dealers, and the service was much used.

The economic situation in regard to purse seine and motor

boats was more complicated than with the provision of ice, since it demanded considerable capital investment or control on the part of entrepreneurs themselves. The capitalization of a purse seine complex in the Perupok area in 1963 included $10,000 to $12,000 (cash price) for the net,[1] $3,000 up to $10,000 for a motor boat, probably a similar sum for a second motor boat, perhaps several hundred dollars for a net boat (non-motorized) and perhaps another $4,000 or so for a van to carry the fish to market. (A sign of the times was that whereas the boats were often allowed to go unpainted, these vans, new, were often taste-fully painted and labelled with the owner's name and address !) The total, for new equipment, could well reach between $30,000 and $40,000.

Moreover, successful operation demanded some financial stability since overheads were great and it was said the rate of depreciation of the net was rapid. Hence, as was pointed out to me by various fishermen, the risks were considerable ; if the net did not get fish the losses were heavy. Finance of this order is not only completely out of reach of the average Malay fisherman ; it is also very many times greater than the level of investment of an entrepreneur in lift-net fishing in 1940, even making allowance for changes in the value of money in the interim.

At a very conservative estimate the capital represented by the nets, boats and other equipment of the fifteen or so purse seine units in the Perupok area in August 1963 must have been at least a quarter of a million dollars and was probably con-siderably more. Where did the finance come from ?

In the first place it was emphasized by entrepreneurs and by fishermen alike that the purse seine were financed by Malay money. No government capital was involved ; nor had Chinese put up the money.[2] With a qualification to be explained later, it did seem to me that a considerable portion of the capital had actually been found by local Malay entrepreneurs. These were men who had made money by trading or who had inherited rice or rubber land or other wealth from their fathers and increased

[1] Some nets were said to cost more. But the price then included the cost of 1,000 plastic floats ($1,000), about 130 brass sinker rings weighing 1½ lb. apiece ($150) and net ropes ($800) according to information given me. A net for taking fish is more expensive than one for taking *bilis* (anchovy), the cost of which is about $8,500. In 1930 the cost of a Chinese purse seine net at Pangkor was only $2,000 (*Annual Report Fisheries Department* (1931), p. 6).
[2] One net in the area was run by a Chinese *tauke* and apparently bought primarily by Chinese capital, but even here there seemed to be a genuine Malay partner.

their assets by prudent investment. Some were *juruselam* fishing experts, who by skill and energy had built up substantial resources ; most were middlemen, particularly fish-dealers, who themselves did not go to sea but let the net be taken under the command of a hired *juruselam*, who then received a special share of the proceeds. The tendency of Malay middlemen to move into the finance of production seemed more marked in 1963 than in 1940.

Nets and other equipment could be bought from the supplying merchants either for cash (*kes*) or on the instalment plan (*asoran, angsuran*). To pay cash, which gave a substantial discount (about 20 per cent.) seemed not uncommon. In this case the purchaser sometimes went to Penang or Singapore to buy, as being cheaper than through an agent in Kota Bharu. But whether purchase for cash or by instalments, it seemed unusual for a single entrepreneur to find all the capital ; he entered a combine (*konsi*) with one or more other Malays. The sum each put up might differ ; I got little information on this point, but was told by various men that they had $1,000, $9,000, $12,000 respectively in separate net *konsi*. Success of such a *konsi* did not depend on one of the owners being a fishing expert ; the most successful net in the Perupok area, run by the *juruselam* Yusoh Po' Ni' San (Pa' Nik Hassan), was owned by a *konsi* said to be of five Malay siblings (or cousins) living in Besut. They were said to have paid cash down for the equipment, buying first the net and boats and later a van for carriage of the fish, and the *juruselam* was in effect an employee.

But in many cases the intending net owners could not put up all the required capital. Hence they used partly their own funds and partly borrowed capital. Actual borrowing of cash from a third party to pay over to the seller of the net seemed uncommon ; the balance of the purchase price appeared usually to be found by the seller, either as a merchant supplier willing to be paid in instalments over time or perhaps, more commonly, through a finance company, which insured the seller against loss. I was told of at least one Malay who was in the net supply business, and from whom a group of owners bought a net by paying half the cash down and borrowing half on *asoran*, presumably through him. It seemed to be the normal procedure to operate with these moiety proportions. The price charged for instalment buying was higher naturally than for cash ; $16,000 as against $12,000 were figures quoted to me, with payment spread over twelve months. In such case the seller of the net was recouped from the monthly takings.[1]

It was in this respect that Chinese capital had entered the purse seine field. For the suppliers of these nets were normally Chinese merchants, perhaps linked with Chinese finance companies, based in Singapore or Penang—or Kuala Lumpur for the sale of motor transport—with local agents in Kota Bharu. They did not function as owners of the equipment and were differentiated from the Chinese *daganang* (pp. 60–1, perhaps better written in dialect form *dagangang*) by not being concerned with the marketing of fish. Hence they did not enter the fishing industry in any active way, and their rôle was restricted to ensuring that they got paid on ordinary—if substantial—financial terms for the equipment they supplied. So their relationship to the fishermen was an ordinary commercial one, and the claim of the Malays that the industry was in Malay hands may be admitted as correct.

On the other hand they exercised some control in that where the sale of equipment was not for cash they presumably had the legal right to recall the equipment if the agreed instalments were not kept up. On the whole the purse seine fishing was too new in the Perupok area in mid-1963 to know just how far the creditors were prepared to push their claims. The local impression was that they were usually reluctant to do this because of the trouble of finding another buyer. As local Malays said, the merchant was not a fisherman, nor could he use the net himself if he took it back. " What the *tauke* wants is money ; he leaves the net with the owner and waits to recoup. When the owner does well again, the *tauke* puts the pressure on." But the creditor might threaten action to reclaim the equipment. One night an old friend of mine, a well-to-do fish-dealer whom we had known since his small beginnings over twenty years before, and who now was part-owner of a purse seine and other property, came to me with a letter in English which he wanted me to translate for him. It related to his purchase of a van for $4,000 on which he had paid $1,000 down. It was an intimation in printed style from a Chinese motor company of Kuala Lumpur that they had

---

[1] It was said that the custom was for two-thirds of the net's share to go to the owners of the net and one-third to the provider of the capital. But this would not seem to give payment of the instalments at a high enough rate, and it is possible that this proportional principle applied only when actual funds were borrowed from local Malays, and not to a fixed instalment system. In the time available it was not possible for me to get completely clear the detailed working of this aspect of net finance.

not heard from him, and presumed he might be in financial difficulties. They would like to help, the letter continued, but couldn't do so without hearing from him. If he didn't write soon they would have to " take action " (unspecified). A typed addendum stated laconically " Two instalments, each $132.50." The fish-dealer showed me a receipt for $110 dated after the despatch of the letter. Obviously there had been some confusion of communications. But it seemed also as if there was incomplete comprehension on the part of this buyer as to just what his liabilities were ; with less businesslike Malays this may happen more frequently.

Two characteristics of the economy at this point may be emphasized. The first is that Malay entrepreneurs had shown initiative and energy in appreciating the advantages of the new technology, and financial acumen in organizing their capital and credit resources to this end. This offers an interesting comment on an observation made by Yamey [1] that a topic of considerable practical importance is the response in a peasant economy to a new source of finance, and the type of calculation made by investors.

The second point is that in so doing they had widened the gap between themselves and the ordinary Malay fishermen. This will be discussed later in the context of distribution of earnings from purse seine and comparative incomes. But here it may be taken up in reference to the organization of *Sharikat*.

SHARIKAT AND COÖPERATIVE ORGANIZATION

The term *Sharikat*, from the Arabic, means an association or partnership, and is broadly equivalent to the term *konsi*, from the Chinese *kongsi*, a combine, a guild (pp. 59, 155, 254). The former term, from its Arabic derivation, has a slightly superior Malay status and is in general official use for " company ". The expression *Sharikat Berkerja Samasama* (association working together) is the term for a Coöperative Society, often called *Sharikat* for short in this context. The Government of Malaya had for long been concerned about the financial position of the fishermen and among various ways had looked to formation of coöperative societies as a means of improving their command of capital and removing them from the burden of indebtedness. [2]

[1] Firth and Yamey (1964), p. 384.
[2] See e.g. Federation of Malaya, *Year Book* (1962), p. 239. By a scheme of 1957 for the East Coast a sum of $1¼ million had been made available to Fishermen's Coöperative Credit and Marketing Societies in the area.

The situation as regards *Sharikat* in the Perupok area was a little puzzling to me at first. Knowing that the Government had had elaborate plans for the development of fishermen's Coöperative Societies on the East Coast, I assumed rather hastily when I was trying to get a preliminary idea of the general pattern that references to a local *Sharikat* controlling a net were to such a society, and that it operated under Government sponsorship, perhaps even with some Government financial help. It then appeared, however, that most of the purse seine nets were controlled by *konsi* of several men—Malays—contributing to the cost, and that while there was a *Sharikat* the fishermen on the whole did not like joining it—whether from conservatism or individualism was not then clear. Later I was told that the local Malay entrepreneurs did not use Government money ; they preferred to work individually, to sell their fish themselves and reap the profit. " Why should they share in a *Sharikat* ? " it was asked. For poor men it would be different—the wealthy or the energetic prefer to use their own capital, not share with others. Then it turned out that from the net of the *Sharikat* the division of yield was done in the same way as with the other purse seine. The crew were no better off than those of other nets and did not share in the specific net-earnings. If this was so, I asked, why did crew members join the *Sharikat* ? The reply was that crew go to *juruselam* and *jurugan* (boat-captains) who don't get angry with them—if their leaders get angry with them the crew don't like them and don't join. From this it became clear that in this context the *Sharikat* was not a full coöperative organization, and the major benefits were reaped only by the net owners.

The puzzle was resolved when I discovered that two types of *Sharikat* were in fact under discussion at different times. Apart from an abortive Coöperative Society, which was what the fishermen were said not to like, there were two other *Sharikat* which were simply associations of net-owners in the ordinary way, who had taken to themselves the title of *Sharikat*—Company. As Awang Lung pointed out, these *Sharikat* were *konsi* like any other—they consisted simply of a number of people who had put money into a net enterprise, and there was no Government finance involved at all. To my question as to why some *konsi* were called *Sharikat* and others not, someone else replied, " Because the men are many—most *konsi* have two, three or four men." The *Sharikat Perupok*, in which our friend of former days Semain Kote, the ex-shadow-play expert, had an interest, had a membership of ten men.

The situation was explained further by a few historical details, given me partly by Government officials and partly by the local fishermen. There was no Coöperative Society for purse seine ownership in the Perupok area. The first *Sharikat* was what was known as the " *Sharikat Asli* ", " Original Company " or *Sharikat Perupok* ; this was started by a State Senator who lived in Kota Bharu, with the help of money from a few local men. (According to the fishermen, he bought a net in Singapore, using in part a Government loan, but this may be incorrect.) This was not a Coöperative Society in the technical sense of the term, centrally registered and borrowing from the central organization. Nor was the later *Sharikat Kubang Kawoh*, a *konsi* of 5 men who bought their net for $16,000 from a Malay *tauke*, paying half in cash and the other half by instalments. By the senior *juruselam* of this net, who gave me these details, I was told that so far—in about a year—the capital had been recouped (*modal balek*, cf. p. 128) to the extent of $12,500 and the *tauke* had already been paid off. This *Sharikat* was described to me as " working together ", but this merely indicated that the net owners were coöperating and not members of a Coöperative group in any official sense. They were not registered with the Government, and their organization did not get finance or other assistance from the Government.

The basis for the Perupok fishermen's rejection of the Government Coöperatives came out very clearly in discussion. One group said that they had asked the Government not for money but for a net, others that they had asked for loans. But all had been refused, hence they ceased to wish to have anything to do with Government organization of their net groups. As my ex-shadow-play expert put it pithily, " *Kerajaan, dia mainmain saja ; dia ta' bagi duit.*" " The Government was just fooling about ; it didn't hand over any money." In justification of the Government it had had some rather unfortunate experiences. I gathered that in Pantai Damat a Coöperative had been formed for line fishermen, to assist them to buy motor boats. Originally these were meant to tow the hand-line boats, but they were used by the members of the Coöperative for other (? more lucrative) towing jobs, with seines, etc. Apparently also some fishermen defaulted on their payments to the Coöperative. So though there was still one small motor boat, Government financed, there was

no Government money put into a purse seine. It appeared too
that some loans had been given without adequate enquiry into
the credentials of the applicants, and their capacity to repay.
Some loans seem to have been made to men who " became "
juruselam for the purpose. As fishermen themselves admitted,
" The Government is afraid that the people here won't pay."
They alleged that the fishermen of Kuala Besut had borrowed
much from the Government and didn't repay.

Granted that there was some basis for Government caution
in making advances of either equipment or money to the Perupok
fishermen, it seems rather regrettable that Government interest
in assisting the fishermen financially was not pursued, especially
since there appeared to be a body of enthusiastic and energetic
trained local officials in various departments whose services could
have been drawn upon for advice. A pilot scheme in which the
snags were worked out till reasonable success was obtained,
accompanied by more systematic and sustained extension work,
would probably have been worth while. As it turned out, it did
seem as if a policy of promise and refusal had discouraged the
fishermen from any form of association under Government
auspices. As a result the more able and energetic of the Malay
entrepreneurs had forged ahead on their own initiative, apparently
with considerable economic benefit to themselves and to productivity. But, as will be indicated later, the question of how
far the poverty of the ordinary fisherman has been affected thereby
still remains.

DISTRIBUTION OF PURSE SEINE TAKINGS

How were the benefits from this new improved form of fishing
shared out ? The principles followed were of the same general
order as obtained traditionally all along the Kelantan and Trengganu coast for large nets, namely a successive fractionating of the
sums involved according to particular functions. This general
procedure, which was described for the lift-net in 1940, and which
is familiar to all fishermen, may serve as a model (pp. 236–43).
But the differences in the technology of a purse seine necessitated
certain modifications.

The scheme though simple in essence was complicated in
practice and was not easy to identify in detail. I had many
talks with *juruselam* and other fishermen on this subject, and,
with the owner's permission, I attended the share-out from a

net one night.   Even then, however, it took me a considerable
time to understand the scheme.   The main principles were as
follows :

Expenses

   (i) Out of the gross takings of a month's fishing (*bĕlum chuchi*
       —" not yet cleaned ") the total amount of expenses for
       fuel, etc. (*belanja*) is subtracted.

Motor boat

   (ii) From the balance now " cleaned " of expenses 20 per
       cent. (" one-fifth ") is first subtracted for the motor
       boats ; if more than one motor, they divide this sum.

Juruselam

   (iii) From the same balance a 5 per cent. share (" 50 out of
       1,000 ") then goes to the *juruselam*.   (If he is assisted by
       a junior expert, this man gets one-third and the senior
       takes two-thirds of the share.)

Fish-handlers

   (iv) From the same balance still a similar 5 per cent. share
       goes to the men who have the work of selling the fish
       (*orang penjual ikan*).

Net and Boat

   (v) The remaining 70 per cent. of the balance is then divided
       into two equal portions.   One, *bagian dalam* (inner share),
       is the share of net and net-boat.

Crew

   (vi) The other, *bagian luar* (outer share), is divided among the
       crew, proportionate to the nights each man was out
       fishing.

   (vii) Within the *bagian luar*, allowance is made for a number
       of special functions : boat captains ; bow and stern
       men ; lamp tender ; head- and foot-rope tenders, etc.
       who each get extra shares.   Hence a crew (*awok*) of, say,
       40 men is calculated as equivalent to, say, 50 persons,
       (*orang*) in terms of shares for division.

Taking as an example the scheme worked out for me by several
juruselam on the basis of a month's gross takings of $10,000, the
broad outline of distribution would look as shown in Table 15
overleaf.

These main principles, however, concealed a considerable
amount of variation in detail since practice can vary from net
to net as regards, e.g., what " expenses " are incurred for crew

TABLE 15

Division of Earnings from Purse Seine 1963

Gross takings	$10,000
i. " Expenses " (approx.)	4,000
	6,000
ii. Motor boat(s) (@ one-fifth)	1,200
iii. *Juruselam* (@ 50 per 1,000)	300
iv. Fish-handlers (@ 50 per 1,000)	300
	4,200
v. Net and Net-boat (*bagian dalam*)	2,100
vi. Crew (*bagian luar*)	2,100
Hence : Individual share (40 men = 50 shares) . . .	$42

maintenance, and how far allowance is made for the fact of a good or poor catch on any night. Moreover, the exact method of division is a matter for each *tauke*; there is no fixed rule, but only a general consensus backed by ethical sanction and the opinion of the crew. As far as the observer is concerned, his understanding of the matter is apt to be obscured rather than clarified by the ponderously elaborate discussions that go on, with much handing to and fro of cash, lasting perhaps for four hours or so.

Some of the complications may be seen by considering those major items such as the *belanja* and *bagian luar*, where most variation is likely to occur. The " expenses " item is that of recurrent overheads and tends to be often the largest single sector of the account as well as the most open to variation. As such, it is the item looked upon most suspiciously by the crew, who commonly grumble about its size without apparently much precision in their charges. The " expenses " normally include : diesel fuel, 3 tins or so per engine per day @ $3.30 per tin ; lamp mantles, chimneys and cleaning needles (a lamp often needs a new mantle every night owing to the banging about at sea, at a cost of $4 or so for a score) ; kerosene for the lamps, 2 tins per night @ $4.50 per tin ; ice, say 5 blocks a day @ $1 per block ; expense allowance to the crew of $1 per day each when fish is caught, $2 or $3 when an extra good catch is taken, and (some nets only) 50 cents per day even when no catch is obtained ; and the cost of food and cigarettes for the crew engaged in maintaining and repairing the net when not at sea (an item of $100 or so). For a conservative total of 20 nights at sea with the average run of luck, a minimal total of expenses would be in the region of

$1,500. The amount was usually very much more, especially when two or three motor boats were attached to the net, some perhaps hired. For takings of $10,000, two local *juruselam* separately estimated that the expenses would be in the region of $4,000. The *belanja* are the first charge upon the earnings. Thus one *juruselam* told me that the month before our talk the total earnings of his net had been $850 and the *belanja* had been $970 ; with this loss of $120 there was no division of proceeds at all to the crew at the end of the month.

Bagian luar. There are three main points concerned in the division of shares among the crew. The first is the allocation of extra shares for special duties such as boat captaincy, or lamp tending. The number and size of these varies with different nets, but commonly an extra single share, or a share-and-a-half is given in addition to the ordinary crew-man's share. This means as a rule about ten or a dozen extra shares. The second point is that the system of payment is by number of nights out. The maximum dividend per man is declared. A man who has gone to sea with the net less than the full number of nights get proportionately less money than the maximum. Everyone is paid according to his nights out. The remainder after they have been paid is put into a pool and divided only among those who went out on all the nights. This gives an incentive bonus for full participation. (If, e.g., the maximum share per man is $40, for 20 nights out, then, in a boat of 10 men if 5 went out every night and 5 on only 18 of these nights, the former would first get $40 and the latter $36 apiece ; the $20 left out would be divided among the former 5 men, giving them a total receipt of $44 each.) The third point is that while most nets differentiated crew payments by number of nights out, some also took into consideration the yield per night. One way of doing this was to pay not monthly but more frequently, say the day after a good catch, in effect fractionating a series of smaller totals instead of one large total at the end of the month. (At least one net followed this system in the Perupok area.)

A brief note on the division I saw will illustrate the operations and the attitude of the crew towards them. The *bagian* was announced the day before, for 9 o'clock the following evening, then altered to 7 o'clock, then postponed till 9 o'clock again so that I could attend. By 9.30 the crew were all assembled, but the *tauke* had not yet appeared—he had gone to Bachok to fetch the money. He finally arrived with large wads of bank notes.

The division took place on the verandah of his house, the crew sitting round with their backs to the wall, with the *tauke* and his partner in chairs facing them. A clerk (a regular feature in these groups) sat at the table also. The scene was lit by three pressure lamps. Drinks were served to the company, a bottle of cordial to each man.

There was much discussion about the principles of division, probably because the net was new and so also was the clerk. Both his book and the main record book of the *tauke* were kept in romanized Malay. The *tauke* made known his calculations from the clerk's books that out of 22 days at sea, 18 had " got " fish and that " before cleaning " off the expenses the gross takings were $6,002. When the expenses, motor boat's share, percentage of the *juruselam*, etc., had been taken off (the actual cash was taken out of the central pile), the sum left to divide was $2,020, giving a *bagian luar* of $1,010. From this sum $130 was taken for distribution to the crew at large for their services in carrying fish to the motor truck. The sum remaining for division was announced by the clerk as $980. There was a crew of 44 men, and it was agreed that with the extra shares for special work this was equal to 57 man-shares. The junior *juruselam* (a son of Awang Lung) was acting as actual divider and assembler of crew shares, calculating together with the clerk. He took off his own share and put it in his jacket pocket. Then he carefully counted out $17 per man-share, making $969 in all. During the actual counting out of the crew's money there was comparative silence. The *tauke* and his partner said nothing ; their main job was done earlier and they could leave it to the clerk and crew representative to do the low-grade job. But " What's $17 for a month's work ? " commented a crew man. " Oh ! that's coffee money," said the partner ; " money to buy rice has already been given " (in the day-by-day expense allowance). " Oh no ! " came a comment from an onlooker outside the verandah rails, " it was coffee money that has already been given ; this is for rice." This remark was a not too subtle criticism—that this share, which should have been the major distribution, was in fact only a small sum. Meanwhile I was waiting to see what would happen. Suddenly someone observed, " The clerk's slipped up a hundred dollars (*silap*, mistake, means also that the quickness of the hand deceives the eye), the $980 was wrong ; it should have been $880." " Throw out $2 from each man's share," was the answer, and rather more than the required $100 was thus removed. This the crew took quite quietly.   The distributor seemed worried that $25 was now over and did a recount, setting the *bagian* out in rows of fives until he had settled what to do with it.   Then with nearly a thousand dollars of cash all displayed in piles, " How many men in each boat ? " asked the *juruselam*, and each boat captain came up to decide with his men exactly how much each was entitled to.   At this point the crew, hitherto rather apathetic, woke up and took a keen interest.   This was something they really could check !   When I left at midnight the division was still proceeding, in a very good-humoured atmosphere despite arguments.

RELATIVE EARNINGS OF CAPITAL AND LABOUR

How now do entrepreneur and ordinary fishermen stand relative to each other in this scheme ?

I cannot speak with complete assurance on this matter.   But from general consideration of all my data I would estimate the allocation of running costs (of fuel, ice, repairs, etc.) at just over 25 per cent. of total earnings, and the distributive share of the entrepreneur for return on capital (principally net and boats), earnings of management and profits, as about 30 per cent. of the total earnings.   The earnings of labour, including advances to the crew and their prior allocation of fish, amount to little more than 40 per cent. of the total.

It is interesting to compare these proportions with the corresponding figures for lift-net fishing in 1940, then the pre-dominant mode of livelihood.   The running expenses of the (non-mechanized) equipment were then a small proportion of the total, even if the fish lures be included under this head at 10 per cent.   The return to capital (boats and net) with which was incorporated any management earnings and profit, was in the region of 27 per cent., whereas the returns to crew were over 60 per cent.   Much greater returns to fishing in modern conditions, accompanied by or resulting from much greater capitalization, has resulted in a marked drop in the percentage of earnings going to labour.   A prediction I had made twenty years earlier (1st ed., p. 304), that a higher level of producers' capital would almost certainly involve adjustments in the traditional system of distributing earnings was thus borne out.

There is also the question of where the pinch comes in a period of poor fishing.   What costs are a first charge upon earnings ? It is clear that in a period of low returns it is labour as well as capital which suffers. The first charge upon earnings is running expenses : when the net cited earlier took $850 as against $970 of expenses, no distribution was made to the crew at the end of the month ; when another net took $1,790 and expenses came to $2,000 the crew received only $1 as a token payment. This is partly a function of the low level of capital backing at the command of the entrepreneur—he probably cannot afford to forgo the cash takings available. But it is part of the conventions which govern labour in the fishing industry generally. Although the crew members of a purse seine are in essence labour contributors only, they are not regarded as wage labourers, giving their services irrespective of the return to the enterprise. They are treated in a sense as coöperators, profit sharers who partake in its good or bad fortune. The result is that at the present time the ordinary Malay fisherman working in a net combine is in an intermediate and somewhat parlous state. He is in effect an employed labourer, contributing only manual skills, but he is treated as a participant in a common enterprise and hence is not put on a regular wage basis. Economic security to the crew is then not treated as a prime charge against the gross takings.

The justification for this could be that when earnings of the net are high the crew share. But highly variable yields in fishing and the lack of any minimum wage undoubtedly give rise to hardship when the earnings of the net are low.

It will be clear that a small pay packet at the end of the month may not reflect at all any lack of hard work or skill on the part of the individual crewman—the apparently chance behaviour of fish shoals, the skill of his *juruselam*, the vagaries of the weather may be responsible. It can be understood then that there is grumbling by the crewmen when their share of the proceeds is small, and some may leave their particular net to join another. There is also an opinion fairly generally expressed that in hard times it is not the *tauke* who suffers most. For example, " The *tauke* people are all right, only the *tauke*. A *tauke*, he has a net, he has money, he buys a truck ; the *tauke* of Kota Bharu are ' in with ' him. He gives out only a little cash to the crew, only a very little."

But in economic terms why do not these purse seine fishermen turn to other forms of fishing or to other forms of employment entirely ? To answer this let us first turn to other fishing.

RETURNS TO OTHER FORMS OF FISHING

How now have other modes of fishing, and the fishermen who use them, fared in modern conditions ? There was no doubt that in the Perupok area other types of net fishing had been affected. Small drift nets were still in use—I saw at least half a dozen—by fishermen who went out singly at night in small boats, set their nets, slept at sea and hauled the nets in the early morning. I recorded an occasion on which about ten men used small scoop nets to take crabs and prawns. But gilling nets (*pukat dalam*) for mackerel seemed to have largely lost their function since the purse seine performed it so much more efficiently. What was most obvious, however, was the radical alteration in the position of lift-net (*pukat tangkul*) fishing, formerly the principal occupation of the area. On the way up the coast I had been told of lift-nets operating much as before from Paka, Dungun, Losong and other *kampongs* in the Trengganu estuary, and Besut. But at Perupok the lift-nets were said to be " finished ". Instead of the score or so of 1940 and the 15 or so of 1947, there were in mid-1963 only three lift-nets in the Perupok–Kubang Kawoh –Pantai Damat area, and even these went to sea only infrequently. The local fishermen said that when the lift-nets put to sea they lost money, and that the purse seine took all the fish.[1] Nearly all the local experts had got rid of their nets ; they had sold them to men of Besut, Melawi and Sabah, who were still using this type of gear.[2]

Taking advantage of the new technology to some extent, and to help meet the new competition, the remaining operators of lift-nets had turned to motor boats. Each net had a motor boat which towed the other four or five boats, especially when the wind was adverse, moving the string of them at fair speed, perhaps five or six knots in a calm sea. The lift-net technique was otherwise the same as that of twenty or so years earlier (cf. pp. 97–100, *supra*) ; with coconut frond lures, the expert listening for shoals of fish, etc. But the Perupok lift-nets now tended to put

[1] To my question as to how this could be since *selar kuning*, the staple catch of the lift-nets, were said not to rise to the lamps of the purse seines, the fishermen argued that the lures of the purse seines, further out to sea, kept the fish of the lift-net from coming closer inshore. But one purse seine *juruselam* at least rejected the idea that there was overfishing.

[2] Possibly these buyers were poorer and could not afford the heavy capital investment in purse seines. They seem to have bought the second-hand *pukat takur* fairly cheaply—e.g. paying $700 (for a nearly new net), $300 (for a three-year-old net which had cost $1,500). But I was told that there were wealthy men in Melawi, etc., just as in Perupok, and that the Melawi men just preferred lift-nets.

out only at times of " bright moon ", that is, on days other than those when the purse seine went out at night. This was in order to avoid a possible glut of fish and get a better price. But this meant only intermittent use of net equipment, and involved the crew at least in using the lift-net as only a secondary employment.

Since the lift-nets of the Perupok area were out so rarely while I was there I could make too few observations to get any very clear view of their takings. But they appeared to have very mediocre success. I recorded one catch of about $200 worth, but others of $20, and one even of $4.50 only, while on another occasion the net was not shot at all because of strong currents. If one were to guess at an average yield, it might be perhaps about $50 for a day's work, and this was certainly only a fraction of that from the purse seine.

The method of marketing the lift-net fish and distributing the earnings had also tended to be affected by the pattern set up by the purse seine. Whereas formerly the catch was sold on the beach (v. pp. 189–91, supra), and the proceeds distributed to the crew weekly (v. pp. 236–49, supra), nowadays the major catch is packed in ice and taken by the tuan pukat (net owner) for sale in inland markets. He keeps the proceeds—" the net is his ". He gives the crew approximately one box of fish in ten for themselves, to sell as they wish. In more detail, the mode of distribution, I was told, was as follows :

(i) The crew of the fishing expert's boat (perahu sampan) get a share out of the general takings after the owner has sold the catch.

(ii) The crew of the boat which carries in the fish (perahu ikan) get their share from the box or so of fish given initially by the net owner.

(iii) The crew of the net-boat (perahu pukat) likewise get their share out of this initial gift of fish.

(iv) The crews of the other two boats get no fish given them for cash sales, but only have their makan lauk, i.e. casual fish for personal sale and domestic consumption. They do not share in the initial apportionment by the net owner when the net first gets fish ; all they do is help (tolong sahaja). But next time out the rôles are reversed and the " helping boats " are those which divide the approximate one-tenth share of saleable fish allocated by the net owner. " What happens if the net rarely

goes to sea ? " I asked. " Ah ! That is the trouble."
" Then why do people continue to go out with the net ? "
" Because as poor people, if they can't go out, they can't
eat."

Because of lack of time for more enquiry, I am not satisfied
that I got the full details of this distributive scheme. But what
seems in essence to have been happening is that led by the example
of the purse seine the lift-net owners have taken to marketing the
bulk of the catch themselves and dispensing with the services of
both the " sea-dealer " and the " land-dealer " (*peraih laut* and
peraih darat, cf. pp. 111–14, 186). Moreover, it appears that they
were in effect handing over as payment to their crews what they
regarded as " fair shares " of the catch. But the conception of
what are regarded as " fair shares " may vary considerably according to what position one occupies in the economic structure.
What seemed fairly clear was that these lift-net crew men were
in 1963 in a less advantageous position as far as income was
concerned than in 1940 ; an inducement for them to go out,
apart from the always present prospect of a spectacular catch,
may have been that it was a day-time not a night-time job.

From a more general economic point of view the competition
of the purse seine had meant that some of the lift-net experts,
especially the older ones and the less successful ones, had been
driven quite out of business by the new type of fishing. They
sold their nets and their large boats, buying instead a small sailing
boat with hand-lining equipment or a small individually operated
drift net. Other boat owners in lift-net groups too had been
similarly affected since there was no longer the same rôle for so
many of their large craft. Many of them too had to dispose of
these and be reduced to hand-lining or to serving as crew men
with the purse seine.

In this non-industrial context, then, we see the familiar
industrial spectacle of displacement of skilled labour by introduction of a new technique, and of the economic structure of one
branch of production being modified in the direction of another,
more efficient, branch. This has a more general bearing on the
character of the peasant social and economic system as a whole
(see later, p. 346).

The other major form of fishing in the Perupok area which
occupied the daily energies of a considerable number of men was
hand-lining. For the most part, this was practised in 1963 in

much the same way as in 1940 (*v.* p. 86, *supra*). Small sailing boats were used, each carrying one, two or occasionally three men, and taking a range of fish, including Spanish mackerel, large horse mackerel, dolphin, squid, dogfish, catfish, sea bream. Division of earnings from these boats, where there was a two- or three-man crew, apparently followed the same principles as before (*v.* p. 253, *supra*). But a few hand-line fishermen in 1963 were using motor boats, which greatly increased their fishing range, much beyond the 10 miles or so where the sailing craft went for squid or the 20 miles (4 hours' sailing) where they went for large fish such as Spanish mackerel. Sometimes, when the wind was strongly adverse or a storm was feared, line fishermen in sailing boats would combine to hire a motor boat to tow them in—at 50 cents per boat from the near and $1 from the far fishing grounds.

The hand-line fishermen came in as a rule in the late afternoon, with the sea breeze, lowering the sail and mast and paddling in the last few yards. A few people, including the fisherman's wife, his particular dealer, a neighbour or a would-be buyer of fish for the evening meal all helped to pull the boat up the beach, on skids of coconut leaf rib, often tucking up their skirts and going out knee-deep in the water to do so. Some fish were kept back for meals at home (*makan lau'*) and the rest of the catch was sold, either in bulk to the dealer with whom the fisherman had a special arrangement, or if of poorer quality in lots of 20 cents or so to fellow-villagers and casual passers-by seeking fish for supper. If the catch contained a fine fish such as a Spanish mackerel it was the prerogative of the particular dealer to bid for and ultimately take it off.

The language of achievement was concerned with these fine fish, especially with these Spanish mackerel, though also with squid in numbers. The ordinary expression for success in fishing is *dapat*—" got ", or if very successful *dapat banyak*—" got in plenty ". All that will be added by way of explanation on the beach at this time is " *Dua ekor* "—" Two tails "—or whatever the number of Spanish mackerel caught may be. Now *ekor*, " tail ", is the general classifier for all fish, as well as for animals, and in selling any kind of fish on the beach they are numbered in " tails ". But the context of situation is such for hand-lining at this time of the year that " two tails " without specification means only two Spanish mackerel ; two of any other fish would have the name of the fish added. The price of a Spanish mackerel is from about forty to more than a hundred times that of most ordinary small fish, so there is reason to single it out in this way. Similarly the expression " *ta' ada dapat* ", literally meaning " haven't got ", does not necessarily mean that no catch has been obtained ; it probably means only that no Spanish mackerel has been caught. The analogous expression *kosong*, " empty ", is of the same order. So also " *ta' ada sa(tu) ekor* ", though it *may* mean just what it purports to say " not a single tail " may mean only of Spanish mackerel ; if the fisherman was entirely unsuccessful then more usually " *ta' ada sa(tu) ekor ikan* ", " not a single tail of fish ", is the expression. All this is linked with conceptions of status by the fishermen and others of the community, as well as with income-getting ; a man who gets three or four good-sized Spanish mackerel (2 ft. to 2½ ft. long) has reason to be proud of himself.

INCOME OF FISHERMEN

I turn now to consider the actual incomes obtained by the Perupok fishermen, insofar as they can be estimated from the data I collected. I would emphasize here the point made again in Appendix I—that the only way to get an adequate idea of fishermen's incomes is to take systematic day-to-day records on the spot. This is demanding work, but general impressions are no substitute for it. So even though our stay in the area in 1963 was short, we tried to get as much concrete evidence as possible.

As regards purse seine fishing, in August 1963 I was able to get nearly complete records of the operation of 12 nets consecutively over a period of three weeks' fishing, equivalent to about a full month's use. (Since the technique involves attracting fish by light, there is no fishing during the period of full moon, when the day-time is utilized for mending the net, etc.) During this period two of these nets were taken away by their *juruselam*, one up coast and the other to islands off the coast, so these were eliminated from my calculations. Details of the remaining ten are given in the accompanying Table 16.[1] The total catch for the ten nets for this three-week period was about 5,000 boxes (about 350 tons) of fish, or roughly 1 ton of fish per

[1] I owe many of the details of this table to Che' Nur, a son of the house where we lived; he collected daily figures of purse seine catches for me, and as I checked, methodically went to the landing place of each net to ascertain the results. I had also other sources of information, e.g. a clerk's record for one net.

man of crew. This seems very substantial. But returns from the purse seine, as from any form of sea fishing, are highly variable. Not only do they depend on the phase of the moon, but also on

TABLE 16

Purse Seine Catches

August 1963—Perupok Area

(Catches in boxes of fish : * indicates fish of better quality ; m.l. indicates no fish for bulk sale but *makan lauk*, for home consumption and personal sale only ; n.m., the net was not shot ; n.o., the net did not put to sea ; O, fished, no catch ; dot only, no record.)

Night	Net									
	A	B	C	D	E	F	G	H	I	K
1 . . .	n.m.	30*	70*	50*	7*	16*	.	.	.	.
2 . . .	15	15	160	n.m.	n.m.	35	n.o.	n.m.	m.l.	.
3 . . .	10	35*	50*	n.m.	90	0	.	11	15	.
4 . . .	n.m.	n.m.	n.m.	n.m.	n.m.	n.m.	n.m.	n.m.	n.m.	.
5 . . .	n.m.	m.l.	85*	m.l.	7	m.l.	n.m.	16	n.m.	n.m.
6 . . .	12	85	100	30	70	35	55	100	50	100
7 . . .	n.m.	n.m.	10	n.m.	45	n.m.	n.m.	7	n.m.	m.l.
8 . . .	100	35	n.m.	n.m.	45	20	n.m.	m.l.	45	40
9 . . .	n.o.	70*	30	7	100	15	13	n.o.	n.m.	m.l.
10 . . .	n.m.	35	100	50	n.o.	35	50	30	n.m.	15
11 . . .	n.o.	n.o.	n.o.	n.o.	45	n.o.	n.o.	n.o.	n.o.	n.o.
12 . . .	n.o.	30	40	n.m.	80	15	n.o.	10	70	20
13 . . .	n.o.	50	n.o.	20	n.o.	15	30	8	25	n.m.
14 . . .	16	70	100	20	n.o.	20	m.l.	n.o.	n.o.	100
15 . . .	10	80	70	30	n.m.	25	15	n.o.	n.m.	20
16 . . .	25	60	40	70*	20	35	40	100	30	50
17 . . .	m.l.	n.m.	m.l.	100	n.o.	30	100	n.o.	45	40
18 . . .	40	70	80*	40*	n.m.	25	n.m.	n.o.	20	50
19 . . .	n.o.	n.o.	50	n.o.	n.o.	n.o.	m.l.	n.o.	n.o.	n.o.
20 . . .	n.m.	50	50	30	n.o.	20*	40	n.o.	35	90*
21 . . .	0	40*	n.o.	15	10	30	35	n.o.	m.l.	50
Nights out .	16	19	18	19	15	19	16	11	17	15
Nights fished .	10	16	16	13	11	17	11	9	11	13
Fished, no catch . .	1	0	0	0	0	1	0	0	0	0
Makan lauk only . .	1	1	1	1	0	1	2	1	2	2
Days catch sold . . .	8	15	15	12	11	15	9	8	9	11
Total boxes .	228	755	1035	462	519	371	378	282	335	575
Average boxes per night†	11	36	50	22	25	18	20	14	17	34

† Adjusted in G, H, I, K to a period of 21 nights record.

the state of the currents, the wind and the weather, apart from the presence or absence of shoals of fish of desired saleable quality.

It will be seen that of the 202 nights available for fishing by all the nets, they put to sea on 165, or 75 per cent. of these. They did not shoot the nets—because of bad weather, strong current or lack of fish—on 38 or 23 per cent. of the nights out. Of the other 77 per cent. of nights out they got no fish or they obtained only fish for the crew with none for bulk sale (*makan lau'*) on 9 per cent. So on only 68 per cent. of the nights out did the net get saleable bulk catches, their average per night for the whole three weeks' period being about 25 boxes apiece.[1] But only a small proportion—perhaps one-quarter—of the catches were of any appreciable quantity of relatively fine fish such as *kembong* (mackerel), and the price therefore was relatively low. *Kembong*, according to my informants, always fetch a good price —up to $50 and even $70 per box, and never as low as $20– $30 per box of about 500 or so fish, whereas herring (*beluru*) may fetch $15–$20 per box and *selayang* only $2 or $3 per box. So a moderate sized catch of 50 boxes (3½ tons) of mixed *kembong* and *lolong* (mackerel and small horse mackerel) may fetch $1,000, where a catch of double the bulk of *selayang* (another small horse mackerel) might fetch only a quarter the price. The monthly taking of a net therefore tended to vary greatly. But skill and energy of different *juruselam* showed itself in the quality as well as the volume of the catches, and over the 3-week period of observation the range was from about 250 boxes of fish to well over 1,000 boxes, which may have well fetched over $10,000. The average would seem to have been in the region of 500 boxes of fish, giving perhaps $5,000 for a month's work.

The principles of division already set out indicate that an individual member of a purse seine crew might expect to get at the end of a month between $15 and $20 cash from a $5,000 gross intake. But the total income of such a man was composed of four sectors : his monthly cash share ; his daily cash distribution of a dollar or so when the net got fish ; his daily allocation of 100 fish or so for personal consumption and sale, worth another $1, when the net got fish ; and the occasional Spanish mackerel or other fine fish caught by hand-lining on the way to and from the fishing grounds.

[1] For the month before the period of observation, from the clerk's record for one purse seine, out of 24 days out, 6 were " empty " and the total catch listed was 250 boxes, or only 10 boxes per night.

It is difficult to estimate on the basis of my short period of study just what the total earnings of a crew man in 1963 were on the average, if only because of the variable amount of daily fish from the last two sources which he got to sell. But estimating the value of this fish at roughly the equivalent of the cash advances he received, one can make an approximate calculation. With five nets I was given details of the gross cash takings of each for a month ; I also got some information about advances and final shares for two nets more. All these data indicated a range of crew man's monthly earnings from about $20 or $25 for un-successful nets, through $40, $50, $55, $65 to $90 for men in successful net groups. Considering together with this the figures for the ten nets for which I had records of the volume of fish taken (see Table 16), and reckoning the catch at an average price of $10 per box, an average full month's earnings of an ordinary crew member at this period of the year would seem to have been not much more than about $50. (Cf. p. 278 for 1940.)

Considering the present-day cost of living, for any except the smallest family, this sum could leave little if any margin for saving or even for routine non-food purchases.[1]

A most striking feature of income from hand-lining was its great variability. Men fishing from motor boats, with their greater range, tended to do best. On one day, I was told, a boat with three men took $120 worth of fish ; on the ordinary principles of division, after deducting cost of fuel and one-third or one-half for the share of boat and engine, this gave about $20 or $25 per man. Another three men at the same time took $75 worth of fish. But the next day the latter crew got no fish at all ! In general, catches were at a very modest level. During the month of August 1963 I recorded the results of hand-lining by about forty fishermen in about thirty boats, nearly all using sail. Some of these records were taken rather haphazardly, especially during the early period of my stay in the village, but most of them were made regularly every afternoon on the same stretch of beach, following the fortunes of individual boats and fishermen for periods of between ten days and a month. In all I recorded 360 cases of fish catches, involving 540 days of man effort ; the records covered twenty-seven days of actual fishing.

[1] See Rosemary Firth (1966) for further details. According to official figures for Malay, Chinese and Indian rural households in 1957-8 in the country generally, the average monthly consumption of the lowest income groups—under $150 per month—was $99, of which $67 expenditure was for food alone. Federation of Malaya, *Year Book* (1962), p. 489.

In practically every case I saw the actual catch, in a large number of cases counted the fish (or made an approximate estimate if they were in large numbers), and in very many cases observed the bargaining process of sale to a conclusion. In cases where I was not able to see the sale or learn the exact figure afterwards I estimated the cash takings by comparison with the current price of the particular kind of fish that day. In a few cases I got the sales figure afterwards from the fisherman or his wife.

In all, for these 540 man days the takings were approximately $1,620, an average of almost exactly $3 per man per day in cash. This average figure, however, must be considered in its proper context if it is to be used in interpreting the economic position of these fishermen. (For 1940 comparison, see pp. 273, 274.)

The arithmetical average of $3 conceals a very wide range of variation. The highest average earnings on one day was $10·45 per man, but this was a day when very few boats were out and only fifteen men in all were involved. Because of the scarcity of fish that day prices were very high. Cash earnings were as low as 90 cents per man on the average for another day when fishing was very poor. On this day twenty-two men were concerned. On any given day there was great variation in the cash earnings of different boats and men. On the day of highest average cash earnings one man took $26.75, two men in one boat took $22.75, two men in another boat $21.25 ; but four men in two boats took only 75 cents each. On the day of lowest average cash earnings six men in four boats got no cash earnings but only fish for their households, and two men in two boats got not a single fish. Throughout the whole period of the 360 boat cases of hand-lining there were thirty-eight cases, or more than 10 per cent., where the fishermen caught only fish for their households but got no cash takings. In another thirty-seven cases they got not enough fish for a meal or even no fish at all !

The variation in earnings was not simply a case where the good fishermen consistently got good incomes and the others did poorly. Even the fishermen who on the average did reasonably well also suffered considerable variation in their earnings. One man who took nearly $100 for twenty days out and who got the highest individual earnings of $26.75 in one day, took only $1.50 in cash in the whole of the following week. Two men who earned about $150 in cash over nearly four weeks, and who took over $65 in two successive days, got only $8.50 between

them in cash for the last week of my observation. It is clear
then that hand-lining is a hazardous occupation from the point
of view of earning a steady income.

These hand-line fishermen do not go out on Fridays because
they would not be able to return in time for the midday prayers.
Their average earnings of $3 per day have therefore to be ex-
tended to include four incomeless days per month. In addition,
there are many days in a season when a boat does not go out,
because of the need for repairs or repainting or mending a sail;
or because of bad weather or fear of a storm; or because of
social preoccupations such as a daughter's wedding, a spouse's
funeral, or the violent madness of the woman next door—all
actual instances known to me. But the cash income again has
to cover these days not out.

I cannot give an exact figure for a full month's income for
any considerable number of fishermen. But it would seem that
in a month when fish by hand-lining should be reasonably
plentiful, cash earnings might be expected in 1963 to be about
$50 per month for an ordinary fisherman, ranging up to perhaps
as much as $100 per month for a very skilful or exceptionally
lucky one. To these figures must be added in real income terms
the value of the fish retained for household use, which I would
guess to be the equivalent of between $5 and $10 per month,
depending in part on the size of the household. It was clear
that in 1963, as with the purse seine fishermen, such an income
left little margin for more than routine food expenditure.

There was some occupational mobility in that line fishermen
might turn to work as crew with purse seine. But the attraction
of hand-lining is that it gives greater independence as well as
opportunity for exercise of personal skill and for employment of
a small amount of capital. With the purse seine, any differences
of skill except in the few special roles of bowman and sternman,
lamp tender, boat captain, tend to be evened out; the bulk of
the work is unskilled. Moreover, any special skill which an
ordinary crew man may have can make little difference to either
the total income or to his own share in it. This is not so with
hand-lining, where individual skill counts for a great deal.
Even with motorized hand-lining, where mechanization has
resulted in a larger share of the takings going to the capital
factor, the abilities of any member of the crew of, say, three men
can affect the earnings considerably—whether they pool their
catch or keep it individually. Hence fishermen who feel they

have distinct skill with the line tend to keep to this occupation and not to change to purse seine. Again, purse seine work is very hard and is regarded as primarily young men's work. Though a few old men with no resources do go out with this net, most elderly men with some skill tend to go line fishing. The work is often strenuous—paddling in a calm or against the wind —with long hours—on a bad day, from perhaps 3 a.m. to 7 p.m. or 8 p.m. But a man is more his own master, and he can intermit more easily than if he were a member of a net crew. Such considerations enter into the choices made by fishermen, choices made primarily, but not solely, on income grounds.

That the income level of all these fishermen was low in comparison with the cost of living, and their general economic position was precarious was expressed to my wife and me again and again, both in general complaint and in comment, often wryly humorous, on particular crises. After a succession of bad days they were apt to get very gloomy. I asked Yoh, wife of our friend and neighbour Awang, how he had done one day ; " He hasn't sold ; squid for 40 cents ! How are we to eat ? " (She had a large family.) This was not an idle question—enough fish for supper, but not enough money for rice, spices, oil, she meant. Their neighbour, an old man who with his partner had earned about $2 one day, said to me, " I can't endure it, Tuan ; it hurts here," with his hand to his back. Another two-man boat took only about $1 worth of fish—" Our earnings aren't enough to eat on, today ! " A man who had only a few fish, not enough to sell, described his boat as " Empty ; if one hasn't got anything, how can one drink coffee ? " Of two men, with about 50 cents worth of fish between them, one commented, " We haven't got anything ; my belly hurts "—then, grinning ruefully, " We can't eat rice, can we ? " Such remarks, randomly reported, illustrate the attitude of the fishermen and their families —that they are close to a poverty line where it is not a matter of being unable to afford luxuries, but of having to worry where one's next meal is coming from. This of course has its impact upon family relations. One instance struck me most vividly— a young man who I knew had had poor catches for some days drew up his boat ; when the bottom boards were lifted up it was empty once again. As we were gathered round, an old woman (mother or mother-in-law, probably) said bitterly, " Empty, three days empty ! ", while the fisherman stood there silently and miserably, with head turned away.

INCOMES OF MINOR FISH DEALERS

The economic position of the other ordinary villagers seemed
for the most part to be not very different from that of the fishermen and their families.   This may be illustrated by a brief consideration of the minor fish dealers, who have been regarded at
times as being in a situation of advantage, able to exploit the
primary producers.

The beach fish dealers for the most part aimed to sell fine
fish fresh in the inland markets, especially Kota Bharu.   Whereas
formerly they tended to have quite a trade in dried fish, which
they cured themselves, for primarily technological reasons described earlier, this aspect of the industry had in Perupok largely
passed from their hands.   I did find some commercial fish-drying.
One woman had invested in 6 boxes of small horse mackerel and
a box of herring, which she was drying to take and sell in her
natal village near the Siamese border.   On the herring alone,
allowing for cost of salt, labour for gutting, rail freight and fares,
she reckoned to get a profit of $8 to $10; on the horse mackerel,
a fish of poorer quality, she might make as much.   Without
making a close calculation it seemed that her earnings from this
rather laborious work were in line with those of other small-scale
local fish dealers.   Incidentally, the practice of a woman married
away from her natal home returning there periodically to sell
some product not local there, is not uncommon.   It offers obvious
advantages—commercially, the woman already has contacts
through her kin ties; and socially, it allows her to pay for her
trip to see her relatives and friends.

The minor fish dealers on the Perupok beach concentrated
mainly on Spanish mackerel and squid when these were in season,
buying them in the afternoon, packing them in ice overnight,
and taking them to the inland markets the following morning.
These fish fetched on the beach from about $2 to as much as
$10, according to size and thickness, and a dealer would handle
each day from two or three to about a dozen.   His capital outlay
on these together with other fish varied greatly, but was commonly up to about $50.[1]   If a dealer had only very few fish then
he would either sell them to another dealer, hoping for a slight
profit, or would combine with other dealers to make up a box
of fish which would be worth while taking to market, considering

[1] My records over about six weeks for just over 200 Spanish mackerel show
that they were sold for just over $600, an average of $3 apiece.   When these fish
were plentiful, one of my informants counted 200 on one day in the Kota Bharu
market, handled by six dealers specializing in this fish alone.

the cost of ice and fares. There were in all about half a dozen such dealers on the Kubang Kawoh sector of beach where I was making my observations. It was of interest that two of them, who were women, mother and daughter, operated commerically on exactly the same terms as the men.[1]

The competition of the dealers to secure fine fish had resulted in a series of arrangements which I do not think existed, at least to the same extent, in 1940. By these, not necessarily but usually in return for cash loans—from about $10 to about $30, mainly to aid in the purchase of boats—individual boat owners in regular hand-lining were in contract to sell their main catch to individual dealers. The number varies—one dealer, Awang, had nine such fishermen in contract to him, another, a woman Limoh, had four. The arrangement was not for purchase of the fish at a fixed price; there was keen haggling, of the kind described earlier (v. supra, pp. 193–5, 216–18). Sometimes the fish dealer wouldn't rise to the fisherman's price (or the latter wouldn't accept the dealer's bid), and the pair separated, leaving the transaction to be concluded later. Often the dealer bore off the fish in some haste with the price unresolved, but in effect on his own terms. Payment was made to those regular suppliers at the end of the week, and the dealer might then cut the price (potong) by a dollar or so if he had lost on his own sales. This naturally could lead to argument and recrimination, but in addition to their economic interdependence dealer and fishermen were often kinsfolk, and the arrangement in general seemed to work amicably enough, with a lot of good-humoured banter.

An idea of the scale of operations and level of profit of a fish dealer of this type is given by a series of transactions fairly regularly described to my wife and to me on her own initiative by Limoh (Halimah), whose husband was also a fish-dealer, part-time.[2] Like the other dealers she bought mainly Spanish mackerel, but included squid and crabs as well as other kinds of fish in her deals wherever she saw a chance of profit.

It was from Limoh I first learned of the tied-boat system,

---

[1] One man, Awang, was a son of Yusoh Pa' Nar, in 1940 a fisherman (v. p. 212 supra) and who by 1947 was without boat or net and served as crew man or as fish-dealer. The women were Me' Sung (v. Rosemary Firth (1943), p. 62) and her daughter Limoh (Halimah).

[2] Limoh was the niece of our housekeeper and came to our house frequently. She was an excellent informant, understanding what we wanted. In comment on our talks she once explained to others, " He wants to know all about buying and selling fish ; if we have profit he wants to know ; if we lose he wants to know." I was able to check most of Limoh's purchases on the beach, and believe her account to have been accurate.

when she told me one evening on the beach that there was now only one boat left to come in from which she could buy. I asked why didn't she go up the beach then to Pantai Damat —only a couple of hundred yards ? She replied, because there would be trouble with the dealers up there, just as there would if they came down to Kubang Kawoh. I asked then about the boat of Awang-Yoh, still to come in. She replied, " Limoh can't buy there—Adang (Adam, another dealer) buys." " But," I said, " you are his niece. Why not buy from him ? " " Because Awang uses the money of Adang—he has borrowed it," and she explained that this was what happened when a fisherman wanted to buy a boat and hadn't the capital. Kinship ties may be involved in this transaction, but of " her " four boats only one was that of a close kinsman, a mother's brother. The day-to-day arrangement was fairly rigid. If none of the four boats got fish " then Limoh doesn't get fish ". But there was some element of flexibility in the long-term arrangement since if a fisherman was dissatisfied with his own dealer he could apply to another, borrow the money to repay his debt and so change allegiance. Each party then had a sanction against the other : if the fisherman tried to insist on too high a price for his fish the dealer would " cut " the price when he came to pay ; if the dealer tried to force sales at too low a price the fisherman would transfer to another dealer. Knowledge of these sanctions by both parties helped to keep their arrangement in equilibrium. The dealer, however, though relying mainly on these arrangements did on occasion enlarge his field by buying fine fish from the purse seine crew men. (Cf. *Tangkap*, pp. 60, 377.)

Limoh and her mother (also a fish dealer) coöperated closely but had separate accounts. She and her husband worked together when he was on shore dealing in fish. But when he went to sea as a line fisherman and she bought his fish, they bargained like any other producer and dealer. This was not real haggling, however ; since her husband had a partner and they split the proceeds, clearly there had to be a figure for accounting purposes which would represent the market price of the fish. Limoh told me that unlike some other dealers she did not afterwards cut the price agreed upon with " her " fishermen. (This may not have been always so, but her statement was confirmed by one of them.) Fish dealing is a fairly specialized trade, and according to Limoh there were six dealers in the Kota Bharu market who dealt only in Spanish mackerel ; she sold regularly only to one.

Clearly even a small-scale fish dealer operating on the beach needs capital—for loans to buy boats, for some cash purchases of fish and for running expenses of ice and bus or taxi fares. Fishermen were apt to have a jaundiced view of the wealth of such dealers ; one said $1,000 would be needed and pointed to Limoh's gold bangles as evidence, adding that dealers never lost money ! But in fact the liquid capital they had at command was probably for the most part fairly small, and all their assets may not have been more than a few hundred dollars.

A summary of Limoh's transactions recorded in full for eleven days gave her a gross profit of $33.70 in eight of them, with a loss of $7.50 on two others and another day on which she neither gained nor lost but got her capital back. All this gave a net profit of $26.20 on a capital expenditure of $347.95 for fish, plus $34.70 for ice and fares. On this total outlay of about $400 she made then about $6\frac{1}{2}$ per cent. profit. From my other less systematic data, such transactions seemed to be fairly representative. Allegations then that these beach fish dealers exploited the fishermen and made exorbitant profits were unfounded. Moreover, from the point of view of making a living it is not the level of profit that is significant but the amount. From all these transactions Limoh earned on the average a little less than $2.50 a day, which is rather less if anything than what a fisherman would probably get on the average. But like the fishermen, she was in a hazardous occupation. Apart from the dozen or so days on which I observed her to make substantial transactions, there were at least three on which she did very poorly for lack of fish to buy. On one of these days three of her boats were not out, and from the fourth she got only a small Spanish mackerel for $1.50. On two other occasions she had a completely blank day owing to bad weather and poor catches. She seemed fairly representative of the beach dealers, and the condition of her house and the style in which she, her husband and two sons lived, showed that they were at much the same economic level as the fishermen. (For 1940, see pp. 232-3.)

GENERAL SOCIO-ECONOMIC POSITION OF THE COMMUNITY STUDIED

It is now necessary to look more generally at the economic position of the community which we studied in detail in 1963. The survey we organized was able to cover only about one-half of the area we studied in 1940, but with greater density of settlement

embraced 177 households,[1] comprising 823 persons, as against
331 households with 1,301 people in the earlier survey.

The general pattern of settlement was much the same as
before, though the main road through the area had been greatly
improved and carried almost constant motor traffic. Dwellings
also were of the same general type as twenty years before, but
on average were of better quality, more having tile roofs instead
of thatch, wooden walls instead of bamboo, and concrete bases
to the foundation posts instead of these being merely set in the
ground. The village of Perupok proper (just outside our survey
area) had a modern mosque and a new market, shifted inland
from the original sites (see Fig. 12) and both serving the inhabitants of our area. The beach scene was much less spectacular than before, with the withdrawal of many of the large
gaily painted sailing boats of the lift-net fleet, but was still the
scene of great activity as a retail fish market of continually
shifting pattern.

The community was still one of strong local tradition, with
about 70 per cent. of the adult men and women having been
born there, and another 20 per cent. coming from villages within
about a ten-mile radius. Apart from the few Chinese, most of
whom were Kelantan-born anyway, only six members of the
community came from outside Kelantan—two men from Patani,
one man from the Perhentian islands, two women from Trengganu and one from north Pahang. There were no Malays from
the West coast (cf. pp. 67-8). The community had an elaborate
network of kinship ties, and despite some considerable and
growing differences in wealth and education (cf. pp. 344-5) its
culture was still very uniform, with no marked social stratification.

Occupationally, the pattern was superficially very much the
same as before. If the distribution of primary occupations of
males in the area in 1963 (Table 17) be compared with that
for 1940 (p. 76) it will be seen that the proportion of fishermen
at the later date, at 73 per cent., was almost identical with that
of twenty years before, at 75 per cent.[2]

But the actual type of employment showed a very marked
change. By 1963 lift-net fishing as a primary employment for
about 75 per cent. of the men had been almost completely

[1] Analysis of the data in terms of household size, distribution of population by
sex and " economic stage ", etc., is given by Rosemary Firth, op. cit. (1966).
[2] This may be compared with the situation inland across the Kemassin River,
at Tanjong Pauh, where five-sixths of the households were engaged in rice-planting
(Dobby (1957), pp. 3-42).

TABLE 17

PRIMARY OCCUPATIONS OF POST-PUBERTAL MALES

Occupation	Economic Age Category				Total
	b *(14–19)	c (20–60)	d (60 +)		
Lift-net : Expert		1		1	
Crew		4		4	
					5
Other nets : Purse Seine					
Expert		1		1	
Motormen		2		2	
Crew	28	65	4	97	
					100
Line fishermen	2	56	7	65	65
Total fishermen	30	129	11		170
Dealers : fish		11		11	
Shopkeepers		11		11	
Total middlemen		22			22
Craftsmen, Officials :					
Barber		1		1	
Carpenter		3		3	
Taxi/Van driver		5		5	
Trishaw peddler		1		1	
Caretaker		1		1	
Coconut climber		3		3	
Copra maker		1		1	
Mechanic		1		1	
Rice planter		7		7	
Rubber tapper		1		1	
Soldier		1		1	
Religious official		1		1	
Religious teacher		3		3	
School teacher	1	1		2	
Bomor (medicine man)		1	1	2	
					33
Not Occupied		5	4		9
Total					234

* All ages approximate only.

displaced by purse seine fishing for about 60 per cent. of the men. The proportion of line fishermen had increased quite markedly. Again there had been some increase in the proportion of middle-men, craftsmen and officials, a change consonant with the general trend to modernization of the economy. One significant point should be noticed, the employment of young men who had passed through the puberty rite of *masok jawi*, but who were not fully adult. Whereas in 1940 almost none of these would have been at school, in 1963 about one-third in this category were receiving some form of literary education, and have not been included in the Table. But it is also to be remarked that the remainder, a considerable number, were serving as crew, in an unskilled capacity with the purse seine.[1]

I now turn to consider the economic position of the men who were primarily fishermen. What assets concerned with their employment did they own? From our census it appeared that of the 170 fishermen, only 17 had both boat and net, 33 had a boat but no net, and 9 had a net (or interest in one) but no boat. But 111 men had neither boat nor net, that is, no durable fishing capital. In other words, less than one-third of the fisher-men had a boat which enabled them to go to sea independently —mainly hand-lining, while about two-thirds of the fishermen had no significant capital at all. These figures may be con-trasted with those given for 1940 (pp. 135–6). In 1940 more than one-third of the fishermen owned boats, and about three-fifths of them owned nets or net components. Only about two-fifths had no significant capital at all. The figures are not exactly comparable because in 1940 the area covered included the village of Perupok with rather more relatively prosperous fishermen. But it seems evident that with the entry of the larger-scale capital units of the purse seine complex there was a falling off in investment in smaller capital items. In general, a distinctly higher proportion of the fishermen of 1963 were dependent for their livelihood on the capital resources of others.

ENTREPRENEURS

From this point of view it is of interest to take a backward look and to see the entrepreneurial position of men in 1963 com-pared with that in 1940. In 1940 the entrepreneurs in the

[1] For further examination of problems in this field, see Rosemary Firth (1966).

XVIIIa SIMPLE RURAL HOUSING, 1963
Palm-thatch roof and bamboo walls denote a fairly poor fisherman; the supporting posts are set directly in the ground.

XVIIIb GOOD QUALITY RURAL HOUSING, 1963
Tile roof, walls of the main structure in sawn timber, shuttered windows mark a more prosperous owner. Concrete foundations to the major supporting posts help to avoid destruction by termites.

Perupok area were of two kinds, the lift-net experts and the major fish dealers. The latter had some investment in fishing equipment but on the whole these two sectors were independent. The lift-net expert might, with a little exaggeration, have been described as the economic aristocrat of the fishing community, for the most part possessed of what was then considerable capital, and receiving the respect of his fellows. By 1963 the economic aristocrat was rather the large-scale fish dealer who had tended to take over the ownership of the fishing equipment and employ the fishing expert as technical manager or foreman. The fish dealer tended now to be referred to as *tauke*, a description, as already mentioned, originally applied to Chinese shopkeepers and merchants.

Few of the entrepreneurs of 1940 had survived until 1963. The man who had been probably the wealthiest fish dealer, Po' Che Su (pp. 176, 252, *supra*), was long dead, though his widow had carried on apparently some very profitable investments. Of the score of lift-net experts some, including Japar, the most successful in 1940, were dead (p. 155, etc. *supra*). Others, such as Muso, had retired and were relatively well off. Still others, such as Awang Tokor, Awang-Yoh and Awang Lung, had retired from lift-net work, but still perforce went to sea, as line fishermen. A brief history of two of them illustrates a not uncommon economic cycle.

Awang-Yoh in 1940 was a lift-net expert who might have been looked upon as a " rising man ", although at that time he had only moderate success, and did not seem very efficient. It was rumoured that he was financed in part by the fish dealer Po' Che Su (p. 176). In 1947 after the war he had neither boat nor net. But as time went on he recovered his position and, according to his own account, about 1953 sold a boat for $1,000 and invested half the money in land, which he then lost through failure to register the purchase properly. About 1959–60, probably feeling lift-net work too much for him, he sold his net (and ring net) for $1,000 and the large boat for $500, buying with the proceeds his house from his sister Me' Sung. When we arrived in the middle of 1963 he was line-fishing, an occupation to which he was driven by lack of other resources.

In 1940 Awang Lung was a fairly successful lift-net expert owning boats and nets (see pp. 153–4 *supra*). In 1947 he was still very prosperous, having just acquired a new lift-net ; he had three boats, including one old one and another bought for

$400 with mast and sails about a year before. With a year or two intermission during which he went line-fishing, gill-net fishing and trading to Trengganu for sugar and gambier and to Siam for salt and rice, he had kept with the lift-net work. But in 1947 he claimed that he was a poor man, having nine or ten people to feed in his house. By 1963 he had retired completely from net work. He went to sea line-fishing from time to time, but had ceased completely his work as a lift-net expert diving to " listen for " fish. " The ears no longer hear and the eyes no longer see " was his succinct expression at this point. He had sold his net and one boat, and the other two boats he had given one each to his sons. One of them also ran a motor boat in a purse seine group (cf. p. 322) ; the other son also had a motor boat and was said to be very rich. Awang Lung himself, always of a moralistic frame of mind, had turned much more to religion in his later years. He had begun to study the Koran systematically and went to mosque most regularly. He also took a rather lofty attitude towards shadow play and spirit medium performances, saying that he had heard people tell of them and heard their noise, but had not himself seen them— though he later qualified this by saying that he had done so when young. " Religious men such as Haji and people who are studying the Koran do not go to the shadow play," he said. I commented, " The shadow play expert and those who go are Muslims and Malays." He replied, " Muslims true, but . . ." —and his sentence faded away. He was described by himself and others as a poor man.

Who were the new entrepreneurs ? I can speak with no great assurance of them as a whole, but I did get some idea. A few such as the fish dealer Daud, now become Haji Daud, had survived and expanded from the pre-war days. In 1940 he dealt in fish, cattle, poultry and copra, had a house with a tiled roof and two sheds with fish-curing equipment, and he and his wife had 50 gallons of padi a year, from land given her by her father. He lent a little money, but was not a wealthy man. Twenty years later he had a larger house, also with tiled roof, owned an old sailing boat and had a share in a purse seine and two motor boats, had acquired an acre of rice land and another of rubber land, recently replanted, and got about 600 coconuts per crop from other land near the sea. In 1960 he had gone to Mecca, with his wife, and despite this expenditure was still building up his capital and buying a van on instalments (cf. p. 314). Other entrepreneurs like him were men of ability and energy who had built up their resources by prudent hard work. But a number seemed by repute to be men whose fathers or grandparents had been able to give them a good start in life through inherited wealth. Local opinion seemed divided on how far these people owed their rise to their inheritance or to their capacity for saving. Awang-Yoh said rather wistfully, " Rich people are those who save—but how they save we just don't know."

What was quite clear was that the economic level of these entrepreneurs was very much higher indeed than that of the ordinary fishermen and beach dealers (and also, from such information as I had, of craftsmen such as tile-makers or cement-block makers). These entrepreneurs were spoken of as the *orang kaya*, the rich men, of the community. Not only were they much more wealthy, they were also much more economically powerful than their forerunners a generation earlier. Moreover, though this is only an impression which I did not have the opportunity of checking, it seemed that the social distance between them and the ordinary run of *kampong* folk was greater than before. They were linked with the ordinary *kampong* folk in kinship, but the tie tended to be acknowledged on both sides only to a very limited degree. When a wedding took place in a rich man's house, he might invite his poor kin. But he would not necessarily do so because, as our housekeeper pointed out to us, he would know that the poor kin might well not be able to afford the dollar which is the normal contribution given by the departing guest to his or her host on such an occasion. It is perhaps not going too far to look on this set of people as an incipient mercantile class growing up in a rural environment. Again, this was a development foreseen a generation earlier (pp. 161, 296).

CHANGE IN ECONOMY AND SOCIETY

Prediction is not the primary aim of an anthropological study. But it is of interest to see whether or not predictions made have in fact been borne out, since one may then check the accuracy of analyses and explanations. In 1940 it had appeared to me that mechanization and modernization of the East Coast fishing industry was very probable and that this would lead to certain radical changes in its economic structure. Particularly would this affect the relations of the capitalist entrepreneurs to the main

body of fishermen. I suggested " The introduction of powered
boats, with their greater capital outlay, would tend to change
the existing pattern of economic relationships in the community.
. . . The increased costs would demand a re-arrangement in
the established system of distributing earnings and there would
be more likelihood of the gap between wealthy and poor fishermen being widened. A special group of power-boat owners
with superior economic status to the ordinary fishermen might
even be created. Since in these communities economic relationships are closely bound up with other social relationships, from
kinship to recreation, the structure of the peasant society itself
would be affected " (p. 20, 1st edition). I think that these
observations made in 1940 have been borne out by my findings
in 1963.

What seems to appear quite clearly from this analysis is the
strength of economic forces in making for a new kind of society.
Initially at least these economic forces are not automatic ; they
operate through the choices of individuals. The Malays of the
Perupok area were a long time in mechanizing their fishing
fleets, holding back presumably not merely from lack of capital
but from unfamiliarity with machines and perhaps from lack of
conviction that mechanization was worth their while. When
they did decide to turn from sail to motor-driven craft, they
chose not the less efficient though cheaper outboard type of
engine but the more solid inboard type. That choice once
made, apparently by one man, others followed him completely,
though he was in no sense a leader by hereditary criteria or in
wealth or other status, because they perceived the possibilities
of profit therein. But the new technological innovations were
not made singly. Greater power in the fishing fleet gave greater
range, but this would not have yielded full advantage without
ice to keep the fish fresh at sea. Nor would the much larger
catches have been easy to handle by relying as formerly on
public transport; private vans were advisable, and feasible because of the greater availability of such vehicles and the great
improvements in the road system. But at each stage in the
process of modernization fresh choices had to be made. The
spread of the new technology meant a growing sophistication on
the part of the local Malays. But it also meant that the price
of convenience had to be paid. Hence came an increase of
overheads which tended to alter the structure of income distribution in the major part of the fishing industry. Moreover,

it tended to involve also a more remote form of control of major fishing equipment than in the traditional local system whereby the technical expert was usually also the main financier. An effect of these changes has been to widen the wealth range in the society and accentuate the areas of social non-coöperation.

The issues are not purely economic in any very narrow sense. In theory the fishermen at the bottom of the scale have had the choice of remaining in this employment or seeking more remunerative work in agriculture, rubber-tapping, tile-making, etc. But it is partly a matter of skill. The skills of the fishermen are transmissible and handed on, and from an early age a young man builds up an investment in the special kinds of knowledge of wind and weather, currents and the habits of fish which he may be reluctant to abandon and which is useless to him in another employment, to which he must go as a novice. With this background many young men are attracted to fishing and the sea as a way of life, and say they would not be at home as agriculturalists. Moreover, with moderate success a man may hope to own a boat or net, which gives him a degree of status locally, or to train as a fishing expert. For a fisherman to sell boat or net except by way of trade, or to lose his position as fishing-expert, is a " come-down ", and such status factors may well enter into the decisions of men not to abandon the fisherman's life. Yet I think this is a matter of degree. The more the modernization of the fishing industry proceeds without a corresponding raising of level of income of the ordinary crew fisherman, the more is it likely that the fisherman will turn to occupations which offer a more stable if not greatly higher reward, and respond to such direct economic stimuli as are available to them.

What inferences may now be drawn about the more practical aspects of these problems? The function of an anthropological study such as this is not primarily to suggest remedies for difficulties, but to provide an analysis which is of theoretical or comparative relevance. But in contributing towards a clearer understanding of the situation it may offer a framework of facts and ideas which a more practically-oriented study may use.

It is clear that the economic situation of the mass of the fishermen in the Perupok area, and presumably elsewhere in Kelantan, is relatively poor. In any effort to improve this, where is the sensitive point for economic initiative? Some people will say it lies in the economic attitudes of the fishermen themselves : they do not save ; they have no proper sense of prudent calculation ; they waste their resources in expensive ceremony. In comment on this one can say that these Malay fishermen do display a great deal of prudent calculation. They have a passionate interest in small margins in bargaining and in their domestic affairs they watch keenly their expenses.[1] Moreover, the " wasteful " wedding and other ceremonies are also very carefully calculated in the hope, if not of making a profit, at any rate of breaking even. Those who attend them also calculate in return whether or not they can afford to accept the invitation. Economic incentives certainly get response.

Now in this peasant-type society, often regarded as a tightly-knit community of mutual help, how does the social system affect this economic process ? Do the *social* ties between producers modify the operation of the *economic* process ? Most anthropologists would think so. Yet what is interesting above all in this situation is that these economic processes, which had widened the gap between capitalist entrepreneurs and propertyless fishermen, were not cushioned to any apparent degree by the elaborate network of kinship ties in the local social system. The Malays of Kelantan have no system of descent group of matrilineage type as have those of Negri Sembilan ; but they recognize cognatic kin ties widely and implement them in defined ways. What it may imply is that such a bilateral kin structure, unlike a unilineal descent group structure, tends not to promote corporate group action in the face of economic inequality. Such inequality is regarded as a matter essentially of domestic or elementary family concern, and even siblings do not by any means come to each other's aid. And though Islam enjoins charity on all Believers, this is not the same thing as effective action on any scale to even out inequalities of income. Hence, the kinship ties of these fishermen do not inhibit their economic calculation, though they may soften its intensity. As I have shown, economic competition between kin may be keen. What is clear then is that prudent calculation is not enough for economic development, if what is desired is a general economic advance and not merely that of one sector of the society.

It can of course be argued that the only feasible policy in the Malay fishing situation is to let matters take their course, with increasing mechanization, and let the bulk of the fishermen find

[1] See Rosemary Firth (1966).

their own economic level. Even if this did not mean a lowering of their standards of living, it would probably result in considerable occupational change, possibly also migration. This would mean a change in the structure of their society and probably abandonment by them of some of their consumption values, such as elaborate wedding feasts. However, if this took place it would mean a radical re-alteration of their whole way of life ; it might also involve the removal of some of their most meaningful social relationships. A preferable policy from the social point of view is for responsible agencies to focus upon the more balanced development not only of the fishing industry, but also of other employment avenues, and try to promote a general set of higher and more regular incomes. The provision of some labour-intensive employment alternative to fishing might do much to ameliorate the fishermen's condition. It is true that they regard themselves as sea folk, and say they are unskilled at such occupations as rice planting. One of our neighbours said : " I am a fisherman and my children will be fishermen after me." Yet as experience has shown, they will turn in times of poor fishing to a range of other employments—tile making, rubber tapping, threshing rice—any occupation which will yield cash to keep the household going. But there is little local rubber on the coastal lands, and apparently no great labour shortage for the local rice harvest. Tile-making is a small-scale industry which might be developed, as also cement block and pipe manufacture. A programme of road development and other public works, which would specifically set out to provide another more substantial avenue for employment for local labour, would seem to be promising here. What is essential, however, is that such a policy should be operated steadily, with a firm view that it would be expected to make a contribution to the welfare of a depressed sector of the population as well as to the immediate development of communications.

As far as the fishing community is concerned internally, one development which seems desirable is for the assumption of more social responsibility by the more wealthy entrepreneurial sector. At the present time they seem to be in an intermediate position between upholding the traditional norms and separating themselves off, in a number of institutional respects, from the bulk of the people. There are at present few local institutions of a neutral non-ego-centred kind to enlist their talents. But with the advance of education and the extension of local government organization it may be that they will be attracted by the possibility of status and power. Then the electoral system, with its perception of the power of the common voter, may tend to awaken in them a response to the call of collective betterment.

APPENDIX I

NOTE ON PROBLEMS AND TECHNIQUE IN A FIELD STUDY OF A PEASANT ECONOMY

The study of peasant economic systems has hitherto been carried out largely by economists and economic historians, from documentary evidence. But of recent years geographers, agriculturalists, sociologists and anthropologists have also contributed to the study, working mainly from direct field observation. The anthropologist has entered partly because many of these peasant societies occur in colonial and other regions where his sphere of work normally lies, partly because of the lack of documentation there, and partly also because he is convinced of the need for a more integrative approach to these economic systems, the organization and values of which can only be understood in their general cultural context.

In his field study the anthropologist draws where possible on material used or supplied by the geographer, the agriculturalist and the historian. But he is concerned less with the influence of the natural environment or the details of the technical operations on which the economy is based than are geographer and agriculturalist, and less with the details of past forms and processes than is the historian. His primary interest is the present structure and functioning of social and economic relations, in the full complexity which can be seen only by direct observation. His contribution lies essentially in the breadth of his social inquiry, in his direct field technique and in the analytical detail of his observation—what may be termed " micro-sociology "—applied to the study of a sample community. In a sense his work is of a historical order, since he is studying the community during a given period of time. But it differs from that of a historian in that the anthropologist deals only to a small extent with material already filtered or crystallized by the processes of documentation —he is faced by the raw stuff of men's speech and actions, from which he himself must extract what seems to him most relevant. He is primary recorder of the material as well as secondary collector and analyst.

Drawbacks to Anthropological Technique

This position has both advantages and drawbacks. Let us take the latter first. Since his period of observation is usually short—commonly only a year or two years—there is danger that he may not grasp the full complexity of the phenomena, and even when he does, that he may interpret as "normal" what are really temporary special conditions. The first danger he attempts to avoid by concentrating his work for months at a time in a single community—one of roughly 1,000 people seems from general experience to be a convenient unit. Living in its midst, he brings to bear day after day all his apparatus of intensive inquiry. But from time to time he breaks off to conduct a series of rapid comparative surveys of an extensive kind, to set off his small-scale findings against the general background of the institutions and conditions in the region as a whole. The second danger he attempts to avoid by collecting evidence about the past from documents and from the memories of people both inside and outside the community. Since also the anthropologist has to gather his own materials as well as interpret them, and in gathering them it is manifestly impossible to record the totality of behaviour of the community, questions of bias are always present. One bias may be due to his own personal preferences— as by the overweighting of the rôle or opinions of certain individuals in the society because their personalities are congenial to him. Another may be due to the general theoretical mode of that particular branch of sociology in which he may have been trained. There is also the bias inherent in any process of selection. Items regarded as most significant for the elucidation of general principles may be, in fact, less so than others ; anomalous cases may be overweighted ; the sample of any type of behaviour may be too small for adequate generalization, and so on. The only safeguard here lies in awareness of the problems and the adoption of a system of checks to minimize the possible error. In particular, the collection of a wide range of quantitative data is useful as a check on generalizations formed from other kinds of material.

Its Advantages

But the anthropological method has some advantages of a special kind. The anthropologist is living in the midst of the community studied. After he has passed through the inevitable "probationary period" of three months or so necessary to establish himself in the confidence of the people and to get some working knowledge of the language, he does come to see the community life in all or nearly all its aspects. If he is at all sensitive, he does not merely impose his own criteria on the subject matter ; phenomena and their categories are forced upon his notice by events that take place daily before his eyes. (His ears too serve him. A shouting at the far end of the village ; a drum at night ; a rhythmic wail in a nearby house—any of these can lead to a whole new range of inquiry. If he has lived there for some time, they can tell him at once the story of some social or economic event that is taking place.)   If then his records are incomplete, if he has ignored some significant element in the situation, there is every chance that it will obtrude itself upon him.   Even if it does not, his intensive study of the behaviour of individuals will soon lead him to discover a gap in his explanation to date.   For example, if he is studying the economic balance of goods and services in the community, he may come to a point where he thinks that he has satisfactorily explained how the economic side of a marriage feast works.   But then he sees someone carrying through the village a parcel of goods of which he has no previous record ;   or on checking up his amounts of input and output he finds that they do not match. Resulting inquiry leads him to discover a type of transfer or category of service or mode of capital investment which is new to him.   This ability to follow up a new phenomenon on the spot, and to trace out the various channels through which any given item of wealth passes from its producer to its final consumer gives him a command over his sources and a flexible instrument of research of the greatest value.

A common distinction drawn between the natural sciences and the social sciences is that the former are experimental whereas the latter cannot be.   While broadly speaking this is true, the anthropologist can conduct experiments on a minor scale.   By a provocative question or by deliberately neglecting to follow a custom he can test the reactions of his companions. By participating in an economic or social event he injects a new factor—often, an additional supply of goods—into the workings of an institution, and can study the adaptations which are made. If he buys a boat and goes fishing he learns not merely the technology of handling the craft and of catching fish, but also something of bargaining, " fair " prices, rewards for help given and methods of sharing the catch.   If he is a guest at a feast or gives a feast himself (cf. p. 181) he finds out by experience the rules of hospitality, what it is appropriate to take as a contribution and to receive as a gift, how and when to contract debts and pay them as part of correct social behaviour.   In the early period of his work, before he gets the " feel " of the culture, his experiments may even be involuntary and the results embarrassing, but his very mistakes are instructive.   (Later, he usually has friends who introduce him, tell him beforehand what to do, and stand by him.)

The Value of Direct Observation

This emphasis on the value of direct first-hand observation of behaviour may seem exaggerated.   Granted that in order to apply economic reasoning to the conditions of a particular society it is necessary to know the broad facts, why not get them more easily ? Why not ask some reliable source, such as government officials or other Europeans who know the country, or leading men of the community itself ? Why not collect material on a wider basis by means of a questionnaire ? And why not supplement this, if it seems desirable, by a quick personal review to get the general background ? Why go to the trouble of camping in a village for months on end ; following out precisely what A, B, C and the rest do every day ; keeping a log of what goods, food and money come in and go out of their possession ; unravelling the history of a particular boat, canoe, pig or cow through its obscure fortunes ?

It is true that each of the ways mentioned above has its value in certain conditions. From resident Europeans one may learn a great deal of the general customs and economic conditions in the area. But very few of them have the time, even if they have the interest, to study systematically subjects such as the level of capital in a peasant economy, the organization of credit, or the distribution of earnings. From the leading men of the community much also may be learned. But in peasant communities or primitive communities, where the very idea of objective inquiry is a novelty and not easily understood, one cannot rely alone on sources of this kind. The anthropologist does use local opinions a great deal. But he treats them as data in two ways. In one way he gets from them facts which he verifies by his own observation, or which he can accept on the basis of collateral knowledge. In the other way he uses them as indices to further facts. They reveal to him personal prejudices ; claims (real or assumed) to status in the society or in the economic organization ; ideal concepts of how the speakers think the economic organization works, or should work. From the verbal data also he gets ideas of what people in general think is most significant in an economic situation, and how individuals regard their own participation in the economic process. To give only one example, on the last point. A member of fishing crew may stress the labour of paddling the boat and handling the net, and contrast his small share of the takings with the large share got by the fishing expert and net owner. The latter may emphasize his capital outlay, his loans of food and cash to the crew in the monsoon, and the labour he has in keeping his men up to the mark and the gear in good order, and justify his share on this account. Views of this kind not only help to deepen one's understanding of the economic process, but also show one where to look for possible misunderstanding, friction and even breakdown in production. Such opinions are often contradictory, or seemingly so ; they often are partially correct, but deliberately or unwittingly omit some factor of importance because the speaker is personally involved in the situation he describes. Such inconsistencies and partial accounts have their value when compared ; following his usual technique of checking by further opinion and by direct observation, the anthropologist is often led to a synthesis on a wider basis. (Team research, as for instance when husband and wife are working together on related problems, gives a further useful cross-check here.) Much the same is true of data obtained by questionnaire methods. They are valuable as a check on impressions, and as giving a synoptic view of economic phenomena such as occupational distribution or size of consuming units. But used alone, they are quite inadequate for many purposes, as for instance if they are intended to provide data on ownership of goods. It is common anthropological experience that statements so obtained are found when checked to have been misleading—not as a rule deliberately, but because of the complex forms of ownership, in which individual and group rights are often closely interwoven, and can be disentangled only with a knowledge of the general social structure. My own view, which I think most anthropologists would support, is that a questionnaire is best used towards the end of one's stay in a community. By then one has already accumulated much check data in advance, and one is in a position to frame questions in a way which will best yield the type of material required.

Testing Assumptions

The importance of the observational technique and of its use in studying the general culture of the community may be stressed from another angle. In an analysis of the economic organization of a peasant community the anthropologist is not only collecting material to answer questions of fact and interpreting that material to formulate generalizations ; he is also testing empirically some of the basic assumptions on which the questions themselves are framed.

Though the fundamental principles of economics are of universal application, many of its subsidiary assumptions ordinarily used are taken from generalized behaviour in Western industrial civilization. As such, they must be tried out before being taken as valid for an Oriental peasant society, or still more so, for a primitive society. Examples of such assumptions are that the economic interrelations of people are determined by their individual self-seeking ; that the owners of productive resources, for instance, will find the best use for these resources and place them there because in that way they can get the largest returns for themselves. At the same time, it is conceded that certain types of preferences are irrational—that, say, producers are attached to one market for their goods or services without any special economic advantage accruing to them from it, when they could serve themselves equally well elsewhere. Such assumptions seem self-evident enough, and in a sense are true for any economic system. Yet without further definition they are not flexible enough to be used as tools in analysis, as almost any examination of a small-scale society can show. Man is " self-seeking " ; man is often " irrational ". But the paradox is that in some social contexts both his self-seeking and his irrationality may form part of a single complex of ideas and attitudes. The key themes which emerge here are : preference for future rather than for present gain ; and for gain in the relative imponderables of prestige, reciprocal services and social support rather than in that of material goods alone.

The economist often goes half-way to meet this concept, but is rarely interested enough to see it fully. For instance, F. K. Knight, who has given some attention to these problems, emphasizes that the disposition to spend or to save, to consume income in the present or to store up wealth, is more influenced by motives such as " good form " than by mere time preference in consumption. He points out also that the fact of possessing accumulated goods confers social prestige and power over one's fellows.[1] But while for many primitive and peasant societies the first proposition seems to be correct, the second is only partly so. In such societies the prestige and power often do not lie simply in the accumulation—which by itself would render the possessor liable to a social stigma as an ungenerous person, and might cause him to be deprived of the economic benefit of support from his kinsfolk and others. They are gained by the *dispersal* of the wealth through socially approved channels, such as liberal hospitality, lavish gifts to kinsfolk, loans or charity to those in need, or holding a feast or a public entertainment (cf. Chapter VI). As Knight himself stresses in a wider context, the basic economic magnitude is service, not good. Yet the act of dispersal is commonly regarded as waste, or lack of thrift. Knight, indeed, in an unguarded moment follows Marshall and others in saying " The improvidence of savages is proverbial ". But this improvidence, defined in its context of social obligations and the reciprocity it entails, is one way of securing a lien on services, and of obtaining for oneself other less tangible but equally valuable benefits, according to the norms current in the society. What is improvidence to the economist or business man may in

[1] F. H. Knight, *Risk, Uncertainty and Profit*, pp. 133-4 (New York, 1921 ; 5th impression, London, 1940).

fact be prudence to the savage or to the peasant ; he is investing capital in social and economic insurance. So also with the " irrational preferences " ; they may well be part of a network of kinship and other social obligations which sooner or later will yield a return, or are themselves a return for advantages received in other ways.[1] From one angle, then, both " irrationality " and " improvidence " can be regarded as elements in the whole " self-seeking " process. That they can be so regarded shows the need for a re-definition of them in empirical terms. This re-shaping of assumptions of course does not mean that economic doctrine as such needs re-shaping. But it does mean that the descriptive economic analysis, whether pursued by the anthropologist or by someone else using his results, is based on a surer foundation for that particular society.

Conditions of Work

The description of the anthropologist's technique given so far represents broadly the lines along which my wife and I carried out our research in Malaya in 1939-40. We were able to spend in all nearly twelve months in the country. The period was divided roughly as follows. Nearly two months were spent in a general survey of social and economic conditions in Kelantan and north Trengganu, with particular reference to peasant agriculture and marketing, to Malay craft-work, to the wages and conditions of Malays employed in other occupations, and to such matters as technical education. About eight months were spent in intensive research among fishing villages and associated communities near Perupok, in the Bachok District of Kelantan. About a month was spent in collecting material on types of fishing, amount of equipment and systems of distribution in Trengganu and the north of Pahang, on a tour which covered all the important fishing centres. Finally, nearly a month was spent on a rapid comparative survey of peasant conditions on the western side of the peninsula, mainly in Selangor and Negri Sembilan, but with short visits to Perak and Malacca. The survey work combined the collection of personal impressions with consultation of official documents, and with discussion with Government officials and with Malay peasants. The intensive work was devoted to gaining as intimate knowledge as possible of the organization and life, particularly on the economic side, of the selected community of about 1,300 people. Here I studied mainly the fishing industry and the local agriculture, especially with reference to production, distribution and exchange, while

[1] Further material bearing on the argument here presented will be found in my *Primitive Polynesian Economy*, pp. 241, 314-15, 355-61 (London, 1939).

my wife paid especial attention to consumption and the domestic economy. My work lay mainly among the men, hers among the women—a division of labour of particular value in a Muslim community. But both of us spent much time in observing and taking part in the general social activities of the people—in their houses, in the rice fields, in the markets ; at weddings, funerals and circumcision feasts ; at shadow-plays and mediumistic performances for the cure of the sick ; and above all, up and down the beach, which was a highway, a gathering-place and the scene of much work.

Language

All our research here was done through the vernacular language. Equipped with only a brief preliminary study of standard Malay, we had to acquire the local dialect by daily conversation with the people. After several months we perforce became sufficiently fluent in it to carry out all ordinary inquiries and to take part in general talk. In particular, we acquired a budget of the common technical terms in fishing, agriculture, marketing, etc., which served as keys to unlock a great deal of information. This technical language is of the greatest importance, since it consists of a set of highly specialized symbols for a range of complicated processes and activities which are basic to the daily life of the people. Furnished with the right word, one can get a direct answer to a question, or understand a situation at once ; without it, however correct one's speech may be grammatically, one may often puzzle one's informant or be reduced to giving and receiving laborious explanations which often irritate the person one is talking to. Because he is bored, or too busy to waste his time, he may leave the explanation incomplete, or allow one to go away with a false impression. Moreover, knowledge of the right technical terms is of the greatest value in establishing confidence. Many a time in Trengganu, where I was not known, I have approached a fisherman with an inquiry, and been greeted with suspicion. But as soon as he realized from my Kelantan dialect and from my use of fishing terms that I was already " in the know ", the barriers were broken down and he was ready to exchange comparisons and give local information.

Quantitative Material

In an economic study of this kind ordinary inquiry must be reinforced by quantitative data. One main difficulty in the way of getting such material is that to be of real value it must be systematic. This means the steady daily collection of the same kind of facts over a considerable period ; bluntly put, it means hard work, often of a monotonous kind. Another difficulty is that to be effective the observations must cover the same units each time—comparison of the output of different boats or nets on different days is not conclusive. But this is not easy in a fishing community on the east coast, where a boat may land on the beach at any hour of the afternoon, and not always at the same spot. Moreover, identification of units is not simple. In the early stages of work, especially, all boats of the same type tend to look alike to the uninitiated eye, and I had to have recourse to surreptitious noting down of the registration·numbers —surreptitious, because at that time it might have been thought by the fishermen that I was checking up on unpaid licence fees for the government. Later on, one gets to know individual boats as the fishermen do—by their lines, by the shape and patching of their sails, by the course they take when coming in and by other small professional signs.

From the methodological point of view another difficulty arises in deciding what shall be the indices which will most nearly yield what the investigator wants to know. He is not studying output, income, capital investment or distribution in the abstract ; he can only study them as they are represented by a multitude of specific things and actions. He must therefore observe those things and those actions which will lead him to relevant economic generalizations. These items must be measurable, they must be significant, and they must be selected from a mass of others.

In studying output, for instance, it would be desirable to know the weight of fish brought in by each boat or net each day. But in this area the fish are not weighed, and it is manifestly impossible for a private investigator to divert them to a weighing depot or trundle a weighing machine up and down the beach. What he can do, however, is to use indirect methods of measurement which give him some approximate totals. He can note, by sampling, the number of baskets of fish in a catch ; he can find out in various ways the average weight of a basket. Again, from time to time, as when fish are sold retail, or by number, he can count the fish in a basket ; and on his own scales at home he can weigh individual series of fish of that type, thus getting a cross-check. Some idea of the physical volume of production can be obtained by such round-about means.

Quantitative work of this kind means much counting. But the investigator cannot count every item of economic significance that is present in the whole field at any one time. Just as when he wishes to measure output he cannot count and weigh every fish caught, so if he wants to find the trend of market prices he cannot note every individual transaction in the thousands that take place in a single afternoon. He must decide what facts, out of the bewildering flow that is passing before his eyes, are most likely to give the key to the economic processes and concentrate upon those. He must use sampling methods. But his samples must be large enough and free enough from bias to be statistically significant. It may be noted that the use of a true random sample is rarely possible in the study of such a fishing economy. It is preferable to attempt to cover selected groups or units intensively, over long periods, and set the results against their known circumstances and the general background.

The collection of quantitative data must begin as soon as possible in order to cover seasonal fluctuations. But in its early stages the material often lacks precision because of the investigator's imperfect knowledge of the general principles of the economic organization. Thus if he wishes to study the range of output and income of the individual fishermen it can be misleading to take the obvious course (as I did in the first few days) and note carefully the catch of each boat as it draws up on the beach. He will later discover that many of the boats which come in empty have done so not because they have earned nothing, but because the catch from their net has been taken by a carrier-agent, and that his full hold in turn does not represent the takings of his crew alone. He will find out also that a boat which is half-full has some fish from its own net and some from a neighbouring net ; that another with a small catch has sold the bulk to a middleman at sea or at a village further down the coast—and so on. It is only as he comes to know the complexities of the system and the people concerned that the investigator can give precision and full meaning to his figures.

I have described the physical and methodological difficulties inherent in this kind of study because, unless they are realized, the limitations and the value of such quantitative data may pass unnoticed. It is my conviction that such data are essential for a knowledge of a peasant economy. Analysis of the records and comparison of the results with those obtained from other less systematic observations helps to build up the picture of the economic system ; moreover, the records provide a useful objective check on one's own personal impressions and on the statements and estimates of informants.

Types of Record Collected

As part of the intensive study of the Perupok area, the following systematic observations were made :

1. A complete daily record for six months of the value of the catch from each of 20 lift-nets, operating from a stretch of beach about half a mile long. This type of fishing, the major one in

the area, embraced the activities of about 400 men for most of the time.

2. Daily records over the same period of the value of the catch for the majority of all other types of large nets operating in the area, and similar records for a large sample of line fishermen.

3. A sample census of rice production on 222 contiguous plots, covering a total area of about 57 acres, noting owner and cultivator, type of rice planted, approximate yield, and much subsidiary information about manuring, labour engaged, etc.

4. A sample census of vegetable production on 64 plots covering a total area of about 11 acres in and around the sample rice area, with similar data as for 3.

5. A general economic census of 331 households, comprising a record of the number, sex, estimated age category, occupation and kinship relation of the members of each household ; their ownership of boats and fishing nets ; their coco-nut palms or the annual yield from these ; the amount of rice produced, or obtained from land share-cropped to others ; the presence or absence of a vegetable patch ; and whether the house was a single-roomed or multi-roomed structure. Much subsidiary information, e.g. on furniture, income, etc., was also obtained. The census was taken towards the end of our period of work, when most of the people were already known personally to us, and much of the information given could be checked from other sources or was already on file in other contexts. (At such a late stage in the inquiry, a census or other questionnaire can be of great value ; if taken at an early stage it is often useless and can be quite misleading.) The census area lay immediately behind the stretch of beach where the fishing records were taken, and formed a fairly compact economic and social unit with it ; the agricultural area (see nos. 3 and 4) lay immediately behind again.

6. A sample set of budgets from 10 households within the census area, taken daily for periods varying from one to five months. (A detailed analysis of this material and some of the census data has already been given by my wife, *op. cit.*)

These records, apart from helping to reveal details of the economic organization and showing the relation of norm to variations, give data on output, income and expenditure which cannot be obtained from published sources. They could also assist in providing some basic material for broader studies of the tropical peasant economy—as studies of nutrition, or of certain sections of the national income.

APPENDIX II

SUMMARY OF BOATS AND FISHING EQUIPMENT IN MALAYAN WATERS, 1938–9

The following table shows the approximate number of fishing boats. Figures for Kelantan and Trengganu are for 1939, compiled from official returns consulted in Kota Bharu and Kuala Trengganu. Figures for the other areas are for 1938, as given in the S.S. & F.M.S. *Annual Report of the Fisheries Department* for that year, p. 31. Figures for Johore, Kedah and Perlis are not known, but a rough estimate is given for completeness. Final totals are approximate. (For convenience, Johore boats have been all placed in the third column.)

TABLE 18

Singapore Boats.	West coast Boats.		East coast Boats.		Total Boats.
1,315	Perak . . .	3,247	Kelantan . .	3,588	
	Penang . . .	2,073	Trengganu . .	3,310	
	Selangor . .	1,743	Pahang . . .	905	
	Malacca . .	981	Johore (e) .	1,000	
	Negri Sembilan	227			
	Kedah (e) . .	750			
	Perlis (e) . .	150			
1,315	Total . . .	9,200 (approx.)	Total . . .	8,800 (approx.)	19,500 (approx.)

The following table shows the number of items of fishing equipment, distinguishing some of the main types. The figures, which are approximate, are derived from the S.S. & F.M.S. *Annual Report of the Fisheries Department* for 1938, with data from Kelantan and Trengganu incorporated from official local returns. The table is incomplete, since data from Johore, Kedah and Perlis have not been available.

Type.	Singapore.	West coast.	East coast.	Total.
DRIFT-NETS				
Jaring Kurau	—	232	1	233
Jaring Tamban . . .	—	—	481	481
Jaring Těnggělam . . .	—	399	180	579
Jaring Těnggiri . . .	153	910	71	1,134
Pukat Hanyut	—	—	167	167
Other	45	91	9	145
SEINES				
Pukat Bawal	—	163	27	190
Pukat Dalam	—	—	206	206
Pukat Kisa	—	432	—	432
Pukat Payang . . .	2	13	126	141
Pukat Sudu . . .	—	—	68	68
Pukat Tarek (Malay) .	—	33	518	551
Pukat Tarek (Chinese) .	71	155	—	226
Other	36	299	57	392
LIFT-NETS (GROUND-NETS)				
Pukat Tangkul . . .	—	—	213	213
TRAP-NETS				
Ambai	3	1,077	—	1,080
Pompang	—	738	—	738
Gombang	—	218	—	218
TRAPS				
Bělat	91	368	129	588
Kelong	232	—	8	240
Other	74	818	117	1,009
LONG LINES	31	693	202	926
MISCELLANEOUS NETS . .	85	167	235	487

Despite its incompleteness, the table indicates the regional specialization in the various types of net fishing.

APPENDIX III

SUMMARY OF FISHING-BOATS AND NETS IN KELANTAN AND TRENGGANU, ABOUT 1939

The following table shows the number of fishing-boats in Kelantan in 1939, according to stations of registration. The totals, which include river-boats, were compiled from returns at the office of the Superintendent of Marine and Customs. Details of the principal types of boats at Tumpat and Bachok were extracted by myself from the local station records, but I was not able to examine the records at other stations.

TABLE 20

Station.	Kolek Buatan Barat.	Kolek Lichung	Kueh.	River-boats.	Other.	Total.
Tumpat . . .	140	4	79	540	23	786
Bachok . . .	46	129	325	43	29	572
Kota Bharu . .	—	—	—	—	—	30
Cherang Ruku .	—	—	—	—	—	176
Tabal . . .	—	—	—	—	—	290
Kemassin . .	—	—	—	—	—	254
Sungai Pinang .	—	—	—	—	—	742
Kuala Pa' Amat	—	—	—	—	—	738
Total . .	—	—	—	—	—	3,588

Table 21, for Trengganu fishing-boats in 1939, excludes river-boats. The totals for each station were compiled locally for me at the request of the Principal Officer of Customs.

Table 22 shows the numbers of the major types of nets in Kelantan (about 1937) and Trengganu (in 1940). Their general geographical distribution along the coast is indicated in Map 3. The Kelantan figures were obtained from a census taken by a Malay Revenue Officer ; those from Trengganu were furnished by the Principal Officer of Customs.

In addition to these nets there are a large number of casting nets and scoop-nets, and also fishing-traps (*bĕlat*)—105 in Kelantan alone—small lifting-nets for crabs—62 in Kelantan—as well as some portable traps (*bubu*) and large stake structures (*kelong*).

TABLE 21

Station.	Pĕrahu Payang.	Kolek Bĕsar.	Kolek Tangkol.	Kolek Pĕril.	Kolek Kueh.	Total.
Kuala Trengganu	8	47	285	352	—	692
Kuala Besut . .	100	2	43	25	3	173
Sĕtiu	22	26	204	60	76	388
Marang . . .	1	81	215	—	285	582
Batu Rakit . .	—	26	203	50	207	486
Pulau Redang .	—	2	13	6	26	47
Dungun . . .	6	67	25	17	211	326
Paka	—	22	38	18	48	126
Kĕrteh . . .	—	5	9	11	14	39
Kijal	—	36	65	20	29	150
Kemasik . . .	—	12	16	21	6	55
Kemaman . .	10	24	71	—	141	246
Total . .	147	350	1,187	580	1,046	3,310

TABLE 22

NETS IN KELANTAN AND TRENGGANU

Net Type.	Number in Kelantan.	Number in Trengganu.
Pukat Payang	46	32
Pukat Takur	89	124
Pukat Tarek	157	278
Pukat Dalam	100	86
Pukat Hanyut (i)	85	82
Pukat Jaring	219	57
Pukat Tĕnggĕlam	116	64
Pukat Sudu	8	12
Pukat Likung	9	0
Pukat Tanggut	0	12
Pukat Pĕtaram (ii)	0	20
Pukat Tarek Sungai (ii)	13	0
Pukat Takur Baring	49	40
Pukat Talang	3	6
Pukat Todak	0	5
Pukat Duri	0	3
Pukat Jumpol	10	0
Other nets	—	9
Total	904	830

(i) Includes for Kelantan 43 *pukat murou*, of same general type as *pukat hanyut*.
(ii) Often used interchangeably, and possibly of same type.

APPENDIX IV

VARIATIONS IN THE SCHEME OF DISTRIBUTION IN THE MAJOR FORMS OF FISHING

In the different fishing areas along the Kelantan-Trengganu coast there are significant differences in the scheme of distribution of the returns. These depend to some extent upon variations in the form of organization of the activity, but often seem to be simply a matter of locally-evolved practice. As one fisherman said " Every *kuala* (river-mouth) has its own customs." In my visits to nearly all the more important fishing areas along the coast I noted a great number of these local differences, and give a selection of them here (see Map 3).

A. VARIANTS IN THE SCHEME OF DISTRIBUTION OF RETURNS FROM LIFT-NET (*PUKAT TAKUR*) FISHING IN KELANTAN AND TRENGGANU.

Tumpat (*Kampong Dalam Ru*) :

Here the *pĕraih laut* are not members of the net group, but go out and buy the fish at sea, on credit, making a profit or loss on the sale according to the shore market. If they lose, they are not allowed to " cut " the price, since the bargain is struck in the presence of the whole crew, who, it is said, would not allow " cutting ". Since the *pĕraih laut* is not a member of the group he and his crew do not come and help with the repair and dyeing of the net.

The net group consists then of five boats. The first item in distribution is an allotment of 10 per cent. for the *unjang*, and does not vary whether the *unjang* is large or small. But sometimes a large parcel of fish may be given instead, selling for several dollars—up to $15 if the catch is a large one. There is naturally no splitting of the distribution into *pĕraih* and *juru sĕlam's* share-out. The bulk of the takings are divided into three, one-third going to the net and two-thirds to the crew. Each boat gets twice the share of a man, though it was stated that out of the net's share comes $2 for the *pĕrahu sampan* (this, however, may be only occasional, equivalent to the *duit kayoh sampan* of the Perupok area).

Kemerak :

Here the custom is normally to fish with five boats, one of which carries the catch to shore. Some nets fish with six boats, but there is no division with the *pĕraih* as in Perupok. First comes an allotment to the *unjang*. These are of two types—a small type used only by *takur* nets, with an allotment of one-twentieth of the catch ; and a large type used partly by line fishermen and partly by *takur*, for which an allotment of one-tenth is made, with some fish as *makan lau'*. Then the remainder of the takings are divided by three and shared in the usual way. A boat gets one share, equivalent to that of a man. Both *juru sĕlam* and captain of the net-boat get an extra share for their special functions. The small share given to a boat here is probably linked with the fact that most of the boats are small and old.

Pantai Bharu :

Here also the *peraih laut* is not a member of the lift-net group but a middleman outright, buying the catch at sea and taking no share in the general proceeds of the group. He takes the risks of the shore market. " If there is no profit, he gets nothing." The reason for not giving the *pĕraih* a share in the general distribution was given thus : " He does not go looking for crew for the net ; the *pĕraih* at Perupok look for crew." The scheme of division of the takings is the common one of one-tenth for the *unjang*, followed by a division of one-third to the net and two-thirds to the crew. I did not ascertain what was given to boats.

Kuala Besut :

Here, as in the cases just mentioned, the *pĕraih laut* is a middleman buying at sea. " If the net owner does not accept the price he offers he does not get the fish." If he loses badly he may " cut " the price a little. There appear to be two kinds of *pĕraih laut* here. The *pĕraih yang tĕtap* is attached to the net group, often by having lent cash for the purchase of net-ropes, and comes and helps to repair the net on Fridays ; he has the privilege of buying the fish at a cheaper rate, and is, in fact, a type of *daganang*. The *pĕraih ta' tĕntu* buy fish from whatever nets they can. In the division of the takings from the net an allotment of two-tenths or even three-tenths is made for the *unjang*. This is much higher than the normal, but was justified by the fishermen, who said that the *unjang* are larger than in Kelantan, being often put down by line fishermen for their own purposes and used by the *takur* as well. The remaining division is on the one-third, two-thirds basis, and a boat is counted as equivalent to a man. But here, as in some other Trengganu areas, the custom is not to make a division regularly each week, but to postpone it until the takings have reached a substantial sum, and divide perhaps only every two or three weeks. To meet their need of cash the crew borrow from the *juru sĕlam*. It would seem that this irregularity offers possibilities of exploitation or at least advantage to the *juru sĕlam*.

Ayer Tawar (Besut) :

This is a comparatively poor fishing settlement, using small boats and seond-hand nets. There are no local *pĕraih laut* but they come from Kuala Besut and buy the fish outright, on credit. They are attached to individual nets. Division of the takings is irregular, occurring every month or two months, depending on the accumulation of cash. The allotment for the *unjang* is two-tenths or one-third of the total takings—large because the *unjang* are many, and are put down by line fishermen as well as *takur* fishermen. The common system of division of the remainder of the takings into thirds is followed. The share of the net, however, does not go to a single owner, but is divided among the small combine which has raised or borrowed the capital. If there are five men in the combine the cash is divided into six portions, the lender of the money getting one share, and also a dollar for every dollar that a crew-member gets in the other section of the distribution. In effect, then, the lender of the money gets a dividend on the same basis as a net owner and a crew-member, " though he does not go to sea ".

Penarik (Setiu) :

Here the common principle of distribution is followed, the *unjang* getting one-tenth, with the remainder divided into one-third and two-third shares for net and crew respectively. Here there are no *pĕraih laut*, and most of the fish is bought by Chinese for curing.

Merang :

Here the nets are owned by Chinese who buy the fish. The ordinary scheme of division is followed for the bulk catch. But there is a special practice for secondary shares. When a good catch is taken, a share of the fish is given to one boat as *ikan luan* as a reward for its extra work. " The men who handle the fish get a little more because of the work of cleaning the boat afterwards." A different boat takes the fish on successive days, but the net-boat and the *pĕrahu sampan* are excluded. Hence on such a day the net-boat gets an allotment of $1.25, and the *pĕrahu sampan* an allotment of $1, irrespective, so it was said, of the amount for which the fish of the *ikan luan* sold, but being taken out of this amount. Thus if a boat sells its special share for $10, this is reduced to $7.75 as a final bonus to its crew.

Batu Rakit :

The *pĕraih laut* here come from Kuala Trengganu, and buy the fish outright, having to pay cash on the spot if it is demanded. Otherwise the *juru sĕlam* goes once a week into the town to collect the proceeds. The distribution is made only at long intervals, perhaps once every two or three months. The reason given for this was that if small amounts are handed out, they are soon spent, and the men prefer to let them accumulate, often deposited with a Chinese *tauke*. The need for cash in the meantime is met by the liberal allotment of the special fish of *ikan luan*. This is not taken if the catch is small, but tends to be given every few days, all the boats of the group getting it in order. This special increment is justified as at Merang by the statement that handling the fish means that the boat has to be cleaned out. *Makan lau'* is also given in addition ; if the catch is only small, then the boat which carries the fish gets about 15 fish per man, and the others about 10 fish per man.

There are some variations in the ordinary scheme of division. An allotment of one-tenth is given for *unjang* of line fishermen when they are used by *takur* fishermen, but nothing is given to a *juru sĕlam* of another group whose *unjang* has been used. The normal reciprocity is deemed sufficient. The bulk of the takings are divided on the one-third, two-thirds basis, the latter going to the crew, with a boat counting as equivalent to a man. But within this section of the division there are special shares. The *juru sĕlam* gets two shares for his " diving ", two shares are divided among the men who handle the casting of the net, and two shares are also allotted (to the *juru sĕlam* and a few helpers) for the work of dyeing the net.

Kuala Trengganu :

After a percentage taken by the *daganang* (see Chap. II and Appendix V), the common principle of division is followed, into thirds, with a boat receiving a share equal to that of a man. No allotment is made for *unjang* put down by any lift-net fishermen ; for *unjang* of line fishermen used a tenth share, or a large parcel of fish, is given according to their wish.

Kuala Marang :

Here, the centre of a considerable lift-net fishing industry, essentially the same scheme is in vogue as at Kuala Trengganu. The reason given for the absence of any special appropriation for the *unjang* made by the *takur* fishermen themselves was that these structures are made by the members of the net-combine alone ; and not by the ordinary crew. Should the *unjang* of another group be used, the situation is met by ultimate reciprocal rights. "Among *takur* together nothing is given; they go according to their inclination, and fish."

The *unjang* of the line fishermen, when used by a lift-net group, are recompensed by a large bag-net (*saup*) full of fish, taking three or four men to lift. This is preferred by the owners of the *unjang* to a percentage in cash, since they tend to get more profit this way. If fish are " selling " they get $5 or $6 for it, if the market is poor, about $2 or $3.

Dungun :

This area has much the same principles of distribution as at Kuala Trengganu and Kuala Marang. But there is no initial percentage for *daganang,* who are lacking here, and the use of *unjang* of line fishermen is compensated by a *tanggut* (basket) of fish and not by a bag-net full ; as an alternative to fish an allotment of ten per cent. may be made. Here fishing from the *unjang* of another expert was said not to be allowed unless permission was first obtained. Each group keeps normally to its own *unjang,* of which it has about ten, and only if there are no fish there can fishing from those of another group be done. There is, however, no percentage of the takings given in such case.

Here, too, *ikan luan* and *ikan gandoh* are allotted every day on which there is any catch at all. There are no *pĕraih laut,* and all the boats of the group except the net-boat take it in turns to carry the fish to shore. The cash from the sale of the *ikan luan* is divided among the crew of the carrier-boat, who thus are repaid for their work on the net. The net-boat is said to get *ikan gandoh* each day when fish are taken, but it is possible that this is less in amount than the *ikan luan.*

In this area as elsewhere the crew-members come to the *juru sĕlam* in the monsoon season for support, and ask for cash or rice. A common practice is for the expert to arrange with a Chinese dealer to give any such man a bag of rice, which is charged up to the expert and recouped from the man when the season begins again. If a loan of this kind is not made then the crew apply to other experts and transfer their labour. " I have done it myself" said one of my informants "if the *juru sĕlam* did not give it to me I was not angry ; I went to another *juru sĕlam* ". But if the expert has plenty of crew he may prefer to let some go, and may render them no help.

Paka :

In general the ordinary principles of distribution obtain here. No allotment is made for the *unjang* of the lift-net groups themselves, the reason given being that each net is owned by a large number of men—eight or ten—and they all contribute to the work of preparing the *unjang*. Those of line fishermen are compensated when used by one-tenth of the takings in cash, or by fish.

If the partners in the net are using both their own and borrowed money, the net's share, normally one-third of the total, is subjected to extensive sub-division. It is first halved, and from one half the man who has lent capital for the net gets the equivalent of twice a crew-man's share for every $100 of capital he has put in. Any remainder still in this half net's share is handed to him as repayment of principal. His interest is not taken out of the crew's share of the takings—" or how could they live ? " The other half of the net's share is then divided up among the working partners who have put in capital, in proportion to their stake. If together they have put up $100 the cash is divided into ten shares, a man who has put up $20 getting two shares, one who has put up $30 getting three shares, and so on.

Kemaman :

Here the lift-net boats are owned by Chinese, and the nets by Malays, but using Chinese capital. After an initial percentage taken by the Chinese as *daganang*, the common procedure is followed. Reciprocal fishing obviates any allocation for *unjang*, except those of line fishermen, who are given a parcel of fish normally sold to small Chinese dealers for $5 or $6. " But sometimes when the owner of the *unjang* is not there no fish are given." As at Dungun, *ikan luan* are taken every day, and also a compartment of *ikan bělakang* for home consumption or sale.

B. DISTRIBUTION OF RETURNS FROM *PUKAT PAYANG* FISHING IN TRENGGANU AND NORTH PAHANG

The organization of the *pukat payang* comprises a large boat and net in charge of a *juru sělam*, with normally a crew of about 18 men. The catch consists primarily of shoals of small fish, but mingled with these are individual larger fish, of better quality, known as *ikan molek* (fine fish). The elements involved in the distributive scheme are as follows :

i. *ikan tangkap*, or *ikan rajut*. After the catch is made the members of the crew are at liberty to remove from it any of the individual larger fish they can find. They are known as " caught fish " (*ikan tangkap*) since they are grabbed out by the crew as the catch is emptied into the hold. Each man keeps what he can seize and puts it into a loose string bag termed *rajut*, whence the alternative name for the fish. At the end of each day's fishing the contents of the bags are sold for cash to small dealers on the beach, the amount gained varying from 50 cents or so to about $2 per man. The system is modified slightly at Kemaman, where the *rajut* fish are sold collectively if the net is a small one, and the proceeds divided, but are sold separately by each man if the net is of the large type.

These *rajut* fish fulfil an important function, since they supply the crew with daily cash, and the receipts from this source are often larger than the share of the takings from the sale of the bulk catch itself. At some of the larger centres, as Kuala Trengganu and Kuantan, division of the bulk proceeds takes place weekly ; but elsewhere it is usually less frequent. At Beserah a division takes place every two or three weeks, and at Kuala Besut it may not occur for a couple of months or more. In these latter cases the daily cash from the *rajut* fish is essential in enabling the crew to carry on. The reason for delay here is that the dealers who buy the fish work on a smaller capital and often have to wait till they get paid from Singapore before they can settle with the fishermen.

ii. In the division of the proceeds from bulk sales the first charge upon receipts is the percentage or " commission " of the *daganang* where this function is operative.

iii. Then comes an item known as *chabut chero'*, an allotment of one-tenth of the total receipts to members of the crew who undertake the upkeep of the net. At Kuala Trengganu this is distributed among about 8 or 10 men, who receive 20 or 30 cents apiece each time.

iv. The next item is an allotment for the *jong timba*, titularly in charge of baling, but also responsible for furnishing the craft with water, washing it out after use—a job which may take a couple of hours a day—and waking the crew in the early morning. This last is an important function, and is no sinecure. At Kuala Trengganu the *jong timba* gets $4 or $5 per $100 of the total receipts ; at Kuala Besut I was told that he gets $8 per $100, but here he mends ropes as well ; and at Kuantan that he gets $10 per $100, and that he cooks up the dye and dyes the net in addition. Here it appears as if there is no *chabut chero'*.

v. Then comes the main division, into thirds of the remainder, on the *bagi tiga* principle, which seems to apply in all cases. One-third goes to the net and boat together, and two-thirds to the crew. In the division among the crew allowance is made for a number of special functions, though those which are thus singled out appear to vary in different areas. Commonly, one extra share is allotted to the man who manages the foresail (*layar topang*) ; one each to the two men who manage the mainsail *layar agong*) ; one to the man who pays out the net (*mĕnarang*) ; and one to the *juru sĕlam* for the use of his tiny boat (*sampan*) which is employed in searching for the shoals of fish. Other shares often allotted are : two to the *juru sĕlam* for his " diving " ; one to the captain of the boat ; and one to the owner of the anchor rope (*tali saup*). I was told by one *juru sĕlam* in Kuala Trengganu, however, that he took sums for the men who managed the sails out of the earlier *chabut chero'*.

In cases where a boat and net are run by a *juru sĕlam* other than the owner, their share (the *bagian dalam*) is either retained by the *juru sĕlam*, as often happens when the owner is a Chinese who gets his profit out of the handling of the fish, or divided into three, the owner getting two-thirds and the *juru sĕlam* one-third.

Earnings from work with the *pukat payang* are comparatively high. The major portion is provided by the sales of the *ikan rajut* daily, yielding perhaps $5 or more a week. The crew's share in the sales of the bulk fish (*ikan petoh*) is small, often only 50 cents or so per man per week, due partly to the initial percentages taken for special functions, but mainly to the low prices paid by the Chinese dealers.

The work is hard, and the men of the crew seem to live at a high rate. Pa' Che Mat, who went from Perupok several years in succession to Beserah as *juru sĕlam* for a *pukat payang* owned by the Chinese there, gave me some details of their life. He stayed there about three months, and brought back between $40 and $50, as against the $10 or $15 of an ordinary crew-member. Much of their money had gone in expensive living. They did not take rice with them to sea as ordinary fishermen do, since with a crew of nearly a score there would not be room for all their food-boxes, and if one man took food others would steal it. The result was that they each spent about 40 cents a day on coffee, snacks and cigarettes—20 cents before going out in the morning and the same on returning in the afternoon.

C. DISTRIBUTION OF RETURNS FROM SOME OTHER TYPES OF NET-FISHING IN TRENGGANU AND NORTH PAHANG.

(a) *Pukat Tarek*

    i. A percentage of the total receipts is taken by the *daganang*, if there is one, amounting to 5 per cent. or 4 per cent.

    ii. The main feature of the system is the division into two equal parts (*bagi dua*), one going to the boat and the net, and the other to the crew.

    iii. Extra shares, however, are allotted for special functions, and the size and incidence of these varies in different areas.

A special appropriation is usually made as *chabut chero'*, for the work of repairing the net, assistance in dyeing it, and other jobs in connection with it. Unlike the similar item with *pukat payang*, this comes not out of the total receipts but out of the *bagian dalam*, the net's share. At Kuala Trengganu the amount is ordinarily the same as that taken by the *daganang*. At Batu Rakit, where there are no *daganang*, each of the helpers gets an extra sum exactly equal to that which an ordinary crew-member receives in the *bagian luar*, the crew's division. At Kemaman no specific shares are allotted, but the net owner hands out sums of 50 cents or a dollar from the net's share to the men who have helped in the work. At Kijal there seems to be no *chabut chero'* of any kind. But at Kemasik and apparently at Dungun one-tenth of the total proceeds is given, though the men who receive it include in their work the baling and washing out of the boat, etc., which are the functions of the *jong timba*.

At Kuala Trengganu there is no special recognition of *jong timba* for *pukat tarek*. At Batu Rakit, where the boat receives a share equivalent to that of a man in the *bagian luar*, the crew's division, this share is handed over to the *jong timba*. At Kemasik the return to the *jong timba* is included in the *chabut chero'*. At Kijal no specific appropriation is made, but the net owner hands over a dollar or so to the two men who do the work of washing out the boat. At Paka and Kemaman I was told that the *jong timba* receives an extra share for his work, but I have no details of the amount involved.

For the special work of paying out the net a man may get an extra increment, but this appears to be small; at Merabang, Kuala Trengganu, it was described merely as " a little more ". In some areas this appears to be omitted.

(*b*) *Pukat Dalam*

The principles of distribution here are very similar to those of *pukat tarek* fishing; in Trengganu the two nets are commonly worked alternately from the same boat, and the *pukat dalam*, unlike that of the Perupok area of Kelantan, is often owned by a single man, who is also the boat owner.

i. After a percentage has been taken by the *daganang*, if there is one, the receipts are divided on the *bagi dua* basis. One half, the *bagian dalam*, goes to net and boat, the share of the latter at Kuala Trengganu being equal to that of a member of the crew, though being taken out of a different section of the distribution. The other half is the *bagian luar*, and is divided among the crew.

ii. A special appropriation is made for paying out the net (*mĕnarang*). This comes out of the *bagian dalam*. At Kuala Trengganu the extra increment for each of the two men who do this work is equal to the share of an ordinary member of the crew. At Kuala Marang each of the two men gets half the share of an ordinary crew-member as an additional increment. Another special appropriation is made for the two men in the bow of the boat (*orang luan*), " because they get out of breath ". They have to set the pace in the great amount of paddling that has to be done, and they are also higher out of the water, so the strain on them is greater. At Kuala Trengganu they receive about $1 or $1.50 extra, out of the *bagian dalam*, for every $2 of an ordinary crew-member's share. At Kuala Marang their extra increment is less, being perhaps about 50 cents in the same circumstances.

At Kuala Trengganu no special allowance is made for the work of cleaning out the boat, but there is such at Paka, Dungun and Kemaman in the south. I have no exact details of the amount of this. In no case is any allowance given for the work of managing the sails, as with *pukat payang*.

(*c*) *Pukat Sudu*

These nets occur mainly at Setiu in Trengganu, and at Beserah and Kuantan, in Pahang. They are worked with three boats apiece, each with a crew of two or three men.

i. The main principle of distribution of the takings is that of *bagi tiga*, one third going to the net and two-thirds to the crew. A boat receives a share equivalent to that of a man in the crew's division.

ii. At Beserah, and probably elsewhere, certain special increments are given. Before the main division, one-tenth of the total receipts is allotted as *chabut cheroʻ* for those men who repair and dye the net. In the crew's division the *juru sĕlam* receives one extra share, and an extra share-and-a-half is allotted to the men who have the task of paying out the net. At Beserah, but not in the other areas, the boats are owned by Chinese, who have the right of buying the fish. They do not take the boats' share of the takings, but leave it to the *juru sĕlam*, being content with their profit on the fish. The nets are owned by Malays in each case.

(d) *Pukat Tanggut*

The two areas where these nets are mainly used are Penarik and Kemasik. Each is worked from a single boat, with a crew of three or four men.

The principle of dividing the takings is simple—the total is apportioned among the crew, with one extra share for net and boat together. If net and boat are owned by different men, then this share is divided between them equally.

(e) *Pukat Takur Baring*

This type of net, worked from a single boat with a crew of two, three or four men, occurs along the Trengganu coast from Batu Rakit to Dungun, and is particularly common on the beaches immediately south of Kuala Trengganu.

The principle of division of the takings is the same as for *pukat tanggut*.

Any attempt at rationalization of the fishing industry of Kelantan and Trengganu would probably have to over-ride these local differences in the distributive scheme to some extent. But action should at least be taken not out of ignorance of the complexities, but with knowledge of the local technical and economic conditions with which they are related. Moreover, the possible effects of the change should be visualized and explained in advance. Local custom is not immutable, but it has a meaning for the people who observe it. There is then some point in studying such local variations, if only as a basis for introducing a more uniform system.

APPENDIX V

THE *DAGANANG* (*DAGANGAN*) SYSTEM IN TRENGGANU AND NORTH PAHANG

In the extreme north of Trengganu, at Kuala Besut, there are no *daganang* of the developed type. Nets and boats are owned by Malays, who may buy from other Malays on the time-payment system. The fish dealers, also largely Malay, may be attached to the fishermen by the *tĕtap* (or *tangkap*) system ; those of them who buy at sea have often lent money on the net-ropes, and they come and help repair the net on Fridays. There seems to be no Chinese capital in boats and nets. At Ayer Tawar, a little way down the coast, the situation is the same, though from lack of free capital the fishermen combine and borrow money from Malays in other villages, especially in Kuala Besut. A Chinese dealer has a factory for making shrimp paste there, but does not have any investment in fishing equipment.

At Kuala Trengganu, a main fishing centre farther south, the *daganang* system is common, the lender of money on a net coming down when the catch is brought in and supervising the sale, on which he gets a commission, usually 5 per cent. of the total amount. If the bidding is slow he may take the catch himself, and cure it, or if it is keen he may let the catch go at a high price, and be content with his commission. The *daganang* are usually Malays, often retired fishermen or men whose fathers have left them some capital. The institution appears to be an old one here, since I was told by a well-known pilot that his father had been a *daganang*. An interesting development here is that unlike in some other areas, if the user of a net has his own money in it, not on loan, then he is the *daganang* himself and takes the commission. Interest on the capital is thus paid directly, no matter who is the lender.

Here Chinese have provided some of the capital. Some small boats are owned by them, and they buy the catch by right, at a lower price than the current market value. But Chinese capital is not the rule. *Pukat payang* fishermen strongly denied the suggestion that they were in the hands of the Chinese. They alleged that it was so in Kemaman and the south, but that here it was " altogether Malay cash ". They admitted that of the 15 or so *payang* at the Tanjong, one was owned by a Chinese, but stressed that the others were in Malay hands. The buying for the fresh-fish market is done by Malays, and many Malays also cure fish, which they then sell to the Chinese dealers and exporters. The chief complaint there is that the export trade in cured fish is under the control of the noted Dato' Wi (Lim Wee Cheng, who has received the title of Dato' Maha Kurnia) who overbids if necessary to break his competitors, and who has most of the other Chinese dealers as his agents. His contacts in Singapore give him an advantage, and Malays who have attempted to export in recent years have lost money. The fishermen here also insisted that they did not obtain rice and cloth from the Chinese traders, but paid cash. They contrasted this with the situation at Besut and Setiu, where, they said, because of the infrequent division of the proceedings from fishing, the fisherfolk borrowed a great deal of rice from the Chinese. Dato' Wi was also said to have a large interest in the Kuala Trengganu rice-carrying trade. He does not own the boats, but buys from the carriers, paying cash. This is a strong inducement to them since they can unload and get away again at once.

Outside Kuala Trengganu Chinese influence is stronger. At Batu Rakit the Chinese fish curers, of whom there were said to be 16 in 1940, are lenders of money to Malay net owners, about 8 out of 30 seine nets being maintained by Chinese capital. But they do not take a commission as *daganang* ; they have the right of buying the fish from the nets on which money has been lent, and this suffices. Even when the fish from the lift-nets are sold to dealers from Kuala Trengganu they take no commission, and it was stated, are not angry, since they cannot compete with the fresh-fish dealers. When I raised at Kuala Trengganu the contrast between Batu Rakit, with no *daganang*, and Kuala Trengganu with an extensive *daganang* system, the answer of the fishermen was that at the Kuala the Chinese did not buy the fish direct, while at Batu Rakit they did, getting it cheaply The Batu Rakit folk said that they preferred the large dealers to the small ones, since the latter often recover some of their losses by " cutting " the price agreed upon with the fishermen (see Chap. VII), or by evading payment. Ten years ago a dealer of this type got away with over $1,000 worth of fish of " the sons of Batu Rakit ". Here Malay fish curers have been entirely supplanted by Chinese ; formerly a Malay in this trade lost heavily, and was forced to drop out of business. At the present time the Malay fish dealers are all in a small way, and deal entirely in fresh fish, taking it for sale inland by bicycle or carrying-pole.

At Merang, a little to the north, the situation is much the same as at Batu Rakit, with Chinese control perhaps somewhat intensified. Out of 20 seine nets, 11 were said to be owned by Chinese, and 9 by Malays, and the 2 lift-nets there were both Chinese-owned. Here also there is no special commission for a man as *daganang*, but the Chinese net owner takes the customary shares. All fish are sold to the fish curers, of whom 4 are Chinese and one Malay, and if one of these has lent money on a net he gets his interest through his ability to buy the fish cheaply. Here there is heavy borrowing, especially of rice, from the *tauke* (fish curers). " If he gives money, assuredly he loses once ; if he gives rice, he gets a profit. He wants profit—double profit ; rice once, and fish once," said a fisherman. A common practice is for the Malay net owners to borrow ten sacks or so of rice at a time, which they distribute to their crew.

To the south of Kuala Trengganu the *daganang* system of taking a commission on the sale of a catch is in force. At Kuala Ibai and at Chenering the commission of the *daganang*—there 3 per cent.—is taken irrespective of the amount of cash borrowed from him. He gets his loan repaid by taking all the share of the net each week when the division occurs, till all his capital is back, getting his commission at the same time. Afterwards he continues to receive the commission for collecting the purchase money for the catch and supplying it to the fisherman, with the responsibility to pay up if the purchaser fails.

At Kuala Marang also there are a number of *daganang*. The number of *tauke* curing fish was enumerated to me as follows : 5 Malays, 3 Chinese who had embraced Islam, and 16 non-Mohammedan Chinese. Of these, 2 Malays, 2 Moslem Chinese and about 6 other Chinese had money invested in nets. One of the Malays was *daganang* for 5 lift-nets and several seine nets, and one Chinese Moslem was *daganang* for 6 seine nets, 4 mackerel nets and 1 lift-net. Of about 30 lift-nets 10 were said to be Chinese-owned, the remainder being Malay-owned, and it was emphasized that most of the equipment was Malay-owned, the money being simply lent by the Chinese. The rate of commission to the *daganang* here is 3 per cent. for lift-nets, the reason for the lower rate as compared with Kuala Trengganu being, it was said, that the amount of cash borrowed was in most cases small, from $50 to $100. For the nets of Kuala Trengganu, especially the *payang*, larger sums are usually borrowed. Another reason for the lower rate at Kuala Marang was that the *daganang* here were lenders rather than owners ; they could not lose but got repaid bit by bit and did not have the expense of the net, especially of repairs. If a man is the owner of the net then he takes a higher percentage to recoup himself. And if he is not the user of the net, then his expenses are apt to be higher still. If a net belongs to a *tauke* and not to the man who runs it, I was told, it lasts 2 years instead of 3 years—the fishermen are not willing to mend it but treat it carelessly. This was corroborated quite independently by a man of Perupok, who had worked with the *pukat payang* of the south. He had pointed out how the Chinese by his right to the purchase of the fish " beat " the Malays by paying low prices. Then he added with a twinkle, " But men beat them too ! ". He explained graphically how the fishermen, seeing a hole in the net, took the fabric in both hands and ripped it apart, then went to the Chinese owner and said " *Tauke* ! the net's torn to bits. Give us another ! "

At Kuala Marang the *daganang* need not buy the fish himself. As a woman commented : " If it is cheap he takes it himself ; if it is dear he lets someone else take it." Here the ubiquitous Dato' Wi has his agents also, about 5 Chinese fish curers who buy and weigh the fish ; the Hindu clerk of the Dato' then comes each week and pays them cash for the fish at an agreed figure. After the fish have been sold in Singapore the accounts are revised, a bonus being given to the curers if the price has been good, or a cut made if the exporter has made a loss. I was told by one of the fish curers that they like the system ; if they run out of salt or jars they can get them from the Dato', and can borrow from him if their funds run low. Their relations with him are separate from their activities as *daganang*, which are carried on with their own capital. One of these men, a Moslem Chinese, had a lift-net, a seine net, and a boat of his own, bought at a total cost of $515, and about $150 or $200 invested in 4 seine nets of which he was *daganang* but not owner. His total capital invested in fishing equipment was thus about $700. An interesting feature at Kuala Marang is that some at least of the Malay fish curers export direct to Singapore and do not sell to the Chinese exporter.

At Dungun and Paka the *daganang* system does not operate, though I could not elicit reasons why this should be so. Money is borrowed, either from Chinese or from Malays, to assist in the purchase of fishing equipment, and repaid with interest at rates mentioned earlier. At Dungun there were 4 Chinese and 2 Malay fish curers ; at Paka 5 Chinese and 1 Malay. The catch from a net could be sold to anyone, lender of money or not. At Kemasik also, there are no *daganang* ; the lending of capital on a net gives no pre-emptive right to the fish. Capital for the purchase of a net is obtained by pawning property, or by direct borrowing. But the Chinese here have only small capital ; they " use steamer capital " (*pakai modal kapal*) and have to wait for the return of the vessel which has taken their dried fish export before they are in funds again. As a result, little Chinese money is invested in fishing equipment ; the people with spare cash are mostly Malays. At Kerteh, a small village, where the few nets are mostly Malay-owned, there are no *daganang*. There are only two fish buyers, one a Chinese, who has most of the trade and also owns one net which a Malay uses, and the other a Malay. At Kijal, on a side-road to the south of Kemasik, the *daganang* appears again ; here there are 7 Chinese and 1 Malay who are fish curers ; they lend money in small amounts to the fishermen, but do not own boats and nets.   They take 5 per cent. commission as *daganang* for functions like those of the Kelantan *tangkap*, insuring payment to the seller of the catch.   The fishermen here stated that they liked the system, as it made selling safe.   Formerly, however, not all the *daganang* were fish-curers ; one was a Malay who did not deal in fish, but who got his commission just the same —that is, his " commission " was interest on his loan.

At Kemaman, the major fishing centre in the south, Chinese enter deeply into the fishing industry.   Many of the *payang* boats are Chinese-owned, but held in the names of the Malays who run them, and though the nets are Malay, much of the capital in them is supplied by the Chinese.   All the fish curers and dealers here are Chinese.   They are of two types ; those with small capital, who buy the *luan* fish (see p. 327) which anyone may purchase ;   and those with large capital, who are lenders of money and *daganang*.   Each of the latter has 4 or 5 nets under his control, getting the repayment of his capital from the net's share of the proceeds, and taking the fish—at a low price—by right.   Should the catch be sold to another he is angry.   A commission of 5 per cent. is taken in the usual way.   Some of these Kemaman Chinese have also capital invested in fishing equipment at Kijal and other villages in the vicinity.   With the seine nets, however, it is different.   Several Kemaman fishermen emphasized that only a little Chinese money was invested in these nets—a figure or $20 or $30 per net was mentioned by several men. Asked why the loans were so small, they replied that they were afraid lest the Chinese should become able to take the fish at their own figure—as they do with the *payang* catches.

In addition to borrowing capital for boats and nets, the fishermen also obtain rice in the monsoon season from the Chinese dealers, either through their net owner, or direct. A direct application by an ordinary crew-member, however, may be refused, the security not being so good.   The general level of indebtedness of the fishermen in the Kemaman region is high.   Official inquiries in 1933 and 1934 gave an average indebtedness for about 80 fishermen of just over $100 per head, the total sum calculated to be due to 9 Chinese fish dealers in Kuala Kemaman alone being over $15,000.

In the north of Pahang Chinese investment in fishing equipment is also heavy.   At Kuantan, where all the fish curers are Chinese, the *payang* boats and nets are all Chinese-owned.   The fish are sold cheaply to the owner, who thus gets his profits, taking neither commission nor the specific share of net and boat, which are treated as part of his general overheads.   At Tanjong

Luper, by Kuantan, the few *pukat sudu* were Malay-owned, and the capital was said to be Malay and not Chinese. It was the same with some drift-nets, the capital in each case being comparatively small. The folk in this village are mainly from Kuala Trengganu, Besut and Semerak. They complained of the poor returns for their work, but when asked why they did not go back, said that things were bad at home too. One woman, commenting on the difficulty of getting together capital, said that she had been in the village since she was a girl, and was now grey-haired, with grandchildren, and only now had she been able to save up enough to buy a *pukat sudu*, costing $150. The Chinese owned the *payang* because of their high cost, which Malays could not afford. The people here complained at the low prices their fish fetched from the Chinese, who bought it on a credit note, which could then be exchanged for cash at their offices. Only 2 of the fish dealers in the village were Malay ; all the rest were Chinese. At Kuantan, where the main market has very few Malay sellers, some Malays were selling fresh fish from line fishing at a small market at the side. The retail prices of fresh fish here were high as compared with those in Kelantan or Trengganu. An explanation of this seemed to be that the Chinese, controlling most of the bulk sales at low price-levels, indirectly forced up the price of the reduced supplies of fish available for the fresh market.

At Beserah, at the southern end of the beach leading to Trengganu, the *payang* and lift-nets are all Chinese-owned, as are practically all the drift-nets. Of 28 seine nets, 8 are Malay and 20 Chinese ; of 6 *pukat sudu*, 3 are Malay and 3 Chinese. Even when nets are owned by Malays the boats are often owned by Chinese. One Malay, brother of the headman of the village, was in partnership in a *pukat sudu* group. The cost of the net was $120, two-thirds of which was borrowed capital. But the 3 boats of the group belonged to a Chinese, who had the right of buying the fish. He did not take a cent of the boat's share of the yield but was content with his profit on the fish. When a boat was old and unseaworthy, he told the leading fisherman of the group to buy another—at his expense. Before joining this group this Malay used to run a group of nets with 5 boats, contributed by another Chinese fish dealer. He did this for 20 years, getting enough to eat, but saving little. Then, he said, the Chinese promised him a bag of rice for the monsoon, but did not give it, so he left. It is quite clear that the action of handing over equipment to the fishermen to use, without taking the customary Malay share of the proceeds, though apparently generous, is in reality an effective method of securing the product, and at a price which well repays the generosity.

APPENDIX VI

WEEKLY RECORDS OF SALES OF FISH FROM LIFT-NETS, PERUPOK AREA, FOR THE HALF-YEAR, NOVEMBER 11TH, 1939 TO MAY 9TH, 1940

Net	A	B	C	D	E	F	G	H	I	J	K	L	M	N	O	P	Q	R	S	T	U	V	Comments.
1939 Nov. .	115	70	87	65	x	x	55	46	x	x	3	x	x	118½	31½	29	(groups not yet formed)					O	Rough weather
	96	177	155	198	48	102	75	62	S	S	90	S	57	152	3	46						x	Much gill-net fishing
	32	48	45	57	45	O	34	6	.	.	42	.	O	85	3	5						3½	Rough weather
Dec. .	x	x	x	x	x	x	x	x	.	.	x	.	x	x	x	x	(F)					x	Monsoon, too rough
	x	x	x	56	2½	18	2½	x	.	.	x	.	x	23½	x	x	43					x	Monsoon, rough
	x	x	x	x	x	x	x	x	.	.	x	.	x	x	x	x	x					x	Monsoon, too rough
1940	x	x	x	x	x	x	x	x	.	.	x	.	x	x	x	x	x					x	,, ,, ,,
Jan. .	x	x	x	x	x	x	x	x	.	.	x	.	x	x	x	x	x					x	,, ,, ,,
	x	x	18	x	O	x	x	x	.	.	x	.	x	O	x	x	x	(F)	(F)			x	Rough, strong currents
	192	281½	374	253	85	69	294	81	x	O	19	6	35	120	75	O	75½	x	12	(F)		S	Some nets not ready
	O	45	67	70	45	O	O	O	30	x	O	12	19	O	O	19	O	x	O	O		.	Rough weather, holidays
	63	85	107	30	60	47	33	40	88	x	30	21½	20	122	45	108	28	x	78	85		.	Rough weather
Feb. .	77	194	183½	214	171	45	57½	113	107½	39	39	86	257	45	112	70	112	44	19	110		.	Good conditions
	x	x	x	68	x	x	x	35	35	3½	x	O	x	36	28	x	O	x	x	x		.	Strong N.E. wind
	31	162	111	174	151	130	70	O	113	O	41½	16	104½	79	13	14	26	6	4	93		.	N.E. wind, boats painted
	x	105	O	41	30	x	30	7½	O	x	x	13	O	65	O	x	x	x	x	O		.	Strong N.E. wind
March .	87	127	167	175	45	188	79	86	80	6	21	141	95	247	39	135	17	7	15	17		.	Good conditions
	266	55	129½	147	47	136	x	76	213	39	92	41	93	86	104	41	103	95	O	36	(F)	.	,, ,,
	O	53	O	13	O	12	x	O	50	74	18	O	O	68½	O	O	67	O	O	x	O	.	Fish getting scarcer
	158	48	30	134	O	O	x	x	O	x	O	x	x	O	O	O	52	x	x	O	O	.	Fish scarce
April .	319	79	82	122	28	15	x	x	69	x	20	O	O	x	x	x	108	x	S	x	O	.	Much line fishing
	196	85½	140	210	28	10	x	98	77	O	180	99	18	64	22½	O	61	x	.	S	42	.	Fish moderate
	45	O	16	128	x	O	8	30	O	O	O	O	O	40	O	3½	32	x	.	O	16½	.	Fish scarce
	15	23	O	278	O	x	46	43½	O	114	56	89	10	84	47	54	97	x	.	124	30	.	Conditions improving
May .	251	108	x	76½	30	O	104½	49	21	56	99	x	x	84	O	22½	116	x	.	50	O	.	Fish moderate
	41½	O	x	x	O	x	x	O	x	28	3	x	x	3	x	4	x	x	.	x	x	.	Much gill-net fishing
Six months' Total	$1,984½	1,746	1,712	2,509½	815½	772	888½	773	883½	359½	753	524½	708½	1,522½	523	551	937½	152	128	515	88½	3½	

Note :

O indicates no bulk sale for cash, though fish may have been obtained for petty retail sale by the crew, or home consumption. The symbol is given if the net went out at all during the week.

S indicates that the net was sold, and that fishing stopped until a new net was ready.

x indicates that the net did not put to sea at all that week.

(F) indicates that the net group is newly formed.

Each figure is the total takings of a net in a week, from a Saturday to the following Thursday, inclusive ; the material is summarized from full daily records.

APPENDIX VII

SALES OF OUTPUT FROM MAJORITY OF DEEP GILL-NETS (NIGHT-FISHING), PERUPOK AREA, OCTOBER 20TH, 1939 TO MAY 11TH, 1940

Net	a	b	c	d	e	f	g	h	i	j	k	l	m	n	o	p	q	r	s	t	u	v	w	x
1939																								
20 : X	400 @15 x	x	800 @15 x	x	100 @20 x	1700 @13 x	800 @18 x	2200 b.30 x	1000 @15 x	x	x	600 @18 x	500 @15 x	x	x	500 @15 x	x	1000 @18 x	500 @10 x	2500 @17 m.	2000 @17 x	2000 @17	600 @17 x	700 @17
21 : X	x	x	x	1000 @22½ x	x	x	x	x	x	? o	? o	x	x	x	x	x	x	1000 @22 x	2145 @20 x	m.	x	x	500 @20 x	500 @20 o
22 : X	x	x	x	x	x	x	? x	x	x x	? o	? o	700 @18	m. or nil	x	x	x	x	500 @22 x	200 @15 o	x	x	x	800 @18 x	0
23 : XI	? x	? x	? x	? x	? x ? x	? x ? x	? x	x	x x	m. or x	x	x	m. or nil	x	x	x	0 x	7500 @23 x	0 0	·	2500 @22 x	x	6000 @23	0 x
10 : XI	? x	? x	7500 @20 x	5000 @20 m.	m.	800 @25 x	300 @25 x	x	0	600 b.16	600 b.16 x	200 @25 x	1000 @25 x	x	x	x	500 @22 x	x	0	3000 @15 x	1150 b.26 x	x	x	1500 @22 x
24 : XI	x	x	x	m.	x	x	x	0	x	x	x	x	x	x	x	x	x	x	0	x	x	x	x	@22 x
*5 : IV : 40	x	x	x	x	x	x	x	x	x	x	x	x	x	x	x	1500 @22 x	500 @22 x	x	2500 @22 x	m.	0	x	x	x
6 : IV	m.	2500 @20	x	x	2150 @17 x	2800 @17 x	0	m.	x	2700 b.53 600 @16	x	700 @16	m.	x	x	500 @20 x	100 @20 x	900 @20 x	2450 @20 x	600 @16 x	0	m.	0	700 @20 0
7 : IV	0	0	0	0	x	x	0	·	0	x	0	x	·	x	x	1000 @16 0	0	500 @16 0	500 @17 0	600 @16 x	800 @16 x	·	x	0
8 : IV	x x	x	x x	x x	x 0	x x	x x	0	x x	x x	x x	x x	x	x	x x	x 0	x x	1950 @25 0	x 0	x x	x x	x x	x x	x
29 : IV	x	x	x	x	x	x	x	x	x	x	x	x	x	x	x	x	x	·	x	·	·	·	·	·
30 : IV	x	x	x	x	x	x	x	x	x	x	x	x	x	x	x	x	x	·	·	·	·	·	·	·
3 : V	x	x	x	x	x	x	x	x	x	x	x	x	x	x	x	0	x	·	0	·	·	·	·	·
4 : V	x	x	x	x	2740 @20	600 @20	500 @20	x	300 @20	300 @20	x	x	780 @20	m.	m.	x	0	2525 @25	1200 @22	200 @20	m.	2000 @15	x	4000 @16
5 : V	b.27	b.90	b.50	b.30	b.115	b.70	b.50	6000 @12	b.42	6500 @12	5200 @12	5000 b.66	1500 @12	2700 @12	2000 @12	2800 @10	2200 @13	5000 @12	5000 @22	3300 @15	b.70	2000 @15	x	4000 @16
6 : V	4000 @10	5000 @10	2300 @10	b.52	4940 @10	4000 @10	b.37	2200 @12	3500 @10	b.50	4000 @10	3000 @10	4000 @10	4000 @10	2300 @10	4500 @10	800 @10	5000 @10	1600 @10	6400 @10	2200 @10	700 @10	x	5000 @11
7 : V	200 @12	m.	1800 @10	500 @10	4460 @10	800 @10	500 @14	b.27	400 @11	300 @13	m.	m.	1200 @10	900 @10	500 @10	? x	800 @10	m.	m.	1500 @10	1200 @10	3000 @10	x	1300 @10
8 : V	2200 @12	x	1500 @13	1750 @13	1340 @13	1650 @13	1050 @13	2500 @10	m.	500 @18	2000 @10	2100 @10	750 @13	1500 @13	800 @13	? x	1000 @12	2240 @13	0	1500 @10	x	600 @13	1200 @12	1100 @14
9 : V	0	x	0	x	x	0	0	0	0	0	0	0	0	0	0	? ?	0	0	0	m.	x	1000 @15	x	x
10 : V	0	0	m.	0	6250 @13	1000 @13	200 @13	0	0	0	x	0	0	x	x	? x	340 @15	1700 @13	1950 @15	400 @16	x	x	x	0
11 : V	0	1500 @15	0	0	0	0	0	0	0	0	x	0	0	x	x	? x	0	0	0	0	x	x	0	x
Totals	$ 106	212	276	238	410	254	141	154	100	209	142	168	130	103	64	139	75	525	310	271	234	130	187	219

Note :

Sales given in numbers of fish at price in dollars per 1,000, or as total price received (b.) for the lot (*borong*).

m.—*makan laut'* (no bulk sale); o—no catch.

x—net not out.

• In addition to data listed, group w obtained *nil* on Nov. 12, and groups q, r, and s *nil* on Nov. 25, 1939.

APPENDIX VIII

COMMONER KINDS OF FISH TAKEN IN EAST COAST MALAYAN WATERS [1]

Most of these fish are taken also in the south and west, in some cases by different gear.

1. By Lift-net (*Pukat Takur*)

The fish taken mainly by this net are various species of horse-mackerel (*Caranx*), of which the most important is *sĕlar kuning* (*Caranx leptolepis*), a small fish with a yellow stripe down the side, ranging up to about 8 in. long. Other types of small *Caranx* commonly taken are *sĕlayang, sĕlar gilek, lolong,* and *lechen,* while occasionally *gĕrong* and *bĕrka'* (*bĕrkas*), large species 2 ft. or more in length, are taken. At some seasons *tamban bĕluru* (*Clupea* sp., a small blue herring), and *tamban sisek* (*Sardinella jussieu,* a sprat) are also taken ; and also some *kĕmbong* (*Scomber kanagurta,* a mackerel).

2. By Deep Gill-net (*Pukat Dalam*)

The main fish taken by this net is the *kĕmbong,* a mackerel, with some *sĕlar gilek,* a horse-mackerel very similar in size and appearance ; both are caught mainly at night by this net. By day the catch is of mixed type, similar to that of a drift-net, with many jewfish of various kinds ; but sometimes *ayoh* (*Thynnus* sp., a bonito or tunny) are taken instead.

3. By Seines (*Pukat Tarek*)

One type of seine takes almost solely *bilis* (*Stolephorus,* a small anchovy). The other type, of larger mesh, takes a variety of fish, of which *kikek* (*Leignathus* sp., described by Stead under the name of " Silver Bellies ") is one of the commonest ; other common kinds are *layur* (*Trichiurus* sp., termed by Stead " Scabbard Fish " and by Maxwell " Hair Tail ")—a long thin fish with the dull sheen of aluminium, and *pĕlata* (*Scomber microlepidotus,* termed by Stead " Pigmy Trevally ")—a tiny mackerel.

4. By Heavy Drift-net (*Pukat Hanyut*)

The fish mainly taken by this net vary at different seasons. Commonly taken are : *bawal* (*Stromateus* spp., pomfret) ; *mayung* (*Chilinus* sp., a large catfish) ; *pari* (*Trygon* spp., skates and rays) ;

[1] Scientific names are taken from C. N. Maxwell, D. G. Stead and glossaries in the Fisheries Department *Annual Reports* (see Bibliography).

yu (*Carcharias* etc., sharks) ; *pěrupok* (*Pellona* sp., a large herring) ; and small *parang* (*Chirocentrus dorab*, the dorab or wolf-herring).

5. By Light Drift-net (*Pukat Těgělang*)

A great variety of fish is taken by this net, including : *gělama* (*Scianidae*, jewfish) ; *duri* (*Arius* spp., catfish) ; *lidah* (*Cynoglossus* sp., a sole) ; *sa-bělah* (*Psettodes*, etc., flounder) ; *todak* (*Belone* sp., a garfish) ; *sělangat* (*Dorosoma chacunda*, gizzard shad) ; *nipis* (? *Clupeidae*, a thin silvery fish) ; *landung* (? a sea-bream). Prawns (*udang*) and crabs (*kětam*) are also taken in quantity at certain seasons.

6. By Purse Seine (*Pukat Payang*)

In general the same types of fish are taken by this as by the seine net of large mesh, and also : pomfret ; jewfish ; *sagai* (*Caranx armatus*, a large horse-mackerel) ; *chěncharu* (*Caranx* sp., a large horse-mackerel ; *chermin* (*Caranx gallus*, silvery moon-fish) ; *daun bharu* (*Drepane punctata*, spotted moon-fish) ; and many other kinds.

7. By Fine-meshed Drift-net (*Jaring tamban*)

In east coast waters various types of *tamban* (*Clupea, Clupeidae, Sardinella*, i.e. sprats, pilchards and herrings) are taken by this net. They include *tamban běluru, tamban sisek, tamban bujur, tamban bulat*.

8. By Trolling

The fish mainly taken in this way is *těnggiri* (*Cybium* sp., Spanish mackerel) ; but *talang* (*Chorinemus* sp., a large horse-mackerel), and *ayoh* (*ayer*, tunny) are also caught.

9. By Hand-lining

Among the commoner types taken by ordinary line fishing are : *kěrisi* (*Synagris* sp., a sea-bream) ; *mudin* (*Saurus myops*, allied to the Queensland smelt) ; *bělitong* (*Coryphaena hippurus*, a large blue fish with a supercilious expression) ; *kěrapu* (*Epinephelus* sp., a grouper) ; *bajibaji* (*Platycephalus nematophthalmus*, flathead) ; *biji nangka* (*Upeneus* sp., a goat-fish) ; *ikan merah* (*Lutianus* sp., snapper) ; *gěrong* and *běrka'* (both *Caranx*, large horse-mackerel) ; *seng, těnok* or *kachang* (*Sphyraena* sp., barracouta or sea-pike) ; *kirun* (a small fish barred in brown and white). Other types taken by different techniques are : *parang* (*Chirocentrus dorab*, wolf-herring) ; *sělar gilek* (*Caranx*) ; *sutung* (*sotong, Loligo*, squid).

10. By Scoop-net or Push-net˙ (*Saup*)

The fish taken by this, mainly during the monsoon, is *bělanak* (*Mugil* sp., a grey mullet).

The variety of fish in a single catch with a seine or a drift-net is often considerable, and even a day's hand-lining may yield several different kinds. A small selection from a number of sample counts of drift-net and hand-line catches will illustrate this.

Small drift-net (pukat tĕgĕlang) :

i. 310 *landung*, 50 prawns, 70 crabs, 10 large dogfish, 12 small dogfish, 12 jewfish, 6 *pĕrupok*, 2 soles, about 100 *ikan nipis*— in January, 1940.

ii. 20 prawns, 20 skate, 10 soles, 15 small jewfish, 6 horse-mackerel, 1 small dorab, 1 flathead—in March, 1940.

Large drift-net (pukat hanyut) :

i. 6 Spanish mackerel, 8 small pomfret, 13 large pomfret, 10 dogfish, 3 small dorab, 7 large dorab, 2 small rays— in January, 1940.

ii. 40 black pomfret, 2 white pomfret, 1 large horse-mackerel *(talang)*, 21 dogfish, 16 *pĕrupok*, 1 catfish, 1 ray, 2 small dorab—in February, 1940.

Hand-lining :

i. 7 large horse-mackerel *(talang)*, 1 large horse-mackerel *(gĕrong)*, 2 pike *(seng)*, 1 *Coryphaena*, 3 sea-bream, 2 squid— in April, 1940.

ii. 1 Spanish mackerel, 2 bonito, 1 pike, 3 *Coryphaena*, 12 large squid—in April, 1940.

The above examples also give some indication of the numbers of fish in fairly typical catches by these techniques. The number of fish per catch obtained in mackerel netting *(pukat dalam)* has already been shown in Appendix VII. In lift-netting, moderate catches counted were : 6,900 *sĕlar kuning* ; 6,500 *sĕlar kuning* and *lechen* and 800 *kĕmbong* ; 500 *sĕlar kuning* and 1,000 *sĕlar gilek* ; about 9,500 *tamban sisek*. Large catches vary between 10,000 and 20,000 *sĕlar kuning*. In gill-netting for sprats, hauls commonly vary between 1,000 and 3,000 fish ; as many as 10,000 are rarely taken.

Fish taken in shoals are commonly sold in large baskets weighing, when full, roughly a picul (133⅓ lb.). A basket contains about 1,300 to 2,000 *sĕlar kuning* according to size, 1,300 to 1,500 *tamban sisek*, 800 to 1,000 *tamban bĕluru*, about 400 *kĕmbong* or about 450 *sĕlar gilek*. Of fish actually weighed, *kĕmbong* varied between 4 oz. and 6½ oz. (averaging 6 oz.), while 10 fairly large *sĕlar kuning* made 1 lb.

APPENDIX IX

GLOSSARY

The Kelantan and Trengganu dialects differ in many respects from standard Malay, which is based on the speech of Johore. As yet no system of orthography has been worked out to represent these dialects satisfactorily in writing, nor has any intensive linguistic study been made of them, though valuable contributions have been made towards this by C. C. Brown, M.C.S. (see Bibliography).

The main characteristics of the Kelantan dialect, as compared with pronunciation in standard Malay are :

—nasalization of final syllables -*an* and -*am*, so that they are equated with final -*ang*.

—nasalization of the final syllable of many words ending in -*i*.

—slurring of final -*b*, -*p* and -*t*, which are commonly all assimilated to a glottal stop (', represented in standard Malay by a final *k*) ; final -*s* is sometimes also thus assimilated.

—substitution of final -*h* by -*s*, and of final -*l* by -*r*.

—substitution of final -*a* by -*o*, this sound often being prolonged.

—omission of medial -*m* before -*p*, medial -*n* before -*ch* or -*t*, and medial -*ng* before -*k* or -*s*.

—indefiniteness of medial vowels, -*o* often approximating to -*u*, and -*e* to -*i*.

—tendency to substitute medial -*y* for medial -*j* in some words, and to elide medial -*r* in others.

Most of these points are illustrated in the examples below. One of the difficulties in rendering the Kelantan dialect into writing is the considerable variation in individual pronunciation ; spelling used in this book therefore tends to be a mean approximation. In order to bring out the force of the dialect I have departed in some respects from the practice of standard Malay : in particular, the inverted comma has often been used to indicate the glottal stop. (The indeterminate -*ĕ* is inserted in words such as *pĕrahu* and *pĕraih*, where it is not definitely sounded, to conform with standard spelling ; compare words such as *sĕlam* and *tĕgĕlang*, where it is normally sounded).

An exhaustive list of Kelantan words and expressions used in fishing is not possible here. The following list gives some of those most commonly used, including many of those mentioned in this book. In each case the Kelantan form is given first, followed by the standard Malay form, where known to me, in brackets.

acho' (*anchak*)—plaited bamboo-strip tray, on which fish, etc., are dried.

bagi (*bagi, bahagi*)—to allot shares. *Bagian pěrahu, bagian pukat, bagian tuboh*—share allotted to a boat, a net, a member of a crew, respectively. The share allotted to a net is sometimes termed *bagian dalam*—the inside or central share ; the remainder is *bagian luar*—the outside share.

basar (*bangsal*)—a shed ; as *basar ikan kěring*—a shed for storing dried fish.

baso (*bangsa*)—a race or type ; used *inter alia* for " kinds " of fish.

batar (*bantal*)—a package of yarn ; a pillow ; a section of coco-nut palm trunk used to support the keel of a boat.

bidang (*bidang*)—a numerical coefficient, used elsewhere for mats, sails, etc., but in Kelantan especially for nets (instead of *rentang*).

boleh, buleh (*běroleh, běruleh*)—to get, to obtain. Used commonly by fishermen in referring to their day's takings. *Buleh běrapo ?*—How much did you get ? meaning How much did you sell your catch for ? *Buleh tiga amas*—I got a dollar and a half.

buah (*buah*)—fruit ; numerical coefficient of boats. Hence *dua buah*, two boats.

bui (*běri*)—to give ; used particularly in fish-bargaining as a synonym of *jual*, sell. *Pa' puloh tuan ta' bui*, at forty (dollars) the owner won't give (them).

buteh (*butir*)—numerical coefficient of many kinds of objects, including net-floats, *unjang*, etc. *Kueh buteh ketere*—medium-sized boat with prow shaped like the curved cashew nut (given as *ketiri* by Wilkinson, but not so pronounced.)

chabu' (*chabut*)—to strip off, as fish from the meshes of a net ; an initial share or bonus from total takings before the general distribution begins.

daganang (*dagangan*)—a fish-buyer with pre-emptive rights (see Appendix IV)

iuru sělam (*juru sělam*)—literally, an expert in diving ; used in a wider sense for an expert fisherman with large nets, as *juru sělam takur, juru sělam pukat tarek*, etc. ; often pronounced *jusělang*.

An allied term is *juragan*, a boat captain.

kau' (*kaup*)—to skim off with the hand ; used specifically of taking the fish in bulk from a boat. *Kau' lah, kau' ; kita na' běli sekit ; sapuloh, duopuloh sen*—Take them out, take them out ; I'm going to buy a few ; ten or twenty cents' worth.

kěrochong (kěronchong)—inner part of seine or lift-net ; equivalent to *pěrut.*

kong—rib of a boat ; also section or compartment between two ribs, giving a unit of measurement in buying fish.

kreja, krejo, krija, krijo (kěrja)—work ; also a feast.

kupang—12½ cents (as a unit of reckoning, not a coin). *Aku běli sa-riyal dua kupang ; aku ta' bilang.* I'll buy them as a lot at a dollar twenty-five cents ; I won't count them.

layar duo—two sails (set goose-winged) ; used specifically to indicate the boat of a carrier-agent returning with fish.

likung (lengkong)—a circle, used of a fleet of nets which are shot in a circle, as *pukat dalam.*

likur (lekur)—a score added to a number. " *Běrapo ?* " " *Tigo-puloh.*" " *Duo !* " " *Duopuloh muroh ; dua likur muroh lagi.*" " *Tiga likur lah !* " " *Mati kurang duo.*"—" How much ? " " Thirty." " Twenty." " Twenty is too cheap, and so also is twenty-two." " Twenty-three then ! " " Bedrock is twenty-eight."

makan lau'—literally fish or flesh food ; used for that portion of a catch reserved for the disposal of the crew. *Makan lau' na' jual-koh ?*—Are you selling your crew's fish or not ?

mati (mati)—literally to die, dead ; used elliptically for *hergo (harga) mati*—dead, i.e. bedrock, price. *Tuan bělum mati*—the owner has not stated his bedrock price.

měduo ; měnigo ; měngepa' ; mělimo—second, third, fourth, fifth ; applied to sections of net, from centre outwards (Fig. 16), or strakes of a boat, from keel upwards.

měmuka' (měmukat)—to shoot a net ; often slurred almost to *muka'.* An associated expression is *buang pukat,* to pay out the net.

pěrahu (pěrahu)—boat. (In writing the dialect phonetically this might be spelt *prahu.*)

pěraih (pěraih)—dealer, middleman. *Pěraih darat*—dealer on shore. *Pěraih laut*—dealer at sea ; specifically, carrier-agent. (In a dialect form this might be written *praih.*)

pěrut (pěrut)—belly ; central part of a seine or lift-net.

pitis (pitis)—small change, formerly coins of tin with hole in centre ; also used for money in general, equivalent to *duit* and *wang.*

potong (potong)—to cut ; used for " cutting " a price. *Kira baya: tujoh běrapo potong ?* If we reckon to buy at seven dollar: how much do we cut ?

Riyal—dollar, used more commonly than *ringgit*. *Na' jual lapan riyal ta' leh kurang ; tujoh ta' se.* I'll sell for eight dollars and not less ; I don't want seven.

Se (sir)—want, wish. *Orang lain ta' se koh ?* Don't other people want them ?

So (satu)—one. " *Limopuloh so aku bĕli.*"   " *Ta' bui.*"   " *Limopuloh duo.*"   " I buy at fifty-one."   " Won't give them."   " Fifty-two."

takur (tangkul)—*pukat takur*, lift-net (ground-net).

tĕgĕlang (tĕnggĕlam)—to sink ; *pukat tĕgĕlang*—light drift-net.

unjang (unjam in Pahang)—fish-lure, normally of coco-nut fronds.

uta' (utas)—numerical coefficient of ropes, cords ; section of a net, as of *pukat dalam* or *pukat tĕgĕlang*, made up in a fleet.

For those unacquainted with Malayan currency and measures it may be noted that the Straits dollar (equal to 100 cents) was current everywhere in Malaya, and was worth approximately 2*s*. 4*d*. A *gantang* is equivalent to a gallon ; a *gantang* of unhusked rice weighs from 4 to 5 lb., and a *gantang* of husked rice about 8 lb. Four *chupak* make a *gantang*, and in Kelantan four *chentong* (each the measure of volume of a round cigarette tin) make a *chupak ;* two *chentong* make a *leng*. The ordinary measure of weight for retail selling is the *kati*, equivalent to $1\frac{1}{3}$ lb. ; 100 *kati* make a picul, or $133\frac{1}{3}$ lb. Fish are sold by size individually (by *ekor*, " tail ") ; by the piece (*kĕping*) if large ; or by the hundred (*ratus*) if small. In bulk they are sold by the basket (*raga*, roughly equivalent to a *picul* if full) or by the thousand (*ribu*). In some retail markets they are sold by the *kati*.

BIBLIOGRAPHY

BIRTWISTLE, W. (1929–39), *Annual Report of the Fisheries Department, Straits Settlements and Federated Malay States*—from 1928 to 1938. Singapore.

BROWN, C. C. (1935), " Trengganu Malay ", *passim*, and Appendix on Trengganu Fishing Methods, *Journal of the Malayan Branch of the Royal Asiatic Society*, vol. XIII, part III. Singapore.

BURDON, T. W. (1955), *The Fishing Industry of Singapore*, Singapore.

DALTON, H. GORING (1926), " Some Malay Boats and their Uses ", *Journal of the Malayan Branch of the Royal Asiatic Society*, vol. IV, pp. 192–7. Singapore.

DEW, A. T. (1891), " The Fishing Industry of Krian and Kurau, Perak ", *Journal of the Straits Branch of the Royal Asiatic Society*, no. 23, pp. 95–119. Singapore.

DJAMOUR, JUDITH (1959), *Malay Kinship and Marriage in Singapore*, London School of Economics Monographs on Social Anthropology, no. 21 London.

DOBBY, E. H. G. *et al* (1957), " Padi Landscapes of Malaya ", *Malayan Journal of Tropical Geography*, vol. X, pp. 1–143.

DOWNS, R. E. (1960), " A Rural Community in Kelantan, Malaya ", *Studies of Asia*, pp. 51–62. Lincoln, Nebraska.

FEDERATION OF MALAYA (1962), *Official Year Book*, pp. 281–8. Kuala Lumpur.
    (1963), *Monthly Statistical Bulletin*, July, Kuala Lumpur.

FIRTH, RAYMOND (1941), " Economics of the Malayan Fishing Industry ", *Man*, vol. XLI, pp. 69–73.
    (1943), " The Coastal People of Kelantan and Trengganu, Malaya ", *Geographical Journal*, vol. CI, pp. 193–205.
    (1948), *Report on Social Science Research in Malaya*. Singapore.

FIRTH, RAYMOND & YAMEY, B. S. (1964), *Capital, Saving and Credit in Peasant Societies*. London.

FIRTH, ROSEMARY (1966), *Housekeeping Among Malay Peasants*, London School of Economics Mon. Social Anthropology, no. 7, 2nd ed. London.

FREEDMAN, M. & SWIFT, M. G. (1959), " Rural Sociology in Malaya ", *Current Sociology*, vol. VIII, pp. 1–15.

GERMAN, R. L. (n.d.), *Handbook to British Malaya*, pp. 187–92. London.

GIBSON-HILL, C. A. (1954), " The Boats of Local Origin Employed in the Malayan Fishing Industry ", *Journal of the Malayan Branch of the Royal Asiatic Society*, vol. XXVII, part II, pp. 145–80.

GOPINATH, K. (1950), " The Malayan Purse Seine (*Pukat Jěrut*) Fishery ", *Journal of the Malayan Branch of the Royal Asiatic Society*, vol. XXIII, part III, pp. 75–96.

GREEN, C. F. (1925–8), *Annual Report of the Fisheries Department, Straits Settlements and Federated Malay States*—from 1924 to 1927. Singapore.

GULLICK, J. M. (1958), *Indigenous Political Systems of Western Malaya*. London.

KESTEVEN, G. L. (1949), *Malayan Fisheries*. Singapore.

LEACH, E. R. (1949), " A Melanan (Sarawak) Twine-making device with Notes on Related Apparatus from N.E. Malaya ", *Journal Royal Anthropological Institute*, vol. 79, pp. 79–85.

LE MARE, D. W. (1950), *Report of the Fisheries Department, Malaya, 1949*. Singapore.

MAXWELL, C. N. (1922), *Malayan Fishes*. Singapore.

PARRY, M. L. (1954), " The Fishing Methods of Kelantan and Trengganu ", *Journal of the Malayan Branch of the Royal Asiatic Society*, vol. XXVII, part II, pp. 77–144.

SMYTH, H. WARINGTON (1902), "Boats and Boat Buildings in the Malay Peninsula", *Journal of the Society of Arts*, vol. L, pp. 570–87. London.

SWIFT, M. G. (1957), "The Accumulation of Capital in a Peasant Economy", *Economic Development and Cultural Change*, vol. 5, pp. 325–37. Chicago.

(1965), *Malay Peasant Society in Jelebu*, London School of Economics Monographs on Social Anthropology, no. 29. London.

WARD, MARION W. (1964), "Malayan Fishing Ports and Their Inland Connections", *Tijdschrift voor Econ. en Soc. Geografie*, vol. 55, pp. 113–42.

WINSTEDT, R. O. (1925), "Malay Industries, Part I. Arts and Crafts", *Papers on Malay Subjects*, pp. 8–17. Kuala Lumpur.

INDEX

394

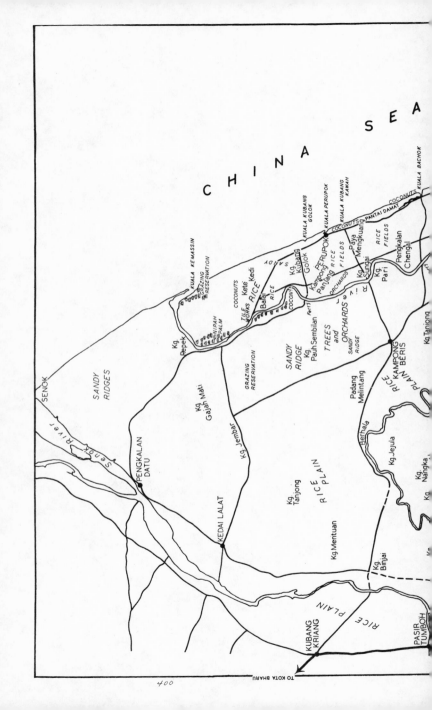

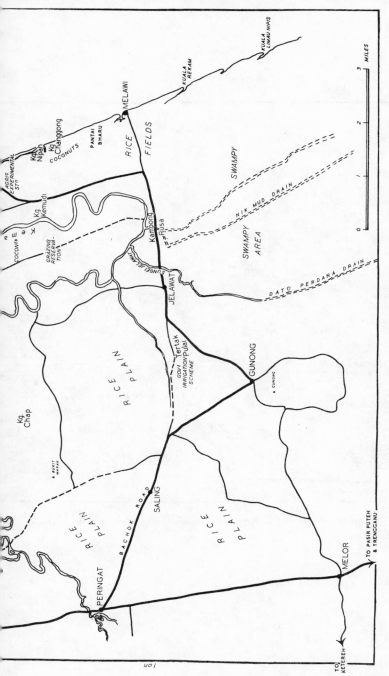

Fig. 11.—Bachok and Surrounding Area.

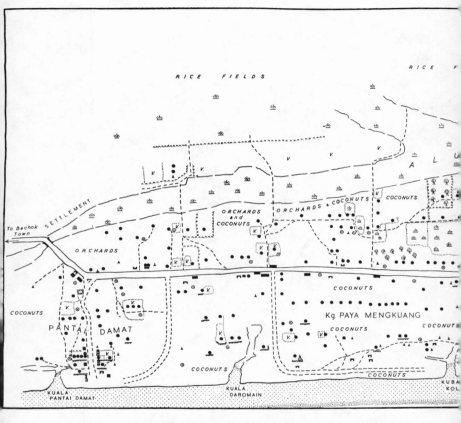

FIG. 12.—Sketch plan of Perupok area, 1940. Looking inland from the beach. The brok

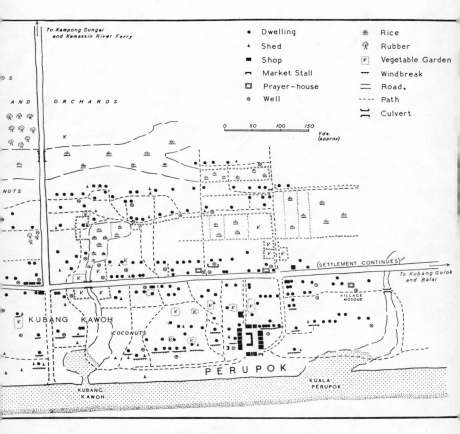

Dwelling
Shed
Shop
Market Stall
Prayer-house
Well

Rice
Rubber
Vegetable Garden
Windbreak
Road.
Path
Culvert

To Kampong Sungai
and Kemassin River Ferry

ORCHARDS

AND

NUTS

0   50   100   150
Yds.
(approx)

(SETTLEMENT CONTINUES)

To Kubang Golok
and Balai

KUBANG   KAWOH

COCONUTS

VILLAGE
MOSQUE

PERUPOK

KUALA
PERUPOK

KUBANG
KAWOH

uble lines show tracks used by buses in fetching fish; only the main footpaths are shown.